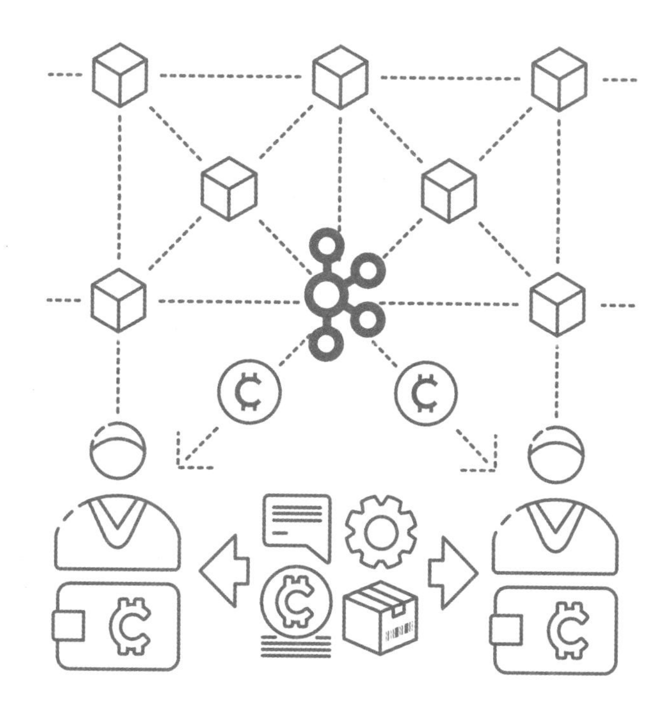

# Kafka进阶

赵渝强 著

電子工業出版社
Publishing House of Electronics Industry
北京•BEIJING

## 内 容 简 介

本书基于作者多年的教学与实践进行编写，重点介绍 Kafka 消息系统的核心原理与架构，内容涉及开发、运维、管理与架构。全书共 11 章，第 1 章，介绍 Kafka 体系架构基础，包括消息系统的基本知识、Kafka 的体系架构与 ZooKeeper；第 2 章，介绍 Kafka 的环境部署，以及基本的应用程序开发；第 3 章，介绍 Kafka 的生产者及其运行机制，包括生产者的创建和执行过程、生产者的消息发送模式和生产者的高级特性等；第 4 章，介绍 Kafka 的消费者及其运行机制，包括消费者的消费模式、消费者组与消费者、消费者的偏移量与提交及消费者的高级特性等；第 5 章，介绍 Kafka 服务器端的核心原理，包括主题与分区、消息的持久性与传输保障、Kafka 配额与日志的管理；第 6 章，介绍 Kafka 的流处理引擎 Kafka Stream；第 7 章，介绍使用不同的工具监控 Kafka，包括 Kafka Manager、Kafka Tool、KafkaOffsetMonitor 和 JConsole；第 8 章至第 11 章，介绍 Kafka 与外部系统的集成，包括集成 Flink、集成 Storm、集成 Spark 和集成 Flume。



**图书在版编目（CIP）数据**

Kafka 进阶 / 赵渝强著. —北京：电子工业出版社，2022.1

ISBN 978-7-121-42653-7

Ⅰ. ①K… Ⅱ. ①赵… Ⅲ. ①分布式操作系统 Ⅳ.①TP316.4

中国版本图书馆 CIP 数据核字（2022）第 015166 号

责任编辑：张月萍　　特约编辑：田学清
印　　刷：三河市双峰印刷装订有限公司
装　　订：三河市双峰印刷装订有限公司
出版发行：电子工业出版社
　　　　　北京市海淀区万寿路 173 信箱　　邮编：100036
开　　本：787×980　1/16　印张：18.75　字数：411 千字
版　　次：2022 年 1 月第 1 版
印　　次：2022 年 1 月第 1 次印刷
定　　价：89.00 元

凡所购买电子工业出版社图书有缺损问题，请向购买书店调换。若书店售缺，请与本社发行部联系，联系及邮购电话：（010）88254888，88258888。

质量投诉请发邮件至 zlts@phei.com.cn，盗版侵权举报请发邮件至 dbqq@phei.com.cn。

本书咨询联系方式：010-51260888-819，faq@phei.com.cn。

# 前 言

## 为什么要写这本书

随着信息技术的不断发展，数据出现爆炸式增长。为了实现对大数据实时高效的分析与处理，Kafka 被广泛应用于大数据实时计算架构中。随着数据的不断增长，Kafka 也得到了不断的发展。各大科技巨头在其自身的大数据平台架构中也大量将 Kafka 用于实时数据的存储与转发，如阿里云大数据平台、腾讯大数据平台、华为大数据平台等。因此，掌握 Kafka 无疑是实现大数据实时计算架构中非常重要的一个组成部分。

笔者拥有消息系统 Kafka 多年的教学与实践经验，并在实际 Kafka 运维和开发工作中积累了一些经验，因此想系统地编写一本 Kafka 方面的书，力求能够完整地介绍消息系统 Kafka。本书一方面总结了笔者在 Kafka 方面的经验，另一方面也希望能够对 Kafka 方向的从业者和学习者有所帮助，同时希望给 Kafka 在国内的发展贡献一份力量。相信通过本书的介绍能够让读者全面并系统地掌握 Kafka，并能够在实际工作中灵活地运用 Kafka。

## 本书有何特点

本书将从 Kafka 的基础理论和体系架构出发，为读者全面系统地介绍每个相关知识点。每个实验步骤都经过笔者验证，力求能够帮助读者在学习过程中搭建学习实验的环境，并将其应用在实际工作中。

本书涵盖了 Kafka 中的各个方面，内容涉及体系架构、管理运维和应用开发，全书共 11 章。如果读者有一定的经验，完全可以不按章节顺序，选择比较关注的章节进行阅读；如果读者是零基础，建议按照本书的顺序进行学习，并根据书中的实验步骤进行环境的搭建，相信读者在阅读本书的过程中能够有很大的收获。

## 适合阅读本书的读者

由于 Kafka 消息系统是基于 Scala 语言编写的，而 Scala 语言又构建在 Java 语言之上，因此本书适合具有一定 Java 编程基础的人员阅读，特别适合以下读者。

- ❑ 平台架构师：平台架构师通过阅读本书能够全面和系统地了解 Kafka 体系，提升系统架构的设计能力。
- ❑ 开发人员：基于 Kafka 进行应用开发的开发人员通过阅读本书能够了解 Kafka 消息系统的核心实现原理和编程模型，提升应用开发的水平。
- ❑ 运维管理人员：初、中级的 Kafka 运维管理人员通过阅读本书在掌握 Kafka 架构的基础上能够提升 Kafka 的运维管理经验。

## 阅读本书的建议

由于本书具有很强的实践性，因此读者在阅读本书时，最好采用理论与实践相结合的方式。在阅读内容的同时，动手搭建实验环境并开发相应的应用程序。只有这样才能更好地理解 Kafka 的工作原理和运行机制。

# 目 录

# 第1章　Kafka 体系架构基础

Apache Kafka 是一个用 Scala 语言编写的开源消息系统，通常用于大数据的流式计算平台。Kafka 最早是由 LinkedIn 公司开发的，后来 LinkedIn 公司将 Kafka 贡献给了 Apache 基金会。Apache Kafka 具备高吞吐量、高容错性、可持久性和可水平扩展的特点。正因为 Kafka 在流数据处理领域中的卓越表现，日前受到越来越多的青睐。

下面举例说明 Kafka 的应用场景。

- 消息系统：Kafka 支持 Topic 广播类型的消息，具备高可靠性和容错机制。同时 Kafka 也可以保证消息的顺序性和可追溯性。传统的消息系统很难实现这些消息系统的特性。同时，Kafka 可以将消息持久化保存到磁盘，从而有效地减少数据丢失的风险。从这个角度看，Kafka 也可以看成是一个数据存储系统。
- 大数据流式计算处理：在大数据计算领域中，计算主要分为大数据离线计算和大数据流式计算。可以将流式数据进行实时采集然后缓存到 Kafka 中，从而进一步通过 Storm、Spark Streaming 和 Flink DataStream 进行数据的实时计算。
- 系统的解耦：由于 Kafka 支持异步的消息机制，可以将 Kafka 用于系统的解耦设计。这样，一个系统的架构变动，就不会影响另一个系统的运行。

## 1.1　什么是消息系统

一般我们把消息的发送者称为生产者（Producer），消息的接收者称为消费者（Consumer）；通常生产者的生产速度和消费者的消费速度是不相等的；如果两个程序始终保持同步沟通，则势必会有一方存在空等时间；如果两个程序持续运行，则消费者的平均速度一定要大于生产者的平均速度，否则消息囤积会越来越多；当然，如果消费者没有时效性需求，也可以把消息囤积在消息系统中，集中对其进行消费。

说到这里，我们再来谈谈队列的分类，根据生产者和消费者的作用范围不同，可以把消息系统分为三类。

（1）第一类是在一个应用程序内部（进程之间或线程之间），相信大家学习多线程时都写过生产者程序和消费者程序，生产者负责生产，将生产的结果放到缓冲区（如共享数组），消费者从缓冲区取出消费，在这里，这个缓冲区就可以称为“消息系统”或“消息队列”。

（2）第二类其实也算是第一类的特例，就像我们喜欢把操作系统和应用程序区别对待来看，操作系统要处理无数繁杂的事物，各进程、线程之间的数据交换少不了消息系统的支持。

（3）第三类是通用意义上的“消息系统”，这类系统主要作用于不同的应用，特别是跨机器、平台，这令数据的交换更加广泛，一般一款独立的消息系统除了实现消息的传递，还提供了相应的可靠性、事务、分布式等特性，将生产者、消费者从中解耦。常见的消费系统产品根据开源与否可分为两类。

- 商业的专用软件：Oracle WebLogic、IBM WebSphere MQ 等。
- 开源的软件：ActiveMQ、RabbitMQ、Apache Kafka 等。

在了解了消息系统的基本概念后，我们通过下面的例子来说明一个消息系统的典型应用架构，如图 1.1 所示。

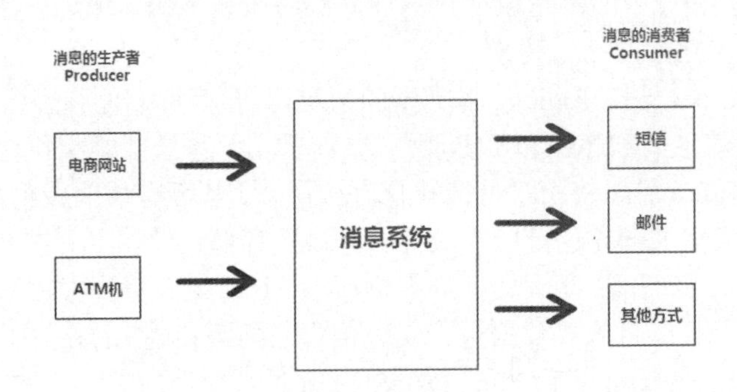


图 1.1　一个典型的消息系统应用

在这个例子中，以一个银行系统为例，当我们在电商网站上进行消费或从银行的 ATM 机上取钱时，银行的消息系统都会发送消息通知我们。这时候，我们就可以把电商网站和 ATM 机看成银行消息系统的消息生产者，当我们消费了银行账户的存款，生产者就会产生一个消息发送到银行的消息系统中，并由该消息系统进行处理，从而通过不同的方式通知我们，例如，短信、邮件或其他方式。我们就可以把这些通知的方式看成银行消息系统的消息消费者，它们负责接收由消息系统转发处理的消息。

一个消息系统的基本组成包括：消息生产者、消息消费者和消息服务器（Broker）。关于它们的具体作用，我们会在后续的章节介绍 Kafka 的体系架构时，再进行详细说明。

## 1.2　消息系统的分类

消息系统的消息通信有两类：具有依时性的同步消息机制及与时间无关的异步消息机制。消息传送中间件有许多不同的类型，它们分别能够支持一类基本方式的消息通信，有时可以支持两类方式。

从消息的传递方式上看，我们又可以把消息系统中的消息划分为队列（Queue）和主题（Topic）。Kafka 支持传递 Topic 类型的消息。

## 1.2.1　同步消息机制与异步消息机制

### 1．同步消息机制

两个通信服务之间必须进行同步，而且两个通信服务必须都是正常的并一直处于运行状态的，随时做好通信准备，发送程序在向接收程序发送消息后，阻塞自身与其他应用的通信进程，等待接收程序返回消息，然后继续执行下一个业务。

图 1.2 展示了一个同步消息系统的典型架构。

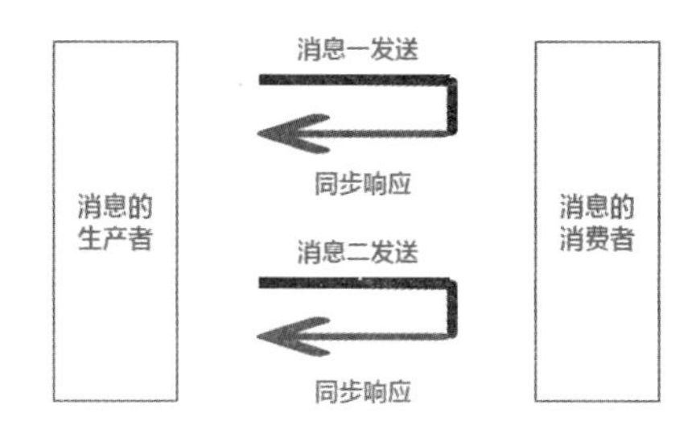


图 1.2　同步消息系统

### 2．异步消息机制

两个通信应用之间可以不用同时在线等待，任何一方只处理自己的业务而不用等待对方的响应。发送程序在向接收程序发送消息后，不用等待接收程序的返回消息，就可以继续执行下一个业务。

图 1.3 展示了一个异步消息系统的典型架构。

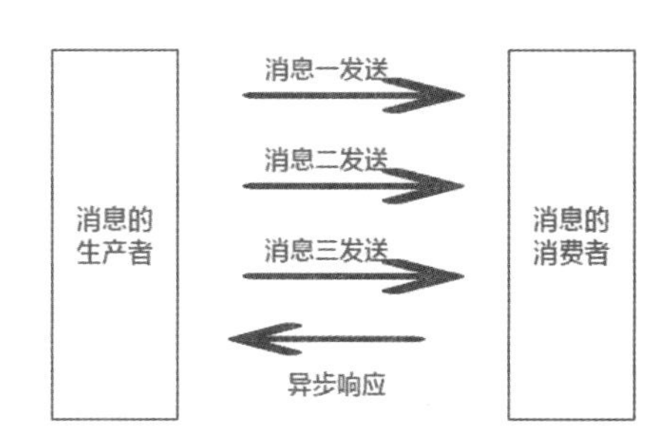


图 1.3　异步消息系统

下面通过具体的例子来说明同步消息机制和异步消息机制的区别。

首先，我们来看同步消息机制的一个例子。在付款的时候，如果 A 已经扫码付款了，没有收到支付成功的状态提示，就会想自己是否已经支付成功了呢？A 就会一直处于等待状态，直到系统反馈一个消息，要么是支付成功要么是支付失败才会进行后续的操作，这就是同步消息机制。

那么什么是异步消息机制呢？比如 A 给 B 发送一封电子邮件，A 不需要知道 B 是否收到了，A 只是把自己的信息传达出去，这样的场景就是异步消息。因为在这个过程中，A 在乎的是把某件事情传达出去，而不必在乎其他人的状态，比如张贴告示也是这样，不需要知道每个人都是否知道这则告示的内容，而是张贴出去就可以了。

在了解了同步消息机制和异步消息机制以后，它们各自有什么样的优点和缺点呢？异步消息传递有一些关键优势。它能够提供灵活性和更高的可用性。系统对信息采取行动的压力较小，或者以某种方式做出响应。另外，一个系统被关闭不会影响另一个系统。例如，你可以发送数千封电子邮件给你的朋友，而不需要他们回复你。

异步消息传递的缺点是缺乏直接性，没有直接的相互作用。思考一下你与你的朋友在即时通话或视频聊天，除非你的朋友及时回复你，否则这不是即时通话或视频聊天，这正是同步消息传递的优点。另一方面，异步消息传递允许更多的并行性。由于进程不阻塞，所以它可以在消息传输时进行一些计算。

## 1.2.2 队列与主题

队列类型的消息也称点对点的消息传递。通过该消息传递模型，一个应用程序（即消息生产者）可以向另外一个应用程序（即消息消费者）发送消息。在此传递模型中，消息目的地类型是队列。消息首先被传送至消息服务器端的特定队列中，然后从此对列中将消息传送至对此队列进行监听的某个消费者。图 1.4 展示了一个典型的队列类型的消息系统。

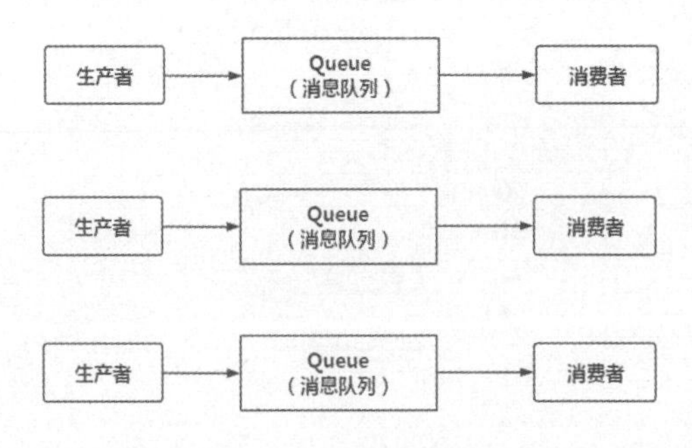


图 1.4　队列消息系统

主题类型的消息也称发布与订阅的消息传递。通过该消息传递模型，应用程序能够将一条消息发送给多个消息消费者。在此传送模型中，消息目的地类型是主题。消息首先由消息生产者发布至消息服务器中的特定主题，然后由消息服务器将消息传送至所有已订阅此主题的消费者。在该模型中，消息会自动广播，消息消费者无须通过主动请求或轮询主题的方法来获得新的消息。Kafka 支持传递 Topic 类型的消息。

图 1.5 展示了一个典型的主题类型的消息系统。

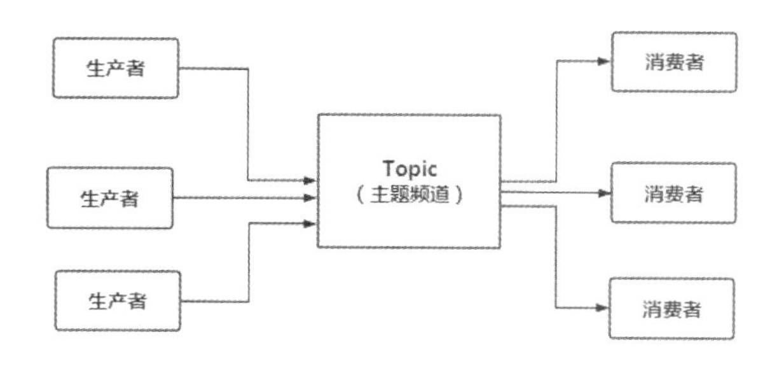


图 1.5　主题消息系统

# 1.3　Kafka 的体系架构

学习 Kafka 消息系统，十分重要的是掌握它的核心架构及其运行机制。本节将详细介绍 Kafka 的核心架构、基本概念，并在此基础上继续介绍 Kafka 的安装与部署。

Kafka 消息系统是一个典型的分布式系统，其组成部分包括：消息生产者、消息消费者、消息服务器及分布式协调服务 ZooKeeper。一个典型的 Kafka 消息系统的集群架构如图 1.6 所示。

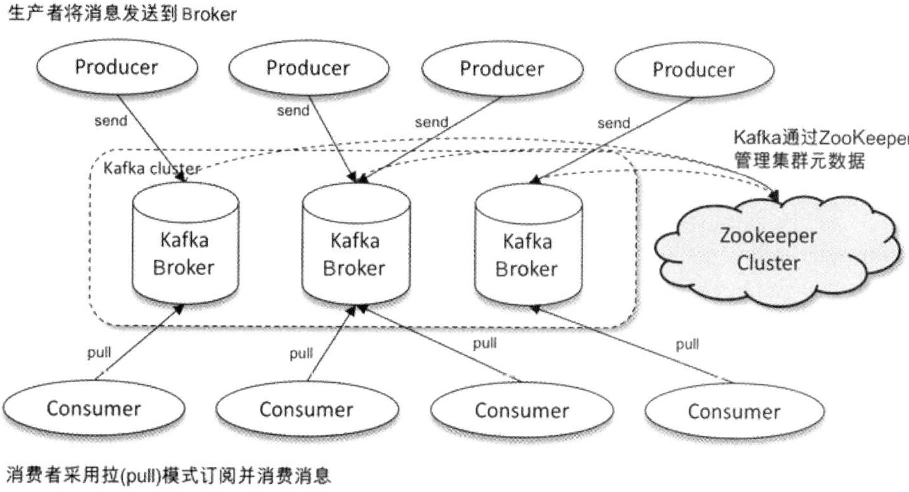

图 1.6　Kafka 消息系统的集群架构

下面我们来列举一些 Kafka 中的术语，这些术语对我们学习并掌握 Kafka 的内容非常重要。

- Broker：Kafka 集群包含一个或多个服务器，这种服务器被称为 Broker。
- Topic：每条发布到 Kafka 集群的消息都有一个类别，这个类别被称为 Topic（可以理解为队列或目录）。物理上不同 Topic 的消息分开存储，逻辑上一个 Topic 的消息虽然保存于一个或多个 Broker，但用户只需指定消息的 Topic 即可生产或消费数据而不必关心数据存于何处。
- Partition（分区）：Partition 是物理上的概念（可以理解为文件夹），每个 Topic 包含一个或多个 Partition，即同一个分区可能存在多个副本。
- Producer：它负责将消息发布到 Kafka Broker。
- Consumer：它向 Kafka Broker 读取消息的客户端。
- Consumer Group（消费者组）：每个 Consumer 属于一个特定的 Consumer Group（可为每个 Consumer 指定 Group Name，若不指定 group name 则属于默认的 Group）。

下面我们分别对这些术语进行介绍。

## 1.3.1 消息服务器

Broker 是消息的代理，Producers 向 Broker 的指定 Topic 写消息，Consumers 从 Broker 拉取指定 Topic 的消息，然后进行业务处理，Broker 在中间起到一个代理保存消息的中转站的作用。

Broker 没有副本机制，一旦 Broker 宕机，该 Broker 的消息将都不可用。消费者可以回溯到任意位置重新从 Broker 进行消息的消费，当消费者发生故障时，可以选择最小的 offset(id)重新读取消费者消息。

关于 Broker 的详细内容，会在后续的章节中进一步介绍。

## 1.3.2 主题、分区与副本

Kafka 中的消息以主题为单位进行归类，生产者负责将消息发送到特定的主题，而消费者负责订阅主题进行消费；主题可以分为多个分区，一个分区只属于单个主题。下面列举一下主题和分区的关系。

（1）同一主题下的不同分区包含的消息不同（发送给主题的消息具体是发送到某一个分区的）。

（2）消息被追加到分区日志文件的时候，会分配一个特定的偏移量（offset），offset 是消息在分区中的唯一标识，Kafka 通过它来保证消息在分区的顺序性。

（3）offset 不跨分区，也就是说 Kafka 保证的是分区有序而不是主题有序。

图 1.7 展示了主题与分区的关系。

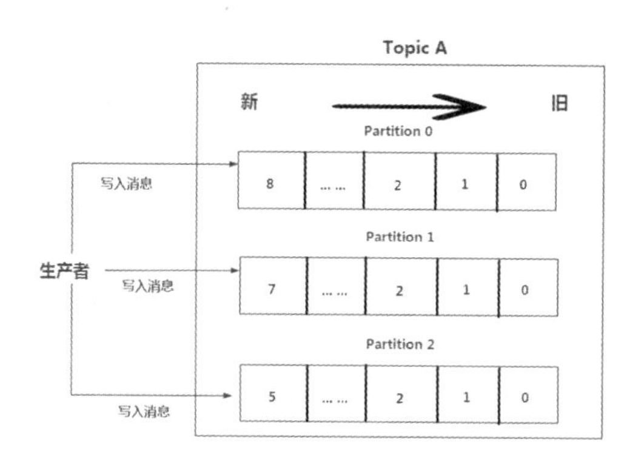


图 1.7　主题与分区的关系

图 1.7 中的 Topic A 有 3 个分区。消息由生产者顺序追加到每个分区日志文件的尾部。Kafka 中的分区可以分布在不同的 Kafka Broker 上，从而支持负载均衡和容错的功能。也就是说，Topic 是一个逻辑单位，它可以横跨在多个 Broker 上。

了解了主题和分区后，再来了解副本。在 Kafka 中，每个主题可以有多个分区，每个分区又可以有多个副本。在这些副本中，只有一个是 Leader 副本，其他都是 Follower 副本。仅有 Leader 副本可以对外提供服务。多个 Follower 副本通常存放在和 Leader 副本不同的 Broker 中。通过这样的机制实现了高可用，当某台机器挂掉后，其他 Follower 副本也能迅速“转正”，开始对外提供服务，这就是 Kafka 的容错功能。

图 1.8 展示了 Kafka 分区的副本机制。

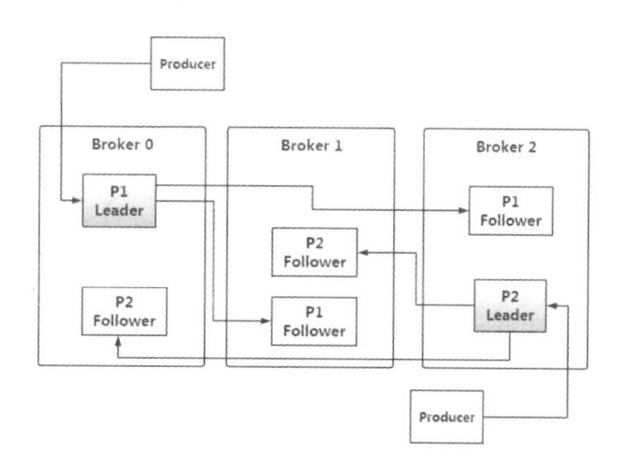


图 1.8　Kafka 分区的副本机制

在图 1.8 中，我们创建一个 Topic，这个 Topic 由两个分区组成：P1 和 P2。可以看出，每个分区有三个副本，由前面的介绍可知，每个分区都将由 Leader 副本负责对外提供服务。

## 1.3.3　生产者

生产者将消息序列化之后，发送到对应 Topic 的指定分区。整个生产者客户端由两个线程协调运行，这两个线程分别为主线程和 Sender 线程（发送线程）。生产者有三种发送消息的方式。实际上，生产者发送的动作都是一致的，不由使用者决定。这三种方式的区别在于是否正常处理到达的消息。生产者的三种消息发送方式如下。

**1. fire-and-forget**

把消息发送给 Broker 之后不关心其是否正常到达。在大多数情况下，消息会正常到达，即使出错了，生产者也会自动重试。但是如果出错了，对于我们的服务而言，是无感知的。这种方式适用于可丢失消息、对吞吐量要求大的场景，比如用户单击日志上报。

**2. 同步发送**

我们使用 send 方法发送一条消息，它会返回一个 Future，调用 get 方法可以阻塞当前的线程，等待返回。这种方式适用于对消息可靠性要求高的场景，比如支付，要求消息不可丢失，如果丢失了则阻断业务（或回滚）。

**3. 异步发送**

使用 send 方法发送一条消息时指定回调函数，在 Broker 返回结果时调用。这个回调函数可以进行错误日志的记录或重试。这种方式牺牲了一部分可靠性，但是吞吐量会比同步发送高很多。可以通过后续的补偿操作弥补业务。

关于生产者，我们会在后续的章节中继续进行详细的介绍。

### 1.3.4 消费者与消费者组

顾名思义，消费者就是从 Kafka 集群消费数据的客户端，展示了一个消费者从一个 Topic 中消费数据的模型。单个消费者模型如图 1.9 所示。

单个消费者模型存在一些问题，如果 Kafka 上游生产的数据很快，超过了单个消费者的消费速度，就会导致数据堆积，那么如何去解决这样的问题呢？我们要加强消费者的消费数据的能力，因此就有了下面要讲的消费者组。

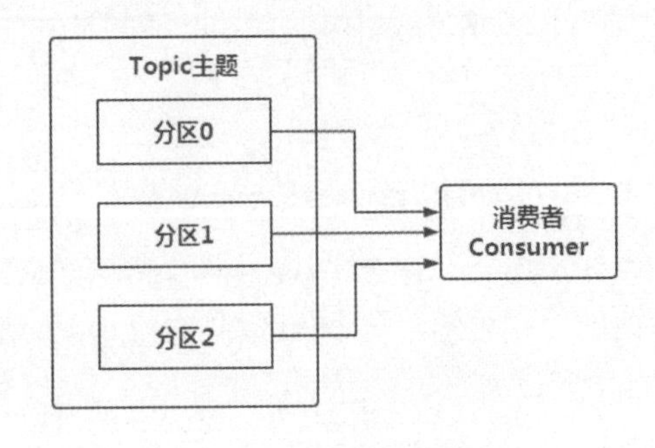


图 1.9　单个消费者模型

所谓消费者组，其实就是一组消费者的集合，图 1.10 解释了什么是消费者组。我们采用了一个消费者组来消费这个 Topic，其消费能力是按倍数递增的，所以一般来说都是采用消费者组来消费数据，而不会采用单个消费者来消费数据。

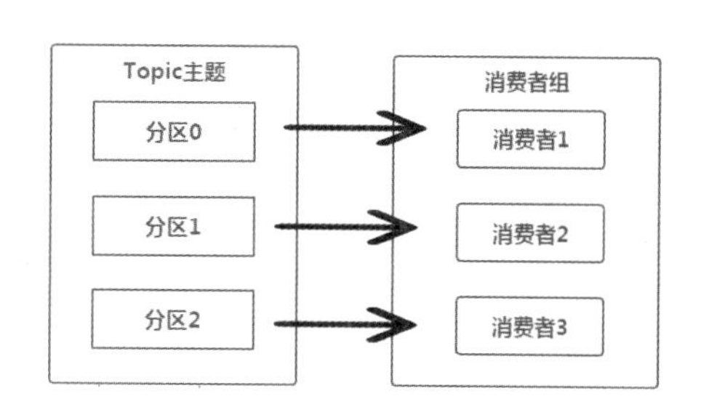


图 1.10　消费者组

# 1.4　分布式协调服务 ZooKeeper

ZooKeeper 是一个开放源码的分布式应用协调服务，是 Google Chubby 的一个开源的实现，是 Hadoop 和 HBase 的重要组件。它是一个为分布式应用提供一致性服务的软件，提供的功能包括：配置维护、域名服务、分布式同步、组服务等。那么 ZooKeeper 能帮我们做什么？

例如，Hadoop 使用 ZooKeeper 的事件处理确保整个集群只有一个活跃的 NameNode 和存储配置信息等，从而实现系统的高可用性；HBase 使用 ZooKeeper 的事件处理确保整个集群只有一个 HMaster，察觉 HRegionServer 联机和宕机，存储访问控制列表等，从而也实现了 HBase 的高可用性。

## 1.4.1　ZooKeeper 集群的架构

图 1.11 展示了一个典型的 ZooKeeper 集群的架构。

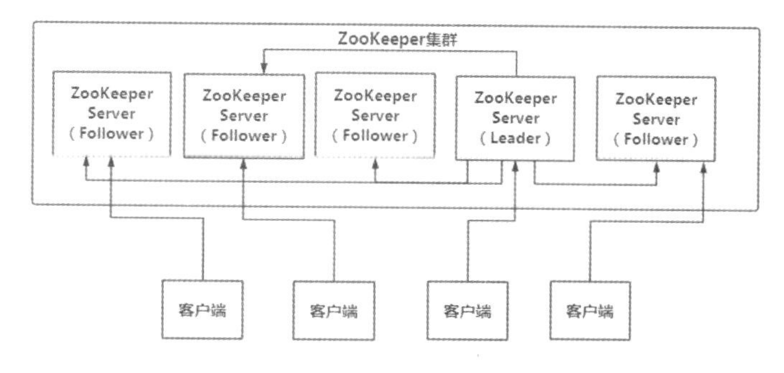


图 1.11　ZooKeeper 集群的架构

客户端可以连接到每个 Server，每个 Server 的数据是完全相同的，每个 Follower 和 Leader 都有连接，接受 Leader 的数据更新操作（并将 Leader 更新的数据同步到 Follower 中），实现数据的同步和一致性。Server 记录事务日志和快照到持久存储的过程。如果 ZooKeeper 集群中过半数的 Server 可用，则整体服务就可以使用，Leader 只有一个，宕机之后，就会重新选择出一个 Leader。

第 2 章将会介绍 ZooKeeper 集群的部署与安装。

### 1.4.2 ZooKeeper 的节点类型

ZooKeeper 节点的类型有两个维度，一个维度是否永久，另一个维度是否有序。两个维度组合成的四种类型如下。

#### 1. PERSISTENT 持久化节点

持久化节点是指在节点创建后就一直存在，直到删除操作主动清除这个节点。否则不会因为创建该节点的客户端会话失效而消失。

#### 2. PERSISTENT_SEQUENTIAL 持久顺序节点

这类节点的基本特性和持久化节点类型是一致的。其额外的特性是，在 ZooKeeper 中，每个父节点会为其第一级子节点维护一份时序，记录每个子节点创建的先后顺序。在创建节点的过程中，ZooKeeper 会自动为给定节点名加上一个数字后缀，作为新的节点名。这个数字后缀的范围是整型的最大值。在创建节点的时候只需要传入节点“/test_”，ZooKeeper 会自动给“test_”后面补充数字。

#### 3. EPHEMERAL 临时节点

和持久化节点不同的是，临时节点的生命周期和客户端会话绑定。也就是说，如果客户端会话失效，那么这个节点就会自动被清除掉。注意，这里提到的是会话失效，而不是连接断开。另外，在临时节点下面不能创建子节点。这里还要注意的就是在客户端会话失效后，所产生的节点也不会立即消失，也要过一段时间，大概是 10 秒以内，可以试一下本机操作生成节点，在服务器端用命令来查看当前的节点数目，你会发现客户端已经停止，但是产生的节点还在。

#### 4. EPHEMERAL_SEQUENTIAL 临时自动编号节点

此节点属于临时节点，并且带有顺序，客户端会话结束节点就消失。

在 Java 程序中，可以搭建 Maven 工程来操作 ZooKeeper，下面给出了 Maven 工程 pom 文件中的依赖信息。

```
01  <dependencies>
02      <dependency>
03          <groupId>junit</groupId>
```

```
04              <artifactId>junit</artifactId>
05              <version>3.8.1</version>
06              <scope>test</scope>
07          </dependency>
08
09          <dependency>
10              <groupId>org.apache.curator</groupId>
11              <artifactId>curator-framework</artifactId>
12              <version>4.0.0</version>
13          </dependency>
14
15          <dependency>
16              <groupId>org.apache.curator</groupId>
17              <artifactId>curator-recipes</artifactId>
18              <version>4.0.0</version>
19          </dependency>
20
21          <dependency>
22              <groupId>org.apache.curator</groupId>
23              <artifactId>curator-client</artifactId>
24              <version>4.0.0</version>
25          </dependency>
26          <dependency>
27              <groupId>org.apache.zookeeper</groupId>
28              <artifactId>zookeeper</artifactId>
29              <version>3.4.6</version>
30          </dependency>
31
32          <dependency>
33              <groupId>com.google.guava</groupId>
34              <artifactId>guava</artifactId>
35              <version>16.0.1</version>
36          </dependency>
37
38      </dependencies>
```

下面的 Java 代码示例创建了不同类型的 ZooKeeper 节点。其中第 12 行代码连接的是 ZooKeeper 集群中的一个节点，这里也可以连接 ZooKeeper 集群。如果要连接 ZooKeeper 集群，将集群中的节点用逗号分隔。

```
01  import org.apache.curator.RetryPolicy;
02  import org.apache.curator.framework.CuratorFramework;
03  import org.apache.curator.framework.CuratorFrameworkFactory;
04  import org.apache.curator.retry.ExponentialBackoffRetry;
05  import org.apache.zookeeper.CreateMode;
```

```
06
07  public class ZooKeeperDemo {
08
09      public static void main(String[] args) throws Exception {
10          RetryPolicy policy = new ExponentialBackoffRetry(1000, 10);
11          CuratorFramework cf = CuratorFrameworkFactory.builder()
12                              .connectString("kafka101:2181")
13                              .retryPolicy(policy)
14                              .build();
15
16          cf.start();
17
18          //PERSISTENT 持久化节点，默认
19          cf.create().forPath("/path01");
20
21          //PERSISTENT_SEQUENTIAL 持久化顺序节点
22          //顺序自动编号持久化节点，这种节点会根据当前已存在的节点数自动加 1
23          cf.create().withMode(CreateMode.PERSISTENT_SEQUENTIAL).
forPath("/node-");
24          cf.create().withMode(CreateMode.PERSISTENT_SEQUENTIAL).
forPath("/node-");
25          cf.create().withMode(CreateMode.PERSISTENT_SEQUENTIAL).
forPath("/node-");
26
27          //EPHEMERAL 临时节点
28          //客户端 session 超时，这类节点就会被自动删除
29          cf.create().withMode(CreateMode.EPHEMERAL).forPath("/path03");
30
31          //EPHEMERAL_SEQUENTIAL 临时自动编号节点
32          cf.create().withMode(CreateMode.EPHEMERAL_SEQUENTIAL).
forPath("/temp-");
33          cf.create().withMode(CreateMode.EPHEMERAL_SEQUENTIAL).
forPath("/temp-");
34          cf.create().withMode(CreateMode.EPHEMERAL_SEQUENTIAL).
forPath("/temp-");
35
36          //由于存在临时节点，这行代码是为了看到效果的
37          Thread.sleep(10000);
38
39          cf.close();
40      }
41  }
```

## 1.4.3 ZooKeeper 的观察机制

ZooKeeper 作为一款成熟的分布式协调框架，其观察机制是很重要的。观察者会订阅一些感兴趣的主题，这些主题一旦发生变化，就会自动通知这些观察者。ZooKeeper 的观察机制是一个轻量级的设计。它采用了一种推拉结合的模式。

一旦服务器端感知主题发生变化，那么只会给关注的客户端发送一个事件类型和节点信息，而不会发送具体的变更内容，所以事件本身是轻量级的，这就是所谓的“推”部分。然后，收到变更通知的客户端需要自己去拉变更的数据，这就是“拉”部分。

Kafka 将集群信息注册到 ZooKeeper 集群中，并由 ZooKeeper 进行观察和监听。当前 Kafka 集群发生了变化，例如，某个 Broker 宕机或出现故障，这些信息都会被 ZooKeeper 感知，从而进行相应的处理。

下面的代码示例展示了 ZooKeeper 的观察机制。在这里的示例中，我们监听 ZooKeeper 节点/testwatcher 的变化。

```
01  import org.apache.curator.RetryPolicy;
02  import org.apache.curator.framework.CuratorFramework;
03  import org.apache.curator.framework.CuratorFrameworkFactory;
04  import org.apache.curator.framework.recipes.cache.NodeCache;
05  import org.apache.curator.framework.recipes.cache.NodeCacheListener;
06  import org.apache.curator.framework.recipes.cache.PathChildrenCache;
07  import org.apache.curator.framework.recipes.cache.PathChildrenCache.StartMode;
08  import org.apache.curator.framework.recipes.cache.PathChildrenCacheEvent;
09  import org.apache.curator.framework.recipes.cache.PathChildrenCacheListener;
10  import org.apache.curator.retry.ExponentialBackoffRetry;
11
12  public class ZKWatcher {
13
14      public static void main(String[] args) throws Exception {
15          RetryPolicy policy = new ExponentialBackoffRetry(1000, 10);
16          CuratorFramework cf = CuratorFrameworkFactory.builder()
17                              .connectString("192.168.157.31:2181")
18                                  .retryPolicy(policy)
19                                .build();
20          cf.start();
21          //监听该目录
22          cf.create().forPath("/testwatcher");
23
24         PathChildrenCache pathChildrenCache =
25                  new PathChildrenCache(cf,"/testwatcher",true);
26
27         //注册监听器
28         pathChildrenCache.getListenable()
```

```
29                        .addListener(new PathChildrenCacheListener() {
30
31           public void childEvent(CuratorFramework client, PathChildrenCacheEvent
event)
32                            throws Exception {
33               switch (event.getType()){
34               case CHILD_ADDED:
35                   System.out.println("新增子节点:" + event.getData().getPath());
36                    break;
37               case CHILD_UPDATED:
38               System.out.println("子节点数据变化: " + event.getData().getPath());
39                    break;
40               case CHILD_REMOVED:
41                   System.out.println("删除子节点:" + event.getData().getPath());
42                    break;
43               default:break;
44               }
45            }
46          });
47          pathChildrenCache.start();
48
49          //测试监听机制
50          Thread.sleep(1000);
51          cf.create().forPath("/testwatcher/childnode");
52          Thread.sleep(1000);
53         cf.setData().forPath("/testwatcher/childnode","Hello World".getBytes());
54          Thread.sleep(1000);
55          cf.delete().forPath("/testwatcher/childnode");
56          Thread.sleep(1000);
57          cf.delete().forPath("/testwatcher");
58
59          pathChildrenCache.close();
60          cf.close();
61      }
62  }
```

### 1.4.4 ZooKeeper 的分布式锁

可以利用 Zookeeper 不能重复创建一个节点的特性来实现一个分布式锁，这看起来和 Redis 实现分布式锁很像，但也是有差异的。因为 ZooKeeper 中分布式锁的本质是一个临时节点。利用 ZooKeeper 的分布式特性，可以实现一个“秒杀”的场景。图 1.12 展示了基于 ZooKeeper 的“秒杀”系统。

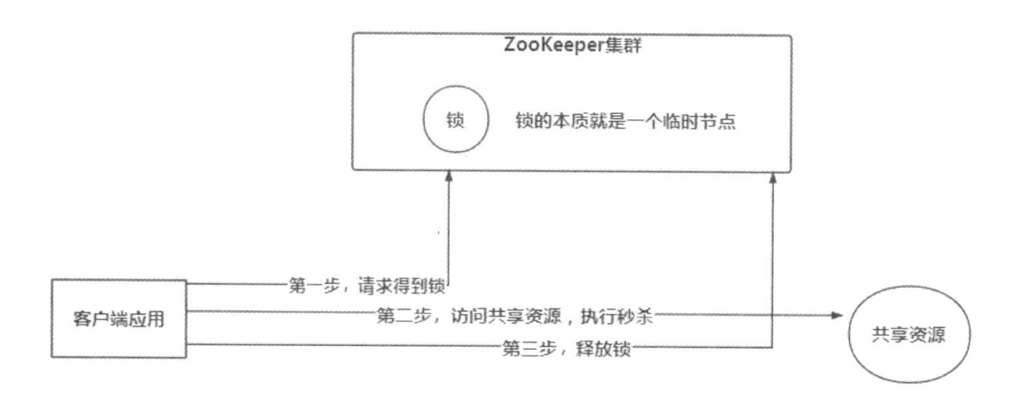


图 1.12　基于 ZooKeeper 的“秒杀”系统

在这个系统中，当客户端应用程序要访问共享的资源并执行“秒杀”时，首先请求 ZooKeeper 集群得到相应的锁；得到锁的客户端应用才能进行第二步操作，访问共享资源，从而执行“秒杀”。如果客户端没有在第一步得到锁的信息，当前线程就会被阻塞，直到成功请求到锁的信息。由于在目前的架构中，在 ZooKeeper 集群中只定义了一把锁，在同一个时刻，只有一个客户端可以成功请求到锁，其他客户端就需要等待。相当于在我们当前的秒杀系统中，同一个时刻只支持一个客户端的“秒杀”。当前客户端执行完“秒杀”后，需要将锁的信息释放，并还给 ZooKeeper 集群，以便后续的客户端能够继续请求锁的信息。

下面的代码完整地展示了这个过程。

```
01  import org.apache.curator.RetryPolicy;
02  import org.apache.curator.framework.CuratorFramework;
03  import org.apache.curator.framework.CuratorFrameworkFactory;
04  import org.apache.curator.framework.recipes.locks.InterProcessMutex;
05  import org.apache.curator.retry.ExponentialBackoffRetry;
06
07  public class TestDistributedLock {
08
09      private static int number = 10;
10      private static void getNumber(){
11          System.out.println("\n\n******* 开始秒杀方法 **************");
12          System.out.println("当前值: " + number);
13          number --;
14
15          try {
16              Thread.sleep(2000);
17          } catch (InterruptedException e) {
18              // TODO Auto-generated catch block
```

```
19              e.printStackTrace();
20          }
21          System.out.println("******* 结束秒杀方法 **************\n\n");
22      }
23
24      public static void main(String[] args) {
25          //定义每次重试的机制
26          RetryPolicy policy = new ExponentialBackoffRetry(1000, 10);
27          CuratorFramework cf = CuratorFrameworkFactory.builder()
28                              .connectString("kafka101:2181")
29                              .retryPolicy(policy)
30                              .build();
31
32          cf.start();
33          //定义 ZooKeeper 的锁
34          final InterProcessMutex lock = new InterProcessMutex(cf, "/mylock");
35
36          //启动 10 个客户端来模拟秒杀的场景
37          for(int i=0;i<10;i++){
38              new Thread(new Runnable() {
39
40                  public void run() {
41
42                      try {
43                          lock.acquire();//请求锁的信息
44
45                          getNumber();//执行秒杀的业务逻辑
46                      } catch (Exception e) {
47                          //打印例外消息
48                          e.printStackTrace();
49                      }finally{
50                          try {
51                              lock.release(); //释放锁的信息
52                          } catch (Exception e) {
53                              // TODO Auto-generated catch block
54                              e.printStackTrace();
55                          }
56                      }
57                  }
58              }).start();
59          }
60      }
61  }
```

### 1.4.5　ZooKeeper 在 Kafka 中的作用

Kafka 使用 ZooKeeper 来实现动态的集群扩展，不需要更改客户端（生产者和消费者）的配置。Broker 会在 ZooKeeper 中注册并保持相关的元数据（Topic、Partition 信息等）更新。客户端会在 ZooKeeper 上注册相关的 Watcher。一旦 ZooKeeper 发生变化，客户端能及时感知并作出相应调整。这样就保证了在添加或去除 Broker 时，各 Broker 间仍能自动实现负载均衡。Broker 端使用 ZooKeeper 来注册 Broker 信息，以及监测 Partition Leader 的存活性。

从生产者和消费者的角度来看，消费者端使用 ZooKeeper 来注册消费者消费的信息，其中包括消费者消费的 Partition 列表等，同时也用来发现 Broker 列表，并和 Partition Leader 建立 socket 连接，并获取消息；ZooKeeper 和生产者之间没有建立关系，只和 Broker、消费者建立关系以实现负载均衡，即同一个消费者组中的消费者可以实现负载均衡。

## 1.5　准备实验环境

我们来了解一下即将用到的实验环境，由于整个环境都需要在 Linux 上进行部署，因此采用 CentOS 作为实验环境，具体如下。

- 操作系统：CentOS 7 64 位。
- JDK 版本：jdk-8u181-linux-x64.tar.gz。
- ZooKeeper 版本：zookeeper-3.4.10.tar.gz。
- Kafka 版本：kafka_2.11-2.4.0.tgz。
- Kafka Manager 版本：kafka-manager-2.0.0.2.zip。
- VMware Workstation。

将部署 3 台 CentOS 的主机，分别是 kafka101、kafka102、kafka103，它们将共同组成一个分布式 Kafka 消息的环境，其架构拓扑图如图 1.13 所示。

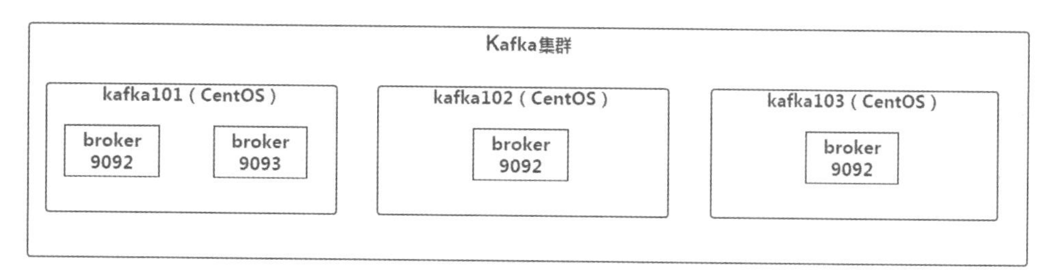


图 1.13　Kafka 集群的部署架构

### 1.5.1　安装 CentOS 操作系统

（1）在 VMware Workstation 对话框中单击“文件”→“新建虚拟机”命令；并选择“自

定义”方式进行安装，如图 1.14 和图 1.15 所示。

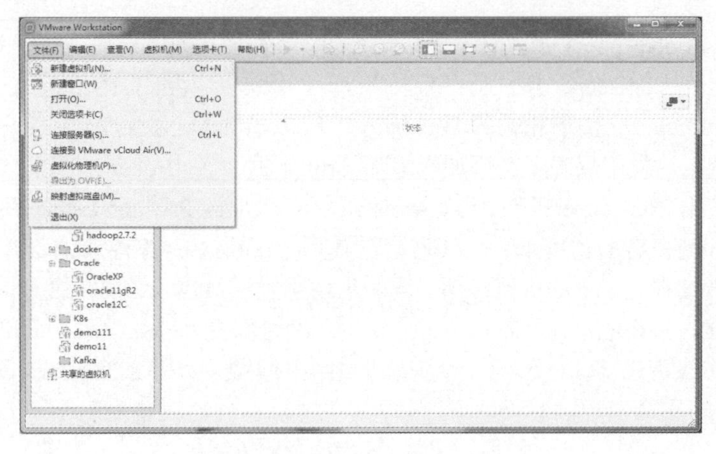


图 1.14　新建虚拟机

（2）在“新建虚拟机向导”界面单击“下一步”按钮；并在“安装客户机操作系统”界面选择“稍后安装操作系统”单选按钮，如图 1.16 所示。

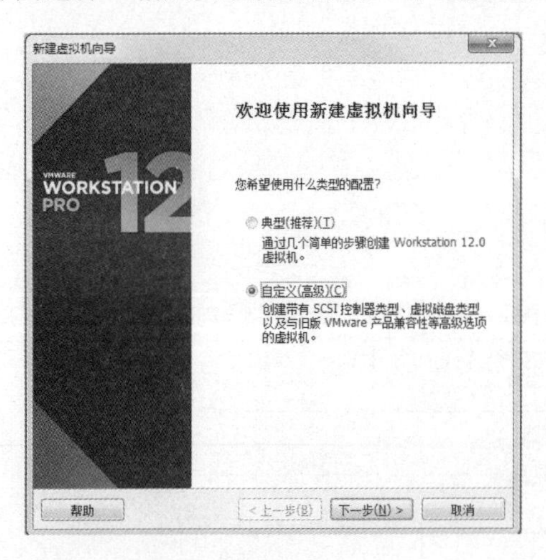


图 1.15　“自定义”虚拟机

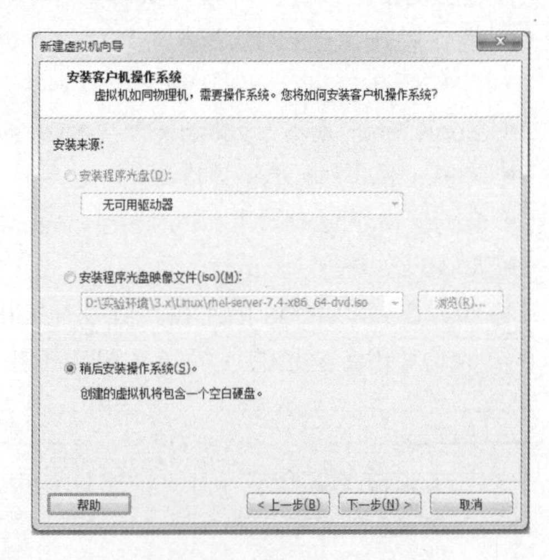


图 1.16　安装客户机操作系统

（3）在“选择客户机操作系统”界面选择“Linux”单选按钮，在“版本”下拉列表中选择“CentOS 64 位”选项，这一步非常重要，如果选择错误，可能造成虚拟机无法正常启动，如图 1.17 所示。

（4）在“命名虚拟机”界面上，输入虚拟机名称，如 kafka101。后续安装的 kafka102 和 kafka103 按照类似的方式设置即可，如图 1.18 所示。

图 1.17　选择客户机操作系统

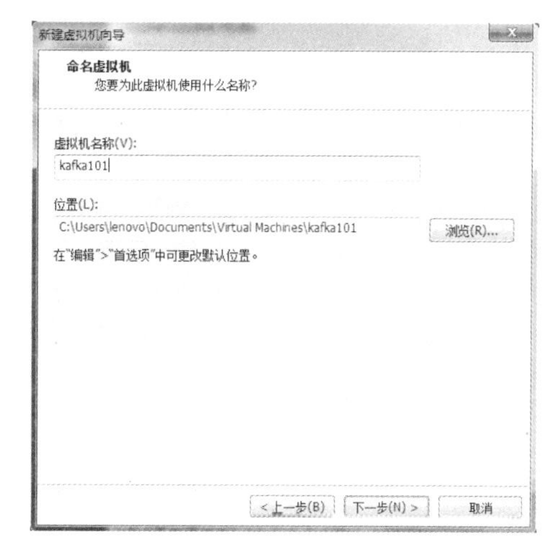


图 1.18　命名虚拟机

（5）连续单击两次“下一步”按钮，直到弹出“此虚拟机的内存”界面。默认的内存设置是 1024MB，即 1GB 内存。这里可以根据自己机器的配置，适当增大虚拟机的内存设置，例如，可以将其修改为 2048MB，即 2GB 内存，如图 1.19 所示。

（6）单击“下一步”按钮，进入“网络类型”界面，这一步非常重要。在实际的生产环境中，通常不能直接访问外网，需要多台主机组成一个集群，集群之间还可以进行相互通信。为了模拟这样一个真实的网络环境，推荐选择“使用仅主机模式网络”单选按钮，如图 1.20 所示。

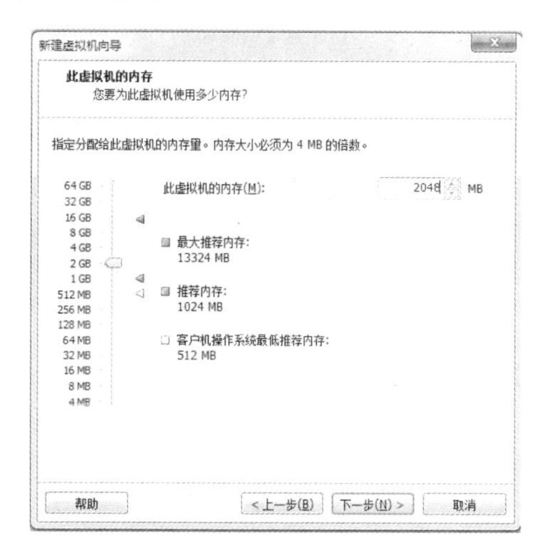


图 1.19　此虚拟机的内存

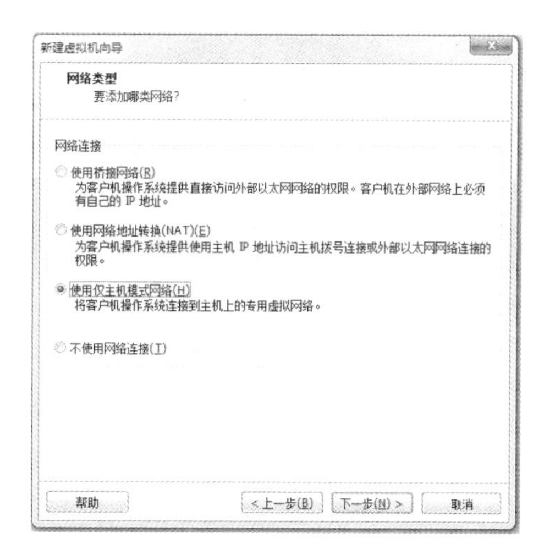


图 1.20　网络类型

选择这样的网络模式后，首先虚拟机不能直接访问外部的网络；其次，如果是一个分布式环境，例如，kafka101、kafka102 和 kafka103 可以使用这样的网络模式保证它们彼此之间可以进行通信。

（7）连续单击四次“下一步”按钮，进入“指定磁盘容量”界面。这里可以根据自己的硬盘的大小进行适当调整，如图 1.21 所示。

（8）连续单击两次“下一步”按钮，进入“已准备好创建虚拟机”界面。单击“完成”按钮，如图 1.22 所示。

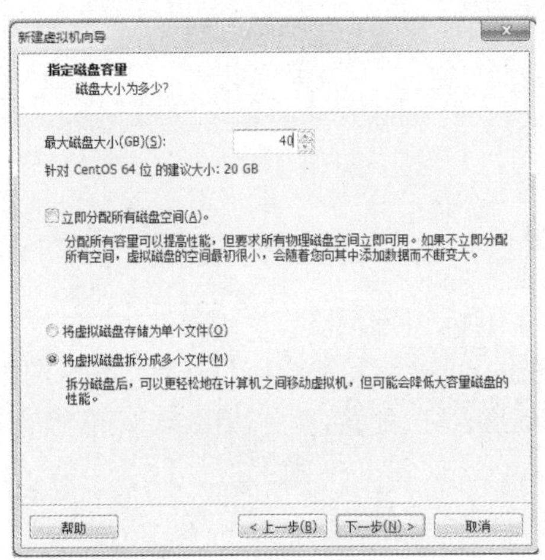


图 1.21　指定磁盘容量

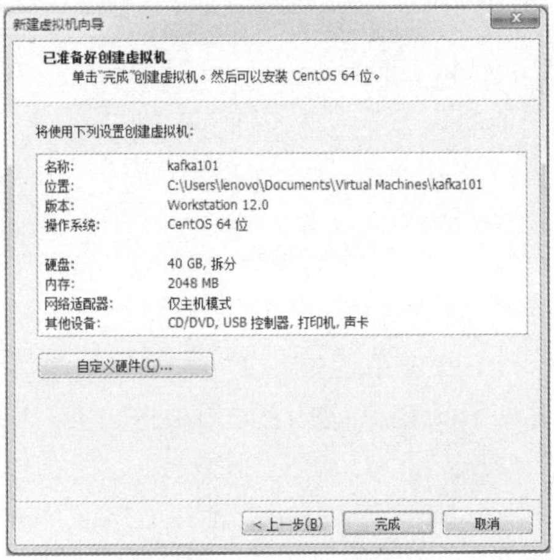

图 1.22　完成虚拟机创建

（9）在 kafka101 的主界面上，选择“编辑虚拟机设置”选项，在“设备”下拉列表中选择“CD/DVD”选项，并将 CentOS 7 的 ISO 介质加载到映像文件的选项中，如图 1.23 和图 1.24 所示。

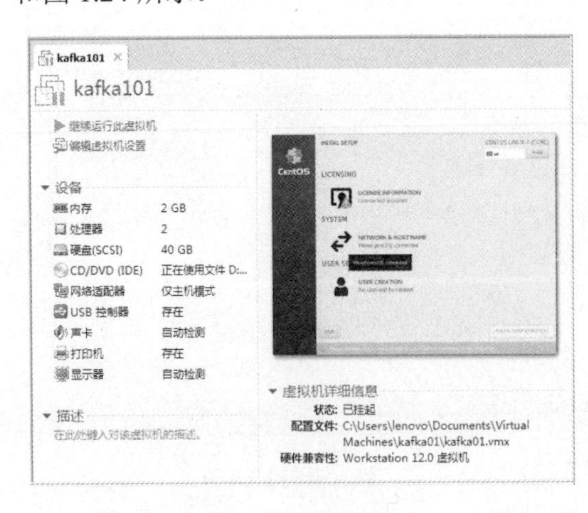


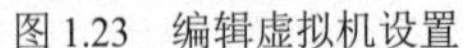
图 1.23　编辑虚拟机设置

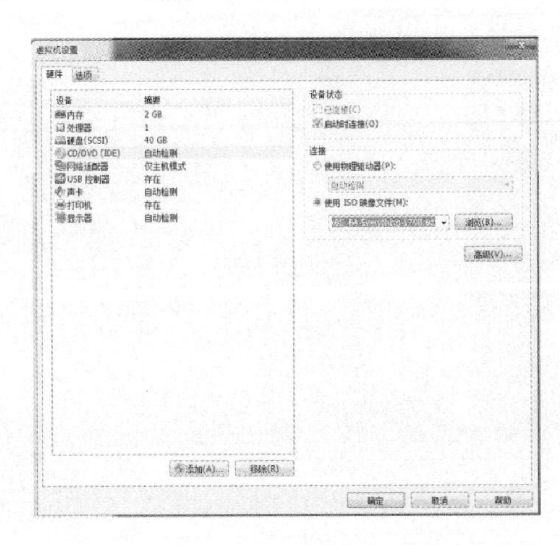

图 1.24　加载 ISO 文件

（10）单击“确定”按钮，并勾选“开启此虚拟机”复选框，等待虚拟机启动，选择“Install CentOS 7”选项，系统开始安装，如图 1.25 所示。

图 1.25　开始安装 CentOS

（11）在欢迎界面单击“Continue”按钮，进入“INSTALLATION SUMMARY”界面，在这个界面进行相关的配置，如图 1.26 和图 1.27 所示。

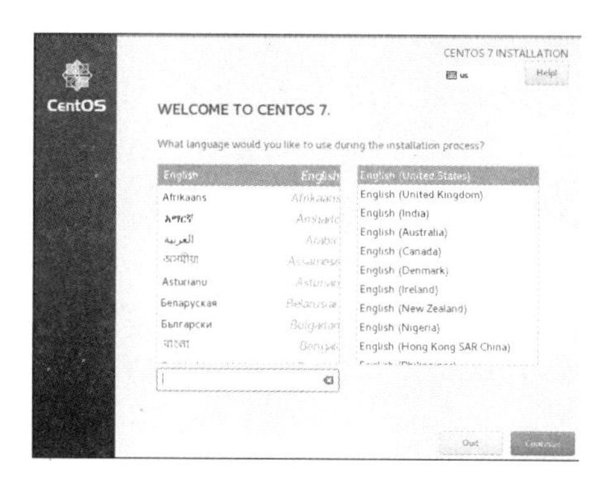


图 1.26　CentOS 欢迎界面

（12）在“SOFTWARE SELECTION”界面选择“Server With GUI”单选按钮和“Development Tools”复选框，如图 1.28 所示。

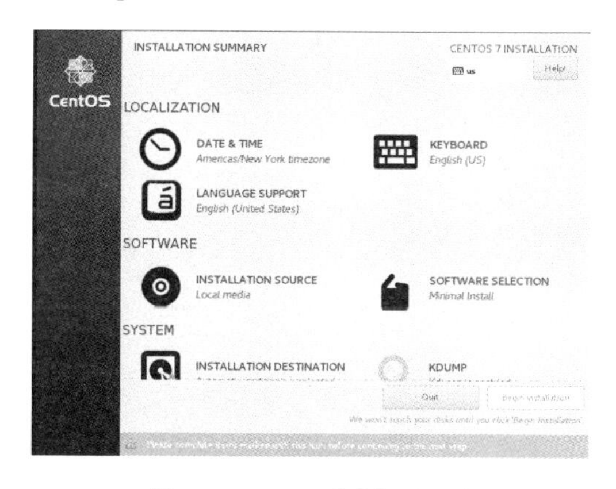


图 1.27　CentOS 安装概要界面

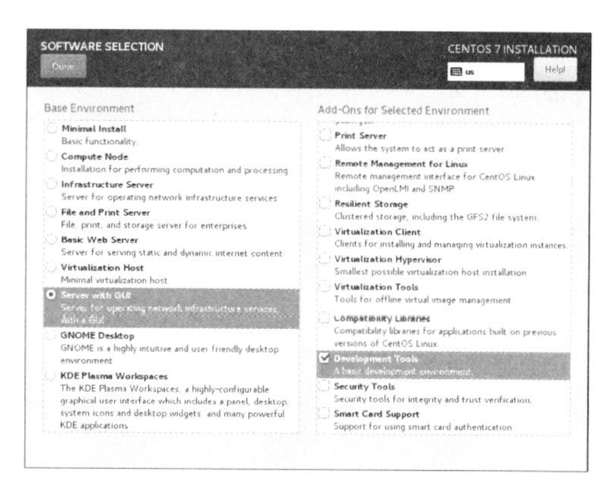


图 1.28　选择需要安装的软件

（13）在“NETWORK & HOST NAME”界面设置主机名和虚拟机的 IP 地址，如图 1.29 所示。

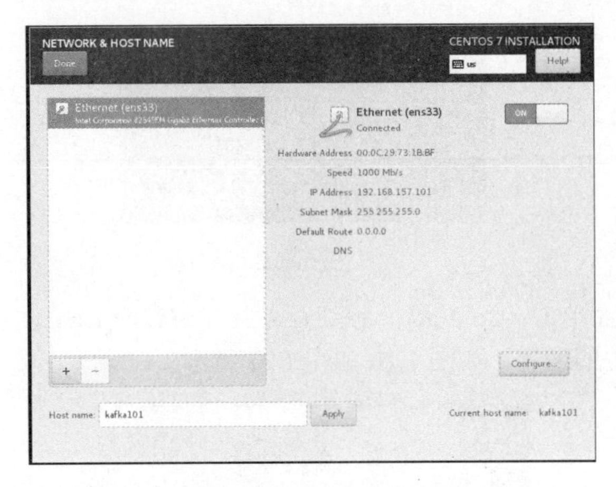


图 1.29　网络与主机名的设置

这里的主机名为 kafka101，IP 地址为 192.168.157.101。CentOS 安装部署完成后，我们就可以通过这个 IP 地址从宿主机连接到 CentOS。后续的 kafka102 和 kafka103，可以根据类似的方式进行安装部署。

（14）完成设置后，直接单击“Begin Installation”按钮进行安装。如果没有特殊说明，只会用到 ROOT 用户，可以设置 ROOT 的密码，如图 1.30 和图 1.31 所示。

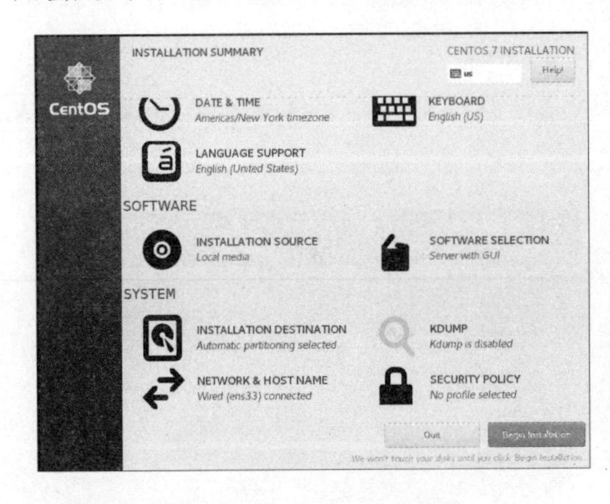


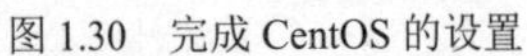
图 1.30　完成 CentOS 的设置

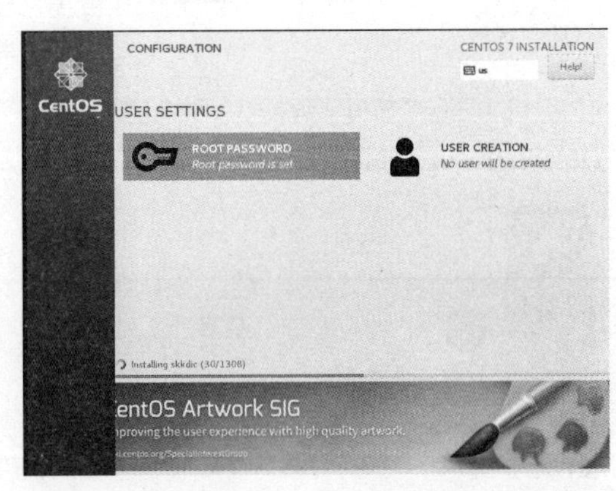


图 1.31　正在安装 CentOS

（15）安装完成后，直接单击“Reboot”按钮重启即可，如图 1.32 所示。

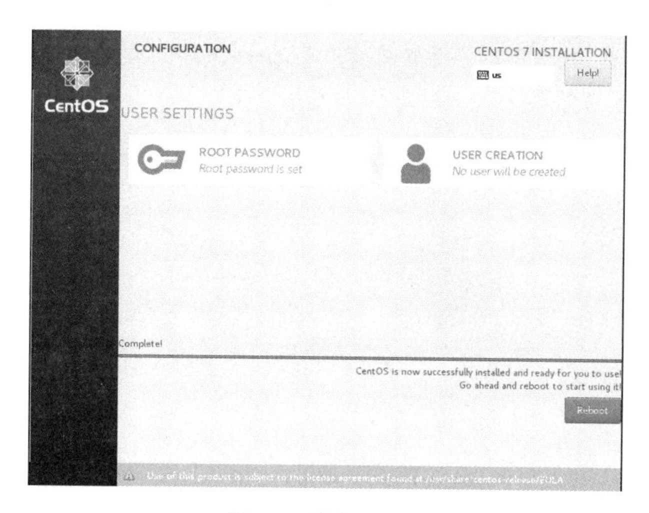


图 1.32　重启 CentOS

（16）接下来就可以运行 XShell，通过 192.168.157.101 的地址从宿主机上登录 CentOS，如图 1.33 所示。

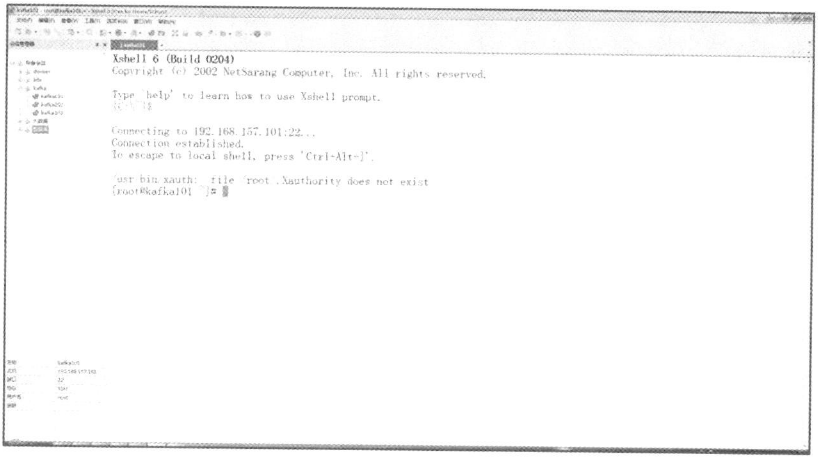

图 1.33　使用 XShell 连接 CentOS

## 1.5.2　配置 CentOS 操作系统

首先需要设置每台 CentOS 的主机名和 IP 地址的映射关系。使用 vi 编辑器编辑 /etc/hosts 文件，将主机名和 IP 地址的映射关系写入，如下所示。

```
192.168.157.101  kafka101
192.168.157.102  kafka102
192.168.157.103  kafka103
```

然后关闭防火墙。如果以后在生产环境中不允许关闭防火墙，那么需要把用到的端口加入防火墙中。这里直接关闭防火墙，如下所示。

```
[root@kafka101 ~]# systemctl stop firewalld.service
[root@kafka101 ~]# systemctl disable firewalld.service
```

## 1.5.3 安装 JDK

为了方便操作，在当前目录下创建两个目录：training 和 tools。我们把所有安装介质都上传到 tools 目录中；最终安装到 training 的目录下，如下所示。

```
[root@kafka101 ~]# pwd
/root
[root@kafka101 ~]# mkdir tools
[root@kafka101 ~]# mkdir training
```

解压 JDK 到/root/training/目录。

```
[root@kafka101 ~]# tar -zxvf tools/jdk-8u181-linux-x64.tar.gz -C ~/training/
```

使用 vi 编辑器编辑~/.bash_profile 文件，设置 JAVA_HOME 和 PATH 的环境变量，并进行如下设置。

```
JAVA_HOME=/root/training/jdk1.8.0_181
export JAVA_HOME

PATH=$JAVA_HOME/bin:$PATH
export PATH
```

环境变量生效。

```
[root@kafka101 ~]# source ~/.bash_profile
```

验证 Java 环境，执行 java -version 命令，如图 1.34 所示。

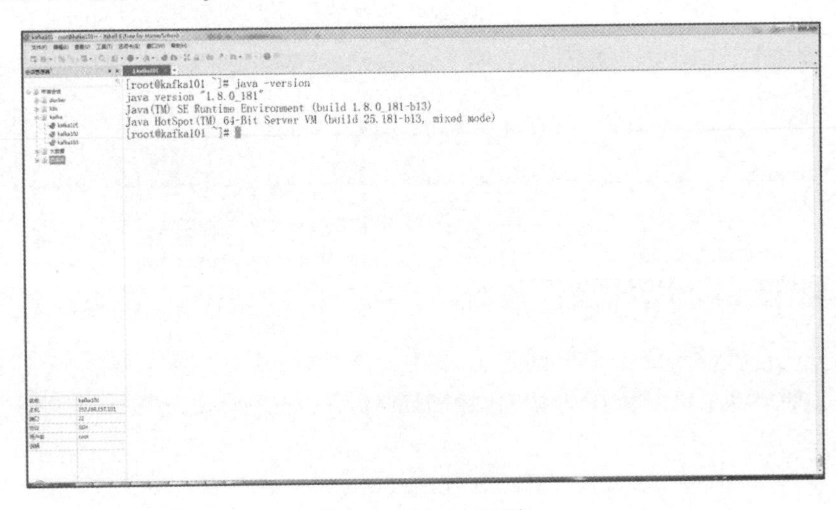


图 1.34　验证 Java 环境

# 第 2 章　部署 Kafka

Kafka 的安装依赖于 ZooKeeper，所以本章首先介绍 ZooKeeper 集群的搭建。当然，Kafka 默认也内置了 ZooKeeper 的启动脚本，在 Kafka 安装路径的 bin 目录下，其名称为 zookeeper-server-start.sh；如果不想独立安装 ZooKeeper，可直接使用该脚本。

通过第 1 章介绍的操作，我们已经准备好 CentOS 的操作系统，并展示了 Kafka 集群的部署架构。由于 Kafka 依赖 ZooKeeper，所以图 2.1 展示了完整的 Kafka 集群的部署架构。

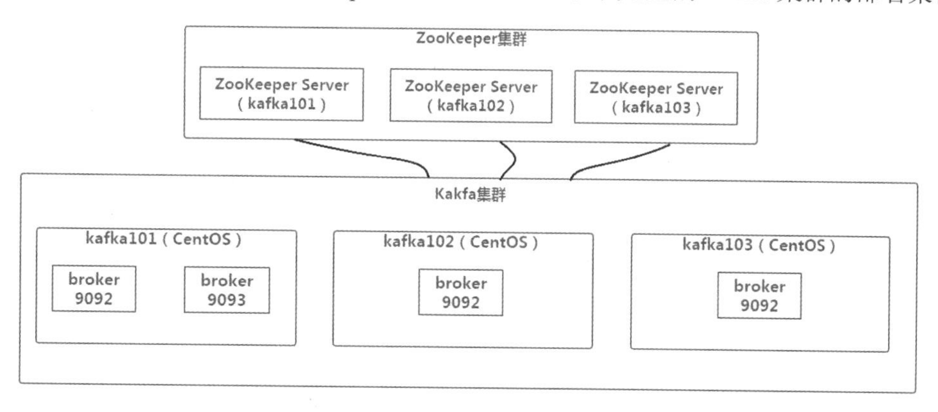


图 2.1　完整的 Kafka 集群的部署架构

在之前部署的三台 CentOS 虚拟机（kafka101、kafka102、kafka103）上部署 ZooKeeper 集群和 Kafka 集群。

## 2.1　部署 ZooKeeper

ZooKeeper 安装部署模式分为 Standalone 模式和集群模式。Standalone 模式比较简单，多用于开发和测试。只需要一个虚拟机就可以完成 Standalone 模式的搭建；如果是生产环境，建议搭建 ZooKeeper 的集群模式，这时候需要三台虚拟机进行搭建。ZooKeeper 集群中的节点具有不同的角色：Leader 和 Follower。

将 ZooKeeper 部署完成后，第 1 章关于 ZooKeeper 功能的示例代码都可以执行。下面我们分别对其进行介绍。

### 2.1.1 ZooKeeper 的核心配置文件

ZooKeeper 的核心配置文件是在其 conf 目录下的 zoo.cfg 文件。需要注意的是，在默认的情况下是没有这个文件的，需要根据 ZooKeeper 的 Sample 文件生成这个文件。

下面我们对这个文件中的配置参数进行相应的说明，以方便后续的配置。下面列出了这个文件的内容。

```
01  # The number of milliseconds of each tick
02  tickTime=2000
03  # The number of ticks that the initial
04  # synchronization phase can take
05  initLimit=10
06  # The number of ticks that can pass between
07  # sending a request and getting an acknowledgement
08  syncLimit=5
09  # the directory where the snapshot is stored.
10  # do not use /tmp for storage, /tmp here is just
11  # example sakes.
12  dataDir=/tmp/zookeeper
13  # the port at which the clients will connect
14  clientPort=2181
15  # the maximum number of client connections.
16  # increase this if you need to handle more clients
17  # maxClientCnxns=60
18  #
19  # Be sure to read the maintenance section of the
20  # administrator guide before turning on autopurge.
21  #
22  # http://zookeeper.apache.org/doc/current/zookeeperAdmin.html#sc_maintenance
23  #
24  # The number of snapshots to retain in dataDir
25  # autopurge.snapRetainCount=3
26  # Purge task interval in hours
27  # Set to "0" to disable auto purge feature
28  # autopurge.purgeInterval=1
```

其中每个参数的具体含义如下。

（1）tickTime。

tickTime 是 ZooKeeper 中的一个时间单元。ZooKeeper 中的所有时间都是以这个时间单元为基础进行整数倍配置的。例如，session 的最小超时时间是 2×tickTime。

（2）initLimit。

Follower 在启动过程中，会从 Leader 同步所有最新数据，然后确定自己能够对外服务的起始状态。Leader 允许 Follower 在 initLimit 时间内完成这个工作。在通常情况下，我们不用太在意这个参数的设置。如果 ZooKeeper 集群的数据量确实很大，Follower 在启动的时候，从 Leader 上同步数据的时间也会相应变长，因此在这种情况下，有必要适当调大这个参数。

（3）syncLimit。

在运行过程中，Leader 负责与 ZooKeeper 集群中的所有机器进行通信，例如，通过一些心跳检测机制来检测机器的存活状态。如果 Leader 在 syncLimit 之后发出心跳包，还没有从 Follower 收到响应，那么就认为这个 Follower 已经不在线了。注意，不要把这个参数设置得过大，否则可能会掩盖一些问题。

（4）dataDir。

dataDir 是 ZooKeeper 存储快照文件 snapshot 的目录。在默认情况下，事务日志也会存储在这里。建议同时配置参数 dataLogDir，事务日志的写性能直接影响 ZooKeeper 性能。可以把这个参数所指向的目录理解为 ZooKeeper 存储数据的目录。注意，这个参数的默认值是/tmp 目录，所以在生产环境中，一定要修改这个参数的值。

（5）clientPort。

clientPort 是客户端连接 ZooKeeper Server 的端口，对外服务器端口一般将其设置为 2181。

（6）maxClientCnxns。

maxClientCnxns 是单个客户端与单台服务器之间的连接数的限制，是 IP 级别的，默认是 60，如果将其设置为 0，那么表明不做任何限制。注意这个限制的使用范围，仅仅是单台客户端机器与单台 ZooKeeper 服务器之间的连接数限制，不是针对指定客户端 IP，不是 ZooKeeper 集群的连接数限制，也不是单台 ZooKeeper 对所有客户端的连接数限制。

（7）autopurge.snapRetainCount。

这个参数指定了需要 ZooKeeper 保留的文件数目，默认保留 3 个。

（8）autopurge.purgeInterval。

从 ZooKeeper 3.4.0 及之后版本，ZooKeeper 提供了自动清理事务日志和快照文件的功能，这个参数指定了清理频率，单位是小时，需要设置为 1 或更大的整数，默认值是 0，表示不开启自动清理功能。

除了上面的配置参数，ZooKeeper 还有一些扩展参数，下面列举出来这些扩展的参数，以及它们的含义。

（1）globalOutstandingLimit。

这个参数默认值是 1000。如果有大量的 client，不仅会造成 ZooKeeper Server 对请求的处理速度小于 client 的提交请求的速度，还会造成服务器端大量请求 queue 滞留而导致内存溢出，此参数能够控制 server 最大持有未处理请求的个数。

（2）preAllocSize。

为了避免大量磁盘检索，ZooKeeper 对 log 文件进行空间预分配，默认为 64MB。每当剩余空间小于 4KB 时，将会再次“预分配”。

（3）snapCount。

snapCount 的默认值为 100 000，在新增的 log 条数达到 snapCount/2 + Random.nextInt(snapCount/2)时，将会对 ZooKeeper 中存储的数据执行快照，将内存中 DataTree 反序为快照文件数据，同时 log 计数置为 0，以此循环。在执行快照的过程中，同时也伴随 log 的新文件创建。snapCount 参数让每个 server 创建快照的时机随时可控，避免所有 server 同时创建快照。

（4）traceFile。

traceFile 用于请求跟踪文件，如果设置了此参数，所有请求将会被记录在 traceFile 文件中。这个参数的配置会带来一定的性能问题。

（5）ClientPortAddress。

这个参数是 ZooKeeper 3.3.0 以后引入的参数，用于指定侦听 clientPort 的地址。此参数是可选的，默认 clientPort 会绑定到所有 IP 上，在物理 server 具有多个网络接口时，可以设置特定的 IP。

（6）minSessionTimeout。

minSessionTimeout 是 ZooKeeper 3.3.0 中引入的新参数，其默认值是 2×tickTime，也是 server 允许的会话超时最小值，如果设置的值过小，将会采用默认值。

（7）maxSessionTimeout。

maxSessionTimeout 是 ZooKeeper 3.3.0 中引入的新参数，其默认值是 20×tickTime，也是 server 允许的会话超时最大值。

### 2.1.2 部署 ZooKeeper 的 Standalone 模式

将在 kafka101 的虚拟机上进行部署，使用的 ZooKeeper 的版本是 zookeeper-3.4.10.tar.gz。

将安装包解压至/root/training 目录。

```
tar -zxvf zookeeper-3.4.10.tar.gz -C ~/training/
```

设置 ZooKeeper 环境变量，编辑文件~/.bash_profile。

```
ZOOKEEPER_HOME=/root/training/zookeeper-3.4.10
export ZOOKEEPER_HOME

PATH=$ZOOKEEPER_HOME/bin:$PATH
```

```
export PATH
```

生成 ZooKeeper 环境变量。

```
source ~/.bash_profile
```

生成 zoo.cfg 文件。

```
cd ~/training/zookeeper-3.4.10/conf/
mv zoo_sample.cfg zoo.cfg
```

编辑 zoo.cfg 文件，修改后的文件内容如下。

```
01  # The number of ticks that can pass between
02  # sending a request and getting an acknowledgement
03  syncLimit=5
04  # the directory where the snapshot is stored.
05  # do not use /tmp for storage, /tmp here is just
06  # example sakes.
07  dataDir=/root/training/zookeeper-3.4.10/tmp
08  # the port at which the clients will connect
09  clientPort=2181
10  # the maximum number of client connections.
11  # increase this if you need to handle more clients
12  # maxClientCnxns=60
13  #
14  # Be sure to read the maintenance section of the
15  # administrator guide before turning on autopurge.
16  #
17  # http://zookeeper.apache.org/doc/current/zookeeperAdmin.html#sc_maintenance
18  #
19  # The number of snapshots to retain in dataDir
20  # autopurge.snapRetainCount=3
21  # Purge task interval in hours
22  # Set to "0" to disable auto purge feature
23  # autopurge.purgeInterval=1
24
25  server.1=kafka101:2888:3888
```

这里主要做了两个修改，一个是 dataDir；另一个是设置了 server.1。参数 dataDir 的含义在前面的内容中已经提到，这个参数用于指定 ZooKeeper 数据存储的路径，在生产环境中需要重新设置。这里我们将其修改为/root/training/zookeeper-3.4.10/tmp。

server.1 用于指定 ZooKeeper 集群中的 server 节点。由于现在部署的是一个 Standalone 的模式，在这种模式下的集群中只存在一个 ZooKeeper Server。因此，这里只设置了一个 server 地址，即 kafka101:2888:3888。kafka101 是我们在前面部署的其中一台 CentOS 虚拟机；端口 2888 表示集群内 Server 节点通信的端口，Leader 将监听此端口；端口 3888 用于选举 Leader 使用。当 ZooKeeper 集群的 Leader 宕机后，ZooKeeper 集群会通过此端口选举一个新的 Leader。

进入参数 dataDir 指定的目录，即/root/training/zookeeper-3.4.10/tmp。创建 myid 文件，并在 myid 文件中输入 1。注意，这个 1 用来表示 server.1 的 ZooKeeper 节点在集群中的哪个主机上。

执行命令 zkServer.sh start，启动 ZooKeeper Server。

```
[root@kafka101 conf]# zkServer.sh start
ZooKeeper JMX enabled by default
Using config: /root/training/zookeeper-3.4.10/bin/../conf/zoo.cfg
Starting zookeeper ... STARTED
```

执行命令 zkServer.sh status，查看 ZooKeeper Server 的状态。

```
[root@kafka101 conf]# zkServer.sh status
ZooKeeper JMX enabled by default
Using config: /root/training/zookeeper-3.4.10/bin/../conf/zoo.cfg
Mode: standalone
```

也可以通过执行 jps 命令查看 Java 的后台进程信息。

```
[root@kafka101 conf]# jps
41894 Jps
41832 QuorumPeerMain
```

这里的 QuorumPeerMain 进程对应的就是 ZooKeeper Server 进程。至此，ZooKeeper Standalone 模式就部署完成了。

可以使用 ZooKeeper 提供的 CLI 命令行工具操作 ZooKeeper。通过下面的命令直接启动 ZooKeeper 的客户端，默认将其连接到本机的 2181 端口上。

```
[root@kafka101 conf]# zkCli.sh
```

也可以通过-server 参数指定连接的主机和端口。

```
zkCli.sh -server kafka101:2181
```

登录 ZooKeeper 客户端后，可以执行 help 命令查看所有可用的操作命令，如下所示。通过具体的示例来演示如何使用 ZooKeeper 的操作命令。

```
[zk: kafka101:2181(CONNECTED) 0] help
ZooKeeper -server host:port cmd args
    stat path [watch]
    set path data [version]
    ls path [watch]
    delquota [-n|-b] path
    ls2 path [watch]
    setAcl path acl
    setquota -n|-b val path
    history
    redo cmdno
    printwatches on|off
    delete path [version]
    sync path
    listquota path
```

```
    rmr path
    get path [watch]
    create [-s] [-e] path data acl
    addauth scheme auth
    quit
    getAcl path
    close
    connect host:port
```

可以通过以下命令来创建 ZooKeeper 的节点。

```
[zk: kafka101:2181(CONNECTED) 33] create /node1 helloworld
Created /node1
[zk: kafka101:2181(CONNECTED) 34]
```

在 ZooKeeper 的根节点下，创建了 node1 节点，node1 节点上的数据是 helloworld。当 ZooKccpcr 的节点创建完成后，通过以下命令查看当前节点列表。

```
[zk: kafka101:2181(CONNECTED) 34] ls /
[zookeeper, node1]
[zk: kafka101:2181(CONNECTED) 35]
```

使用 state 命令可以查看节点状态。

```
[zk: kafka101:2181(CONNECTED) 35] stat /node1
cZxid = 0x14
ctime = Fri Sep 18 06:56:28 EDT 2020
mZxid = 0x14
mtime = Fri Sep 18 06:56:28 EDT 2020
pZxid = 0x14
cversion = 0
dataVersion = 0
aclVersion = 0
ephemeralOwner = 0x0
dataLength = 10
numChildren = 0
[zk: kafka101:2181(CONNECTED) 36]
```

其中，

- cZxid 和 ctime 表示创建时的事务 id 和时间。
- mZxid 和 mtime 表示最后一次更新时的事务 id 和时间。
- pZxid 表示当前节点的子节点列表最后一次被修改时的事务 id。引起子节点列表变化的两种情况是删除子节点和新增子节点。

由于 ZooKeeper 是通过节点来保存数据的，所以可直接通过 get 命令查看节点数据内容，如下所示。

```
[zk: kafka101:2181(CONNECTED) 36] get /node1
helloworld
cZxid = 0x14
```

```
ctime = Fri Sep 18 06:56:28 EDT 2020
mZxid = 0x14
mtime = Fri Sep 18 06:56:28 EDT 2020
pZxid = 0x14
cversion = 0
dataVersion = 0
aclVersion = 0
ephemeralOwner = 0x0
dataLength = 10
numChildren = 0
```

最后，可以使用 quit 命令退出 ZooKeeper 客户端。

```
[zk: kafka101:2181(CONNECTED) 38] quit
Quitting...
2020-09-18  07:01:33,519  [myid:]  - INFO   [main:ZooKeeper@684]  - Session:
0x174a0bdebaa0003 closed
2020-09-18      07:01:33,531      [myid:]       -       INFO              [main-
EventThread:ClientCnxn$EventThread@519] - EventThread shut down for session:
0x174a0bdebaa0003
```

### 2.1.3 部署 ZooKeeper 的集群模式

将在 kafka101、kafka102、kafka103 的虚拟机上部署 ZooKeeper 的集群模式。首先在 kafka101 虚拟机上进行配置，然后通过 scp 命令将配置好的 ZooKeeper 目录复制到 kafka102 和 kafka103 虚拟机上。具体步骤如下所示。

修改 kafka101 的虚拟机上的 zoo.cfg 文件，完整的内容如下。

```
01   # The number of milliseconds of each tick
02   tickTime=2000
03   # The number of ticks that the initial
04   # synchronization phase can take
05   initLimit=10
06   # The number of ticks that can pass between
07   # sending a request and getting an acknowledgement
08   syncLimit=5
09   # the directory where the snapshot is stored.
10   # do not use /tmp for storage, /tmp here is just
11   # example sakes.
12   dataDir=/root/training/zookeeper-3.4.10/tmp
13   # the port at which the clients will connect
14   clientPort=2181
15   # the maximum number of client connections.
16   # increase this if you need to handle more clients
17   # maxClientCnxns=60
18   #
```

```
19   # Be sure to read the maintenance section of the
20   # administrator guide before turning on autopurge.
21   #
22   # http://zookeeper.apache.org/doc/current/zookeeperAdmin.html#sc_maintenance
23   #
24   # The number of snapshots to retain in dataDir
25   # autopurge.snapRetainCount=3
26   # Purge task interval in hours
27   # Set to "0" to disable auto purge feature
28   # autopurge.purgeInterval=1
29
30   server.1=kafka101:2888:3888
31   server.2=kafka102:2888:3888
32   server.3=kafka103:2888:3888
```

这里在配置文件中增加了两个 ZooKeeper 节点，即 server.2 和 server.3。它们分别位于 kafka102 和 kafka103 的虚拟机上。

在 kafka101 的虚拟机上，把配置好的 ZooKeeper 目录复制到 kafka102 和 kafka103 的虚拟机上，执行下面的命令。

```
cd /root/training
scp -r zookeeper-3.4.10/ root@kafka102:/root/training
scp -r zookeeper-3.4.10/ root@kafka103:/root/training
```

在执行 scp 命令的时候，需要允许远程连接主机，并输入远端主机的 root 用户密码，如下所示。

```
[root@kafka101 training]# scp -r zookeeper-3.4.10/ root@kafka103:/root/training
The authenticity of host 'kafka103 (192.168.157.103)' can't be established.
ECDSA key fingerprint is SHA256:9ezjqGdBFdeuu3/hTzuChA8BwGxAYyEQ+mNeyrn5fj4.
ECDSA key fingerprint is MD5:60:3a:71:17:61:fd:5b:81:a1:84:fb:78:78:db:83:8a.
Are you sure you want to continue connecting (yes/no)? yes
Warning: Permanently added 'kafka103,192.168.157.103' (ECDSA) to the list of
known hosts.
root@kafka103's password:
```

在 kafka102 和 kafka103 的虚拟机上设置 ZooKeeper 的环境变量，编辑文件~/.bash_profile。

```
ZOOKEEPER_HOME=/root/training/zookeeper-3.4.10
export ZOOKEEPER_HOME

PATH=$ZOOKEEPER_HOME/bin:$PATH
export PATH
```

在 kafka102 和 kafka103 的虚拟机上生成 ZooKeeper 的环境变量。

```
source ~/.bash_profile
```

将 kafka102 的虚拟机上 myid 文件的内容修改为 2；将 kafka103 的虚拟机上 myid 文

件的内容修改为 3，如图 2.2 所示。

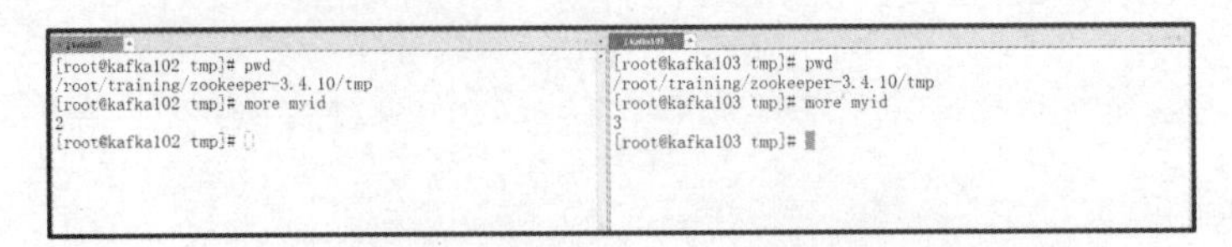


图 2.2　修改 kafka102 和 kafka103 上的 myid 文件

（1）在每台主机上执行 zkServer.sh start 命令，启动 ZooKeeper Server 集群，如图 2.3 所示。

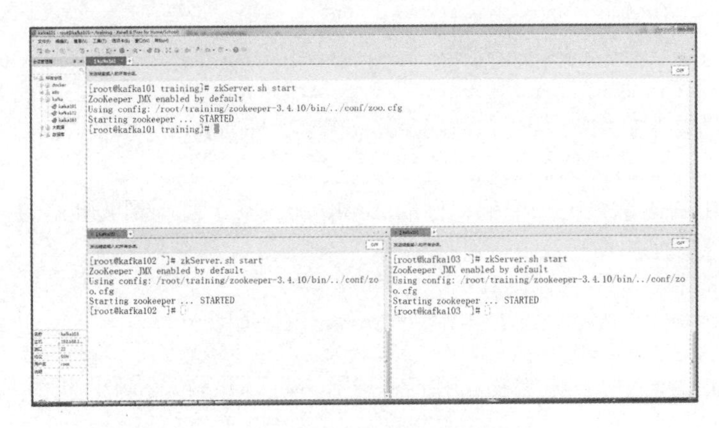


图 2.3　启动 ZooKeeper 集群

（2）在每台主机上执行 zkServer.sh status 命令，查看 ZooKeeper 集群中每个节点的状态，如图 2.4 所示。

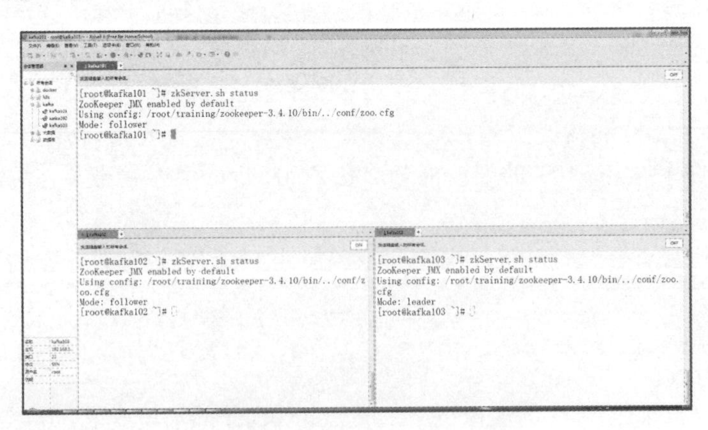


图 2.4　查看 ZooKeeper 集群的节点状态

这里可以看到，ZooKeeper 集群通过选举机制将 kafka103 的虚拟机上的 Server 选举为 Leader；而 kafka101 和 kafka102 的虚拟机上的 Server 选举为 Follower。

ZooKeeper 集群部署完成后，可以使用 ZooKeeper 的命令工具来进行操作，其用法与在 ZooKeeper Standalone 模式下的用法完全一样，这里就不再介绍了。

## 2.1.4　测试 ZooKeeper 集群

ZooKeeper 集群的功能中很重要的就是集群的节点之间数据同步与 Leader 的选举机制。下面进行一个简单的测试。

（1）测试集群的数据同步功能。在每个节点上启动 zkCli.sh 的命令行工具，并在 kafka101 的虚拟机上创建一个节点，试着在 kafka102 和 kafka103 的虚拟机上进行查看，如图 2.5 所示。

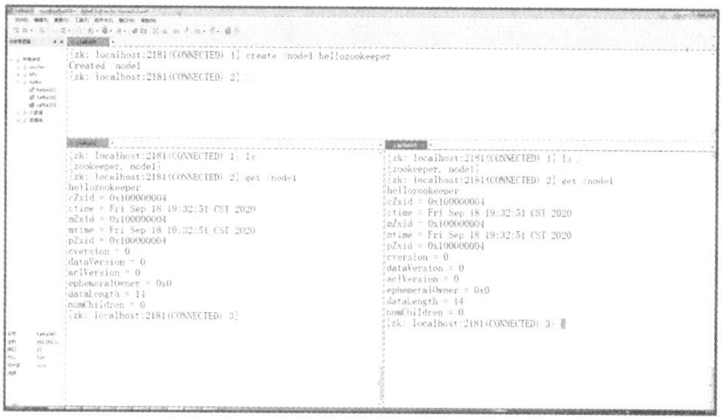

图 2.5　ZooKeeper 集群的数据同步

在 kafka101 上创建一个节点/node1，并保存数据 hellozookeeper。这个新创建的节点和数据将会自动同步到 kafka102 和 kafka103 的虚拟机。

（2）测试集群的选举机制。先来看一下集群当前的 Server 状态，如图 2.6 所示。

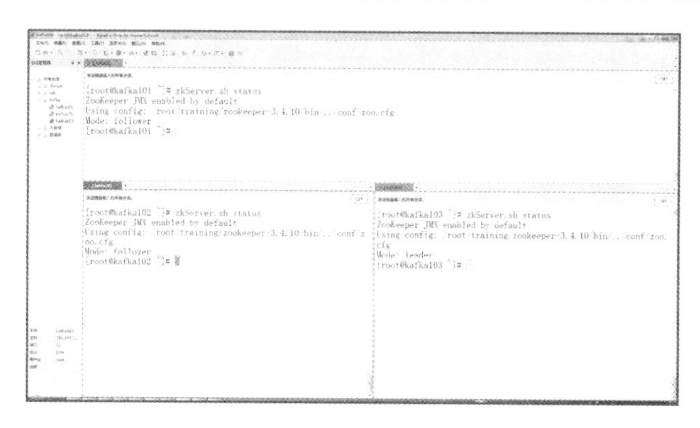

图 2.6　ZooKeeper 集群的 Server 状态

可以看到 kafka103 的虚拟机上的 Server 是 Leader 状态，通过下面的方式可以杀掉 kafka103 的虚拟机上的 Server。

```
[root@kafka103 ~]# jps
1572 QuorumPeerMain
1720 Jps
[root@kafka103 ~]# kill -9 1572
```

其中，1572 是 ZooKeeper Server 的进程号。

（3）重新查看 ZooKeeper 中 Server 的状态，在每个节点上执行 zkServer.sh status，如图 2.7 所示。

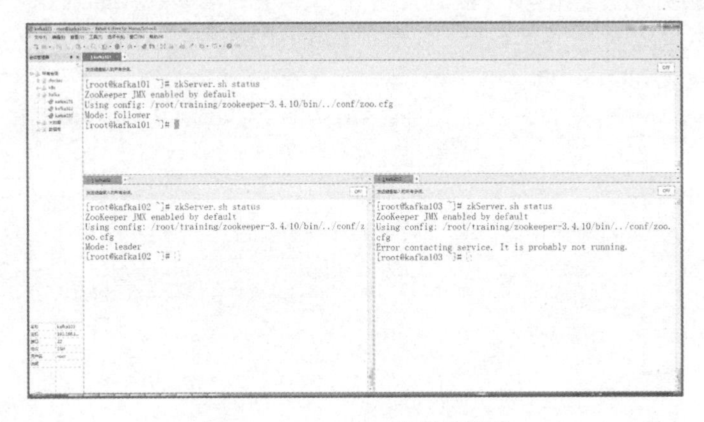


图 2.7　ZooKeeper 集群的选举

通过观察图 2.7 可以看到，kafka103 的虚拟机上的 Kafka Server 已经不能访问了。但是 ZooKeeper 集群利用本身的选举机制将 kafka102 的虚拟机上的 Kafka Server 选举成 Leader，而 kafka101 的虚拟机上的 Kafka Server 依然是 Follower 的状态。

## 2.2　安装部署 Kafka

Kafka 的部署方式分为三种。

- 单机单 Broker 模式。
- 单机多 Broker 模式（Kafka 伪集群模式）。
- 多机多 Broker 模式（Kafka 集群模式）。

前两种模式都是在一台虚拟机的主机上进行搭建的，主要用于开发和测试；在生产环境下，我们应该部署真正的 Kafka 集群模式，即多机多 Broker 模式。

Kafka 的核心配置文件是 config 目录下的 server.properties 文件，下面列出了这个文件的完整内容。后面章节将详细说明每个参数的含义。

```
01  ############################# Server Basics #############################
02
```

```
03  # The id of the broker. This must be set to a unique integer for each broker.
04  broker.id=0
05
06  ############################# Socket Server Settings
07
08  # The address the socket server listens on. It will get the value returned from
09  # java.net.InetAddress.getCanonicalHostName() if not configured.
10  #   FORMAT:
11  #     listeners = listener_name://host_name:port
12  #   EXAMPLE:
13  #     listeners = PLAINTEXT://your.host.name:9092
14  # listeners=PLAINTEXT://:9092
15
16  # Hostname and port the broker will advertise to producers and consumers. If
not set,
17  # it uses the value for "listeners" if configured.  Otherwise, it will use
the value
18  # returned from java.net.InetAddress.getCanonicalHostName().
19  # advertised.listeners=PLAINTEXT://your.host.name:9092
20
21  # Maps listener names to security protocols, the default is for them to be
the same. See the config documentation for more details
22 #  listener.security.protocol.map=PLAINTEXT:PLAINTEXT,SSL:SSL,SASL_PLAINTEXT:
SASL_PLAINTEXT,SASL_SSL:SASL_SSL
23
24  # The number of threads that the server uses for receiving requests from the
network and sending responses to the network
25  num.network.threads=3
26
27  # The number of threads that the server uses for processing requests, which
may include disk I/O
28  num.io.threads=8
29
30  # The send buffer (SO_SNDBUF) used by the socket server
31  socket.send.buffer.bytes=102400
32
33  # The receive buffer (SO_RCVBUF) used by the socket server
34  socket.receive.buffer.bytes=102400
35
36  # The maximum size of a request that the socket server will accept (protection
against OOM)
37  socket.request.max.bytes=104857600
38
39
```

```
40  ############################# Log Basics #############################
41
42  # A comma separated list of directories under which to store log files
43  log.dirs=/tmp/kafka-logs
44
45  # The default number of log partitions per topic. More partitions allow greater
46  # parallelism for consumption, but this will also result in more files across
47  # the brokers.
48  num.partitions=1
49
50  # The number of threads per data directory to be used for log recovery at
startup and flushing at shutdown.
51  # This value is recommended to be increased for installations with data dirs
located in RAID array.
52  num.recovery.threads.per.data.dir=1
53
54  #############################          Internal          Topic          Settings
#############################
55  #  The  replication  factor  for  the  group  metadata  internal  topics
"__consumer_offsets" and "__transaction_state"
56  # For anything other than development testing, a value greater than 1 is
recommended to ensure availability such as 3.
57  offsets.topic.replication.factor=1
58  transaction.state.log.replication.factor=1
59  transaction.state.log.min.isr=1
60
61  ############################# Log Flush Policy #############################
62
63  # Messages are immediately written to the filesystem but by default we only
fsync() to sync
64  # the OS cache lazily. The following configurations control the flush of data
to disk.
65  # There are a few important trade-offs here:
66  #    1. Durability: Unflushed data may be lost if you are not using replication.
67  #    2. Latency: Very large flush intervals may lead to latency spikes when
the flush does occur as there will be a lot of data to flush.
68  #    3. Throughput: The flush is generally the most expensive operation, and
a small flush interval may lead to excessive seeks.
69  # The settings below allow one to configure the flush policy to flush data
after a period of time or
70  # every N messages (or both). This can be done globally and overridden on a
per-topic basis.
71
72  # The number of messages to accept before forcing a flush of data to disk
```

```
73  #log.flush.interval.messages=10000
74
75  # The maximum amount of time a message can sit in a log before we force a flush
76  # log.flush.interval.ms=1000
77
78  ############################# Log Retention Policy #############################
79
80  # The following configurations control the disposal of log segments. The
policy can
81  # be set to delete segments after a period of time, or after a given size
has accumulated.
82  # A segment will be deleted whenever *either* of these criteria are met.
Deletion always happens
83  # from the end of the log.
04
85  # The minimum age of a log file to be eligible for deletion due to age
86  log.retention.hours=168
87
88  # A size-based retention policy for logs. Segments are pruned from the log
unless the remaining
89  # segments drop below log.retention.bytes. Functions independently of
log.retention.hours.
90  #log.retention.bytes=1073741824
91
92  # The maximum size of a log segment file. When this size is reached a new
log segment will be created.
93  log.segment.bytes=1073741824
94
95  # The interval at which log segments are checked to see if they can be deleted
according
96  # to the retention policies
97  log.retention.check.interval.ms=300000
98
99  ############################# Zookeeper #############################
100
101 # Zookeeper connection string (see zookeeper docs for details).
102 # This is a comma separated host:port pairs, each corresponding to a zk
103 # server. e.g. "127.0.0.1:3000,127.0.0.1:3001,127.0.0.1:3002".
104 # You can also append an optional chroot string to the urls to specify the
105 # root directory for all kafka znodes.
106  zookeeper.connect=localhost:2181
107
108 # Timeout in ms for connecting to zookeeper
109  zookeeper.connection.timeout.ms=6000
```

```
110
111
112 ############# Group Coordinator Settings ################
113
114 # The following configuration specifies the time, in milliseconds, that the GroupCoordinator will delay the initial consumer rebalance.
115 # The rebalance will be further delayed by the value of group.initial.rebalance.delay.ms as new members join the group, up to a maximum of max.poll.interval.ms.
116 # The default value for this is 3 seconds.
117 # We override this to 0 here as it makes for a better out-of-the-box experience for development and testing.
118 # However, in production environments the default value of 3 seconds is more suitable as this will help to avoid unnecessary, and potentially expensive, rebalances during application startup.
119  group.initial.rebalance.delay.ms=0
```

下面我们通过具体的步骤演示如何配置 Kafka 的三种不同的部署模式。

## 2.2.1 单机单 Broker 的部署

如图 2.8 所示，在单机单 Broker 的模式下，在 kafka101 的虚拟主机上部署一个 Broker 用于接收和转发生产者发布的消息。在这种模式下，由于只存在一个 Broker，将存在单点故障的问题，即 Broker 本身或 Broker 所在的主机宕机后，都会造成 Kafka 无法正常工作，所以这种模式只能用于开发和测试环境。

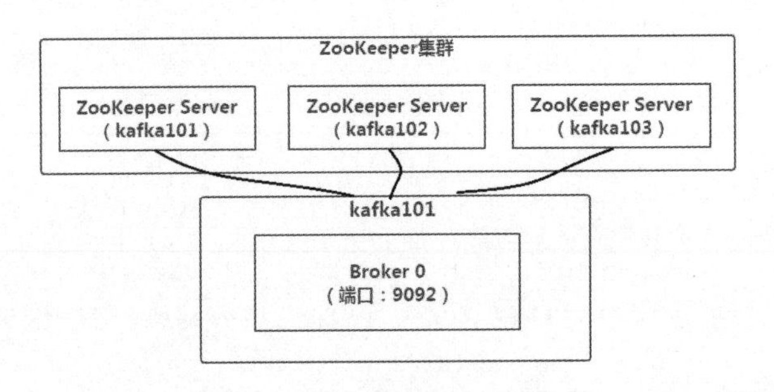


图 2.8　Kafka 单机单 Broker 模式

首先，将压缩包解压至/root/training 目录。

```
tar -zxvf kafka_2.11-2.4.0.tgz -C /root/training/
```

进入 Kafka 的 config 目录，并修改 server.properties 文件。

```
cd /root/training/kafka_2.11-2.4.0/config/
vi server.properties
```

下面列出了需要修改的参数。

```
broker.id=0
log.dirs=/root/training/kafka_2.11-2.4.0/logs/broker0
zookeeper.connect=kafka101:2181,kafka102:2181,kafka103:2181
```

其中，

- broker.id 表示 broker 的 id 号。在一个 Kafka 集群中，不同的 Broker 应该具有不同的 id 号，不能重复。
- log.dirs 表示 Kafka 日志数据的存放地址，多个地址用逗号分隔。这里我们为了方便管理，为每个 Broker 单独创建一个目录来存储这个 Broker 所对应的日志数据。
- zookeeper.connect 表示连接 ZooKeeper 进群的地址。

关于这些 Kafka 配置参数的详细说明，我们会在后续章节中进行介绍。

创建 Broker 0 日志存储的目录。

```
mkdir /root/training/kafka_2.11-2.4.0/logs/broker0
```

启动 Kafka Broker。

```
bin/kafka-server-start.sh config/server.properties &
```

启动成功后，将输出如下日志信息，如图 2.9 所示。

```
[2020-09-19 05:21:09,938] INFO [ProducerId Manager 0]: Acquired new producerId block (brokerId:0,blockStartProducerId:1000,blockEndProducerId:1999) by writing to Zk with path version 2 (kafka.coordinator.transaction.ProducerIdManager)
[2020-09-19 05:21:09,975] INFO [TransactionCoordinator id=0] Starting up. (kafka.coordinator.transaction.TransactionCoordinator)
[2020-09-19 05:21:09,996] INFO [TransactionCoordinator id=0] Startup complete. (kafka.coordinator.transaction.TransactionCoordinator)
[2020-09-19 05:21:10,009] INFO [Transaction Marker Channel Manager 0]: Starting (kafka.coordinator.transaction.TransactionMarkerChannelManager)
[2020-09-19 05:21:10,062] INFO [ExpirationReaper-0-AlterAcls]: Starting (kafka.server.DelayedOperationPurgatory$ExpiredOperationReaper)
[2020-09-19 05:21:10,151] INFO [/config/changes-event-process-thread]: Starting (kafka.common.ZkNodeChangeNotificationListener$ChangeEventProcessThread)
[2020-09-19 05:21:10,288] INFO [SocketServer brokerId=0] Started data-plane processors for 1 acceptors (kafka.network.SocketServer)
[2020-09-19 05:21:10,289] INFO Kafka version: 2.4.0 (org.apache.kafka.common.utils.AppInfoParser)
[2020-09-19 05:21:10,290] INFO Kafka commitId: 77a89fcf8d7fa018 (org.apache.kafka.common.utils.AppInfoParser)
[2020-09-19 05:21:10,290] INFO Kafka startTimeMs: 1600507270288 (org.apache.kafka.common.utils.AppInfoParser)
[2020-09-19 05:21:10,291] INFO [KafkaServer id=0] started (kafka.server.KafkaServer)
```

图 2.9　Broker 启动成功日志

也可以通过 Java 的 jps 命令查看后台的 Java 进程，如图 2.10 所示。

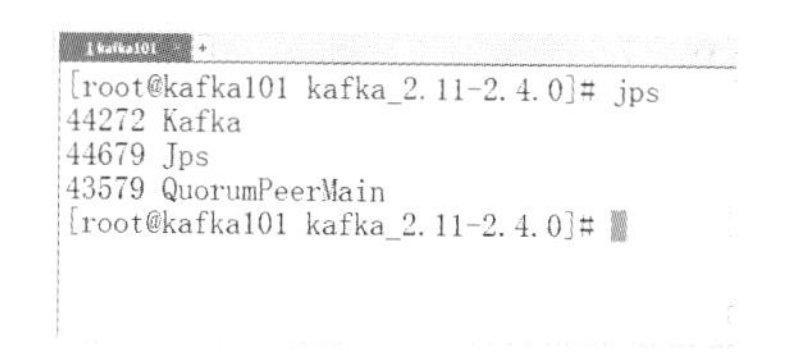


图 2.10　Kafka 的后台进程信息

其中 44272 进程就是 Kafka Broker 对应的进程，而 43579 进程就是我们前面配置好的 ZooKeeper 进程。

## 2.2.2 单机多 Broker 的部署

如图 2.11 所示，在单机多 Broker 的模式下，在 kafka101 的虚拟主机上部署两个 Broker，分别运行在 9092 端口和 9093 端口。在这种模式下，由于只存在一台主机，所以也存在单点故障的问题，即 Broker 所在的主机宕机后，都会造成 Kafka 无法正常工作。如果是两个 Broker 中的其中一个出现了问题，则整个 Kafka 依然可以正常工作。由于这种模式并不是真正的集群，所以也只能用于开发和测试环境。

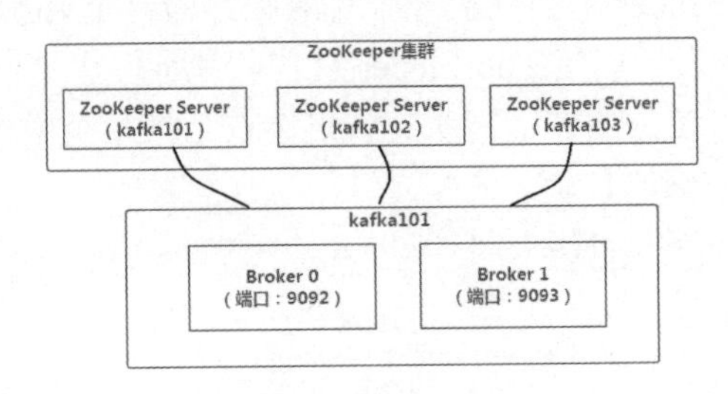


图 2.11　Kafka 单机多 Broker 模式

在 Kafka 的 config 目录下，手动复制一个新的 server.properties 文件。

```
cp server.properties server1.properties
```

创建 Broker1 日志存储的目录。

```
mkdir /root/training/kafka_2.11-2.4.0/logs/broker1
```

修改 server1.properties 文件，内容如下。

```
broker.id=1
port=9093
log.dirs=/root/training/kafka_2.11-2.4.0/logs/broker1
zookeeper.connect=kafka101:2181,kafka102:2181,kafka103:2181
```

注意，参数 port 需要手动添加。

启动 Kafka Broker。

```
bin/kafka-server-start.sh config/server1.properties1 &
```

启动成功后，将输出如下日志信息，如图 2.12 所示。

```
nator.transaction.TransactionCoordinator)
[2020-09-19 05:33:45,949] INFO [TransactionCoordinator id=1] Startup complete. (kafka.c
oordinator.transaction.TransactionCoordinator)
[2020-09-19 05:33:45,961] INFO [Transaction Marker Channel Manager 1]: Starting (kafka.
coordinator.transaction.TransactionMarkerChannelManager)
[2020-09-19 05:33:45,985] INFO [ExpirationReaper-1-AlterAcls]: Starting (kafka.server.D
elayedOperationPurgatory$ExpiredOperationReaper)
[2020-09-19 05:33:46,034] INFO [/config/changes-event-process-thread]: Starting (kafka.
common.ZkNodeChangeNotificationListener$ChangeEventProcessThread)
[2020-09-19 05:33:46,114] INFO [SocketServer brokerId=1] Started data-plane processors
for 1 acceptors (kafka.network.SocketServer)
[2020-09-19 05:33:46,116] INFO Kafka version: 2.4.0 (org.apache.kafka.common.utils.AppI
nfoParser)
[2020-09-19 05:33:46,116] INFO Kafka commitId: 77a89fcf8d7fa018 (org.apache.kafka.commo
n.utils.AppInfoParser)
[2020-09-19 05:33:46,116] INFO Kafka startTimeMs: 1600508026115 (org.apache.kafka.commo
n.utils.AppInfoParser)
[2020-09-19 05:33:46,118] INFO [KafkaServer id=1] started (kafka.server.KafkaServer)
```

图 2.12　单机多 Broker 启动成功日志

也可以通过 Java 的 jps 命令查看后台的 Java 进程，可以看到两个 Kafka Broker 的进程，如图 2.13 所示。

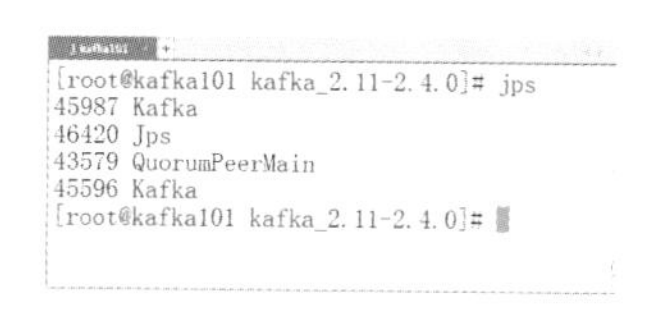


图 2.13　Kafka 的后台进程信息

## 2.2.3　多机多 Broker 的部署

如图 2.14 所示，在多机多 Broker 的模式下，在 kafka101 和 kafka102 的虚拟主机上分别部署 broker。这种模式是真正的 Kafka 集群模式，并提供高可用的特性，可以用于真正的生产环境。

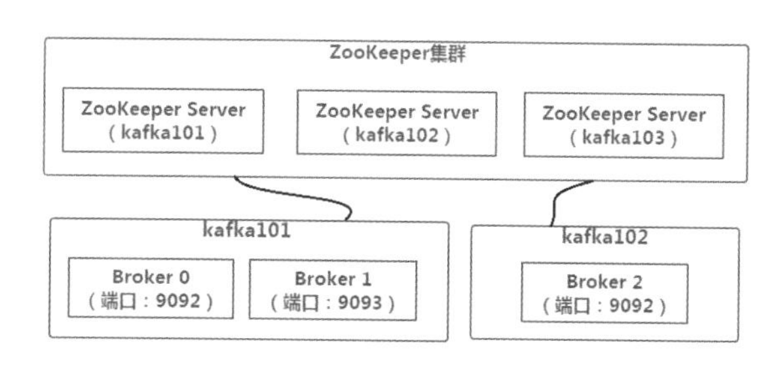


图 2.14　Kafka 多机多 Broker 模式

只需要按照之前单机环境下的方式，在 kafka102 的虚拟机上重新部署一个 kafka 节点，并修改 server.properties 参数即可，下面列出了配置参数的内容。

```
broker.id=2
log.dirs=/root/training/kafka_2.11-2.4.0/logs/broker2
zookeeper.connect=kafka101:2181,kafka102:2181,kafka103:2181
```

图 2.15 展示了最终部署成功的进程信息。

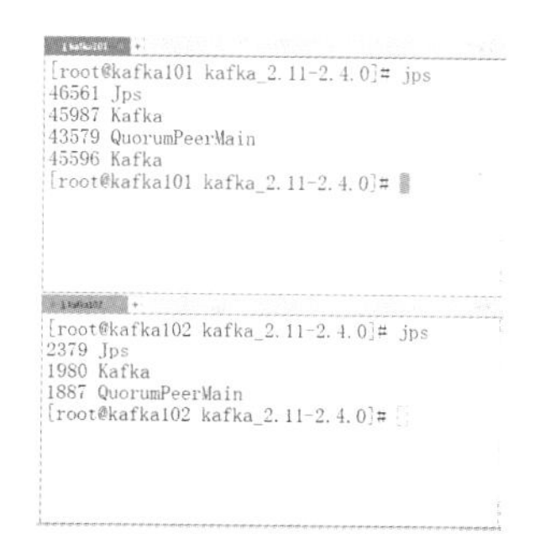


图 2.15　多机多 Broker 模式的进程信息

这里可以看到在 kafka101 的虚拟机上有两个 Kafka 的 Broker；而 kafka102 的虚拟机上只有一个 Kafka 的 Broker。

### 2.2.4 使用命令行测试 Kafka

至此，已经完成了 Kafka 集群的搭建。我们来进行一个简单的测试，创建一个 Topic 主题，并使用 Kafka 提供的命令工具来发送消息和接收消息。

创建一个名称为“mytopic1”的 Topic。

```
bin/kafka-topics.sh --create --zookeeper kafka101:2181 \
--replication-factor 2 --partitions 3 --topic mytopic1
```

其中，

- --zookeeper：用于指定 ZooKeeper 的地址，如果是多个 ZooKeeper 地址可以使用逗号分隔。
- --replication-factor：用于指定分区的副本数。这里我们设置的副本数为 2，表示同一个分区有两个副本。
- --partitions：用于指定该 Topic 包含的分区数。这里我们设置的分区数为 3，表示该 Topic 由三个分区组成。
- --topic：用于指定 Topic 的名称。

使用下面的命令启动 Producer 发送消息。

```
bin/kafka-console-producer.sh --broker-list kafka101:9092 --topic mytopic1
```

使用下面的命令启动 Consumer 接收消息。由于 Kafka 支持的是 Topic 广播类型的消息，可以多启动几个 Consumer，如图 2.16 所示。

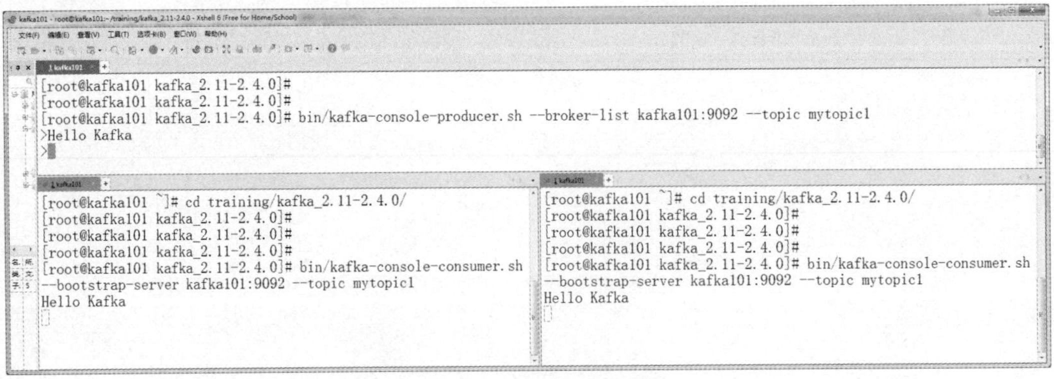

图 2.16 测试 Kafka 的消息发送与接收

这里，我们启动了一个 Producer 和两个 Consumer，并在 Producer 中发送了一条消息 “Hello Kafka”；在两个 Consumer 中，可以看到这条消息被同时接收到了。

下面列出了一些特殊方式的接收命令。

```
从开始位置消费
bin/kafka-console-consumer.sh --bootstrap-server kafka101:9092 \
--from-beginning --topic topicName
显示 key 消费
bin/kafka-console-consumer.sh --bootstrap-server kafka101:9092 \
--property print.key=true --topic mytopic1
```

## 2.3 Kafka 配置参数详解

在 2.2 节配置部署了 Kafka 集群的各种模式。Kafka 的核心配置文件是 server.properties，现在我们对这个配置文件中的每个参数进行详细的说明，具体内容如下。

- broker.id

每一个 broker 在集群中的唯一表示，要求是正数。当该服务器的 IP 地址发生改变时，broker.id 没有变化，则不会影响 Consumers 的消息情况。

- num.network.threads

broker 处理消息的最大线程数，在一般情况下不需要去修改。

- num.io.threads

broker 处理磁盘 IO 的线程数，其数值应该大于硬盘数。

- socket.send.buffer.bytes

socket 的发送缓冲区，socket 的调优参数为 SO_SNDBUFF。

- socket.receive.buffer.bytes

socket 的接收缓冲区，socket 的调优参数为 SO_RCVBUFF。

- socket.request.max.bytes

socket 请求的最大数值，防止服务器端的内存溢出。message.max.bytes 必然要小于 socket.request.max.bytes，会被 topic 创建时的指定参数覆盖。

- log.dirs

Kafka 数据的存放地址，多个地址用逗号分隔，例如，/data/kafka-logs-1，/data/kafka-logs-2。

- num.partitions

是每个 Topic 的分区个数，若在 Topic 创建时没有指定，则会被 Topic 创建时的指定参数覆盖。

- num.recovery.threads.per.data.dir

在启动恢复日志和关闭刷盘日志时每个数据目录的线程的数量，其默认值为 1。

- offsets.topic.replication.factor

副本数或备份因子，其默认值为 3。

- transaction.state.log.replication.factor

事务主题的副本因子。

- log.retention.hours

日志保存时间（hours|minutes），默认为 7 天（168 小时）。超过这个时间会根据 policy 处理数据。无论 bytes 和 minutes 哪个先达到都会触发。

- log.segment.bytes

控制日志 segment 文件的范围，超出该范围则追加到一个新的日志 segment 文件（-1 表示没有限制）。

- log.retention.check.interval.ms

日志片段文件的检查周期，查看它们是否达到了删除策略的设置（log.retention.hours 或 log.retention.bytes）。

- zookeeper.connect

ZooKeeper 集群的地址，可以是多个地址，多个地址之间用逗号分隔。

- zookeeper.connection.timeout.ms

指定消费者将 offset 更新到 ZooKeeper 中的时间。注意，offset 更新时基于的是 time 而不是基于每次获得的消息。一旦更新 ZooKeeper 发生异常并重启，将可能拿到已经拿到的消息，并连接 ZooKeeper 的超时时间。

- group.initial.rebalance.delay.ms

这是一个新增的参数。对用户来说，这个改进最直接的效果之一就是新增了一个 broker 配置：group.initial.rebalance.delay.ms，其默认时间是 3 秒。用户需要在 server.properties 文件中自行将其修改为想要配置的值。这个参数的主要效果就是让 Kafka 推迟空消费组接收到成员加入请求后，立即开启 rebalance。在实际使用时，假设你预估所有 Consumer 组成员加入需要在 10s 内完成，那么就可以将该参数设置为 10 000。

## 2.4 Kafka 在 ZooKeeper 中保存的数据

在前面的内容中，我们介绍了 Kafka 会将集群的配置信息和元信息存储在 ZooKeeper 中。图 2.17 展示了 Kafka 在 ZooKeeper 集群中保存的信息。

这里我们使用了一个 ZooKeeper 的客户端工具：ZooInspector。这是一个利用 Java 开发的 ZooKeeper 客户端工具，利用这个工具可以很方便地管理 ZooKeeper 中的节点信息。

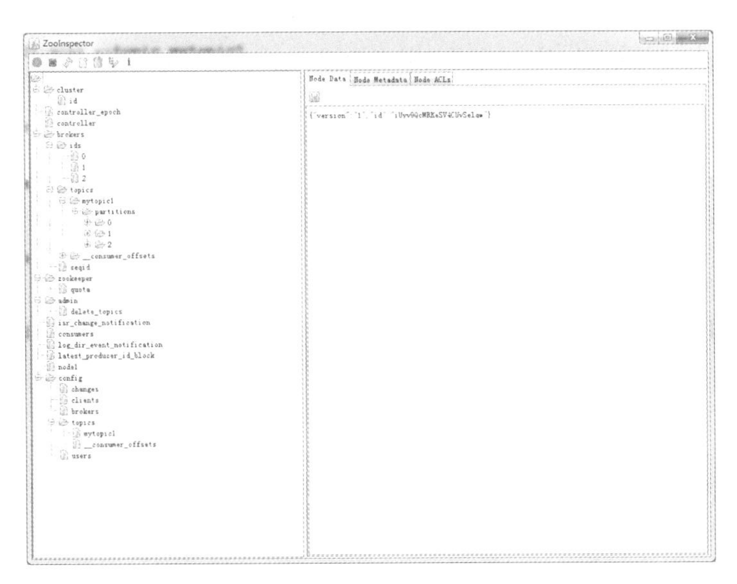

图 2.17 Kafka 在 ZooKeeper 中存储的信息

关于 Kafka 在 ZooKeeper 中存储的信息，我们对其中主要的部分进行如下的说明，如表 2.1 所示。

表 2.1 ZooKeeper 存储信息

| /brokers/ids | Broker 注册信息 |
| --- | --- |
| /brokers/topics/[topic]<br>/brokers/topics/[topic]/partitions/[0...N] | Topic 与分区的注册信息 |
| /controller_epoch | 此值为一个数字，Kafka 集群中第一个 Broker 第一次启动时值为 1，以后只要集群中 center controller（中央控制器）所在的 Broker 变更或挂掉，就会重新选举新的 center controller，每次 center controller 变更，controller_epoch 值就会加 1 |
| /consumers/[groupId]/ids/[consumerIdString] | Consumer 注册信息。每个 Consumer 都有一个唯一的 ID（consumerId 可以通过配置文件指定，也可以由系统生成），此 id 用来标记消费者信息 |

## 2.5 开发客户端程序测试 Kafka

Apache Kafka 是吞吐量巨大的一个消息系统，是用 Scala 语言写的，同时也支持 Java 语言。可以通过搭建 Maven 的工程来开发对应的应用程序，以下是需要在 Maven 的 pom.xml 文件中添加的依赖。

```
01    <dependency>
```

```
02        <groupId>org.apache.kafka</groupId>
03        <artifactId>kafka-clients</artifactId>
04        <version>2.4.0</version>
05    </dependency>
```

## 2.5.1 开发 Java 版本的客户端程序

生产者程序。

```
01  import java.util.Properties;
02  import java.util.Scanner;
03  import org.apache.kafka.clients.producer.KafkaProducer;
04  import org.apache.kafka.clients.producer.Producer;
05  import org.apache.kafka.clients.producer.ProducerRecord;
06
07  public class ProducerDemo {
08
09      public static void main(String[] args) throws InterruptedException {
10          Properties props = new Properties();
11          props.put("bootstrap.servers", "kafka101:9092");
12          props.put("acks", "all");
13
14          props.put("retries", 0);
15          props.put("batch.size", 16384);
16          props.put("linger.ms", 1);
17          props.put("buffer.memory", 33554432);
18
19          props.put("key.serializer",
20                  "org.apache.kafka.common.serialization.StringSerializer");
21          props.put("value.serializer",
22                  "org.apache.kafka.common.serialization.StringSerializer");
23
24          Producer<String,  String>  producer  =  new  KafkaProducer<String,
String>(props);
25          for(int i=0;i<10;i++) {
26              producer.send(new ProducerRecord<String, String>
27                          ("mytopic1", "key"+i, "value"+i));
28              Thread.sleep(1000);
29          }
30          producer.close();
31      }
32  }
```

其中，

- 第 12 行代码，表示服务器端在接收到消息后，生产者需要进行反馈确认的尺度，其主要用于消息的可靠性传输。
  - acks=0 表示生产者不需要来自服务器端的确认。
  - acks=1 表示服务器端将消息保存后即可发送 ack，不需要等到其他 follower 角色都收到该消息。
  - acks=all（或 acks=-1）意味着服务器端将等待所有副本都被接收后才发送确认。
- 第 14 行代码，表示生产者发送失败后重试的次数。
- 第 15 行代码，表示当多条消息发送到同一个 partition 时，该值控制生产者批量发送消息的大小。批量发送可以减少生产者到服务器端的请求数，有助于提高客户端和服务器端的性能。
- 第 16 行代码，表示在默认情况下缓冲区的消息会被立即发送到服务器端，即使缓冲区的空间并没有用完。可以将 linger.ms 设置为大于 0 的值，这样发送者在等待一段时间后，再向服务器端发送请求，以实现每次请求可以尽可能多发送批量消息。
- 第 17 行代码，表示生产者缓冲区的大小，保存的是还未来得及发送到服务器端的消息，如果生产者的发送速度大于消息被提交到服务器端的速度，该缓冲区将被耗尽。
- 第 19～22 行代码，说明了使用何种序列化方式将用户提供的 key 和 value 值序列化成字节。

消费者程序。

```
01  import java.time.Duration;
02  import java.util.Arrays;
03  import java.util.Properties;
04
05  import org.apache.kafka.clients.consumer.ConsumerRecord;
06  import org.apache.kafka.clients.consumer.ConsumerRecords;
07  import org.apache.kafka.clients.consumer.KafkaConsumer;
08
09  public class ConsumerDemo {
10      public static void main(String[] args) {
11           Properties props = new Properties();
12          props.put("bootstrap.servers", "kafka101:9092");
13          props.put("group.id", "mygroup");
14          props.put("enable.auto.commit", "true");
15          props.put("auto.commit.interval.ms", "1000");
16          props.put("key.deserializer",
17                  "org.apache.kafka.common.serialization.StringDeserializer");
18          props.put("value.deserializer",
19                  "org.apache.kafka.common.serialization.StringDeserializer");
20
21          KafkaConsumer<String, String> consumer =
```

```
22                     new KafkaConsumer<String, String>(props);
23
24          consumer.subscribe(Arrays.asList("mytopic1"));
25          while (true) {
26            ConsumerRecords<String, String> records =
27                      consumer.poll(Duration.ofMillis(100));
28
29            for (ConsumerRecord<String, String> record : records)
30              System.out.println("收到消息: "+ record.key() + "\t"
31                                              + record.value());
32          }
33      }
34  }
```

其中,

- 第 13 行代码,表示 Kafka 使用消费者分组的概念来允许多个消费者共同消费和处理同一个 Topic 中的消息。分组中的消费者成员是动态维护的,如果一个消费者处理失败了,那么之前分配给它的 partition 将被重新分配给分组中的其他消费者;同样,如果分组中加入了新的消费者,也将触发整个 partition 重新分配,每个消费者将尽可能地分配到相同数目的 partition,以达到新的均衡状态。
- 第 14 行代码,表示用于配置消费者是否自动提交消费的进度。
- 第 15 行代码,表示用于配置自动提交消费进度的时间。
- 第 16～19 行代码,说明了使用何种序列化方式将用户提供的 key 和 value 值序列化成字节。

## 2.5.2 开发 Scala 版本的客户端程序

生产者应用程序。

```
01  import org.apache.kafka.clients.producer.KafkaProducer
02  import java.util.Properties
03  import org.apache.kafka.clients.producer.ProducerRecord
04  import org.apache.kafka.clients.producer.RecordMetadata
05  import org.apache.kafka.clients.producer.ProducerConfig
06
07  object DemProducer {
08    def main(args: Array[String]): Unit = {
09     val props = new Properties
10     props.put("bootstrap.servers", "kafka101:9092")
11     props.put("acks", "all");
12     props.put("retries", "0")
13          props.put("batch.size", "16384")
14          props.put("linger.ms", "1")
```

```
15          props.put("buffer.memory", "33554432")
16
17          props.put("key.serializer",
18                  "org.apache.kafka.common.serialization.StringSerializer");
19          props.put("value.serializer",
20                  "org.apache.kafka.common.serialization.StringSerializer");
21
22          val producer = new KafkaProducer[String,String](props)
23          var i = 0
24          while(i< 10){
25              producer.send(new ProducerRecord("mytopic1",
26                                          "scala_key"+i,
27                                            "scala_value"+i))
28              i = i+1
29              Thread.sleep(1000)
30          }
31          producer.close()
32    }
33  }
```

其中，参数的含义与 Java 版本的生产者参数设置的含义一致，参考前面章节中的相应参数说明。

消费者应用程序。

```
01  import java.util.Properties
02  import org.apache.kafka.clients.consumer.KafkaConsumer
03  import java.util.Arrays
04  import java.time.Duration
05  import org.apache.kafka.clients.consumer.ConsumerRecord
06
07  object DemoConsumer {
08    def main(args: Array[String]): Unit = {
09      val props = new Properties
10      props.put("bootstrap.servers", "kafka101:9092")
11
12      props.put("group.id", "mygroup");
13      props.put("enable.auto.commit", "true");
14      props.put("auto.commit.interval.ms", "1000");
15      props.put("key.deserializer",
16              "org.apache.kafka.common.serialization.StringDeserializer");
17      props.put("value.deserializer",
18              "org.apache.kafka.common.serialization.StringDeserializer");
19
20      val consumer = new KafkaConsumer(props)
21      consumer.subscribe(Arrays.asList("mytopic1"));
```

```
22      while (true) {
23        val records = consumer.poll(Duration.ofMillis(100));
24        val its = records.iterator()
25
26        while(its.hasNext()){
27          val message = its.next()
28           System.out.println("收到消息: "+ message.key()
29                                + "\t" + message.value())
30        }
31      }
32      consumer.close()
33    }
34  }
```

其中参数的含义与 Java 版本的消费者参数设置的含义一致，参考前面章节中的相应参数说明。

# 第3章　Kafka的生产者

通过学习前两章的内容，我们已经掌握了Kafka的核心体系架构，并部署了Kafka的集群环境。消息的生产者可能存在多个，那么这些生产者产生的消息是怎么传到Kafka应用程序的？发送过程是怎么样的？本章将详细讨论Kafka生产者的执行逻辑。

## 3.1　Kafka生产者的执行过程

生产者产生的消息发送到Kafka应用程序的发送过程，如图3.1所示。

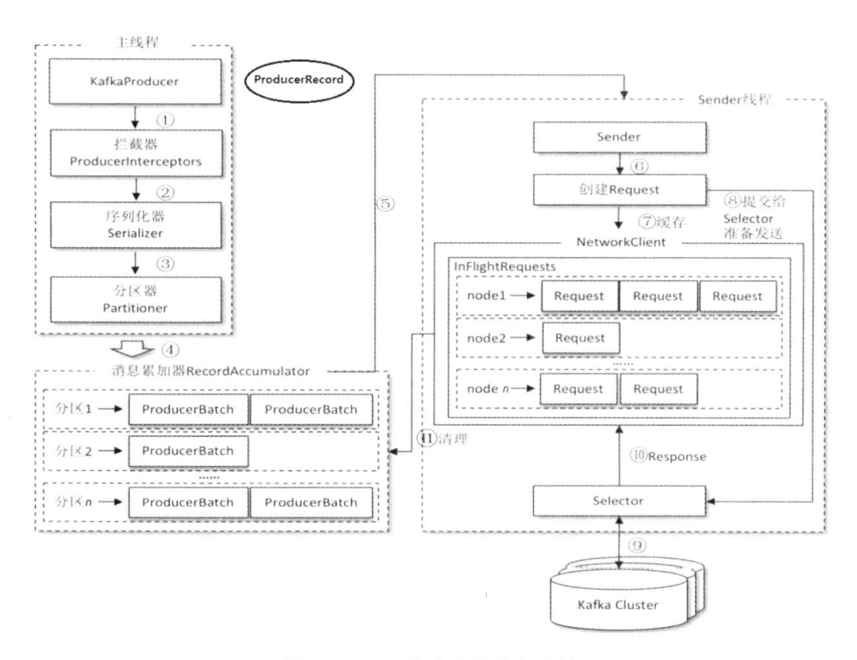


图3.1　Kafka生产者的执行过程

生产者客户端由两个线程协调运行，这两个线程分别为主线程和Sender线程（发送线程），其中，

- 在主线程中由 KafkaProducer 创建消息，然后通过可能的拦截器、序列化器和分区器的作用之后缓存到消息累加器（RecordAccumulator，也称消息收集器）中。
- Sender 线程负责从 RecordAccumulator 中获取消息并将其发送到 Kafka 中。

我们从创建一个 KafkaProducer 对象开始，它将创建一个 ProducerRecord 对象。这个对象是 Kafka 中的一个核心类，它代表生产者发送到 Kafka 服务器端的一个消息对象，即一个 Key-Value 的键值对。在 ProducerRecord 对象中，包含如下信息。

- Kafka 服务器端的主题名称（Topic Name）。
- Topic 中可选的分区号。
- 时间戳。
- 其他 Key-Value 键值对。

其中，十分重要的就是 Kafka 服务器端的主题名称。

ProducerRecord 创建成功后，需要经过拦截器、序列化器将其转换为字节数组，这样它们才能够在网络上传输，然后消息到达分区器。分区器的作用是根据发送过程中指定的有效的分区号，将 ProducerRecord 发送到该分区；如果没有指定 Topic 中的分区号，则会根据 Key 进行 Hash 运算，将 ProducerRecord 映射到一个对应的分区。

ProducerRecord 默认采用当前的时间，用户也在创建 ProducerRecord 的时候提供一个时间戳。Kafka 最终使用的时间戳取决于 Topic 的配置，而 Topic 时间戳的配置主要有以下两种：

- CreateTime 表示使用生产者产生的时间戳作为 Kafka 最终的时间戳。
- LogAppendTime 表示生产者记录中的时间戳在将消息添加到其日志中时，将由 Kafka Broker 重写。

ProducerRecord 在经过主线程后，最终由发送线程发送到 Kafka 服务器端。Kafka Broker 在收到消息时会返回一个响应，如果写入成功，则返回一个 RecordMetaData 对象，它包含主题和分区信息，以及记录在分区里的偏移量，上面两种时间戳类型也会返回给用户。如果写入失败，则返回一个错误。生产者在收到错误之后会尝试重新发送消息，几次之后如果还是写入失败，就返回错误消息。

## 3.2 创建 Kafka 生产者

### 3.2.1 创建基本的消息生产者

要在 Kafka 消息集群中写入消息，首先需要创建一个生产者对象。Kafka 生产者有三个必选的属性，其他属性是可选的。下面列出了这三个必须设置的属性及它们的含义。

（1）bootstrap.servers。

该属性指定 Kafka 集群中 Broker 的地址列表，其地址的格式为 host:port，如果有多个 Broker 地址，可以用逗号进行分隔。当然，该地址列表中不需要包含所有 Broker，因为 Kafka 会将整个集群的元信息和配置信息存储在 ZooKeeper 中，生产者会从给定的 Broker 中，通过 ZooKeeper 查找到其他 Kafka Broker 地址信息。在生产环境中，建议至少要提供两个 Broker 地址信息，这样做的目的是支持容错。一旦其中一个 Broker 出现了宕机，生产者仍然能够通过另一个 Broker 连接到 Kafka 集群上。

（2）key.serializer。

发送到 Kafka 的消息需要经过序列化后，才能实现正常发送与转发。生产者将要发送的消息通过序列化器进行序列化后，生成一个 key/value 的值，并发送到 Broker。当创建 Kafka 生产者的时候，生产者需要知道采用何种方式把消息（即 Java 对象）转换为字节数组。因此通过生产者端的参数 key.serializer 就是这一项配置的工作。它必须实现 org.apache.kafka.common.serialization.Serializer 接口，然后生产者会使用这个类把键对象序列化为字节数组。

（3）value.serializer。

value.serializer 与 key.serializer 一样，用于指定的类会将值序列化。创建一个 Kafka 生产者的代码如下。

```
01  Properties props = new Properties();
02  props.put("bootstrap.servers", "kafka101:9092");
03  props.put("acks", "all");
04
05  props.put("retries", 0);
06  props.put("batch.size", 16384);
07  props.put("linger.ms", 1);
08  props.put("buffer.memory", 33554432);
09
10  props.put("key.serializer",
11          "org.apache.kafka.common.serialization.StringSerializer");
12  props.put("value.serializer",
13          "org.apache.kafka.common.serialization.StringSerializer");
14
15  Producer<String, String> producer = new KafkaProducer<String, String>(props);
```

其中，第 05 行～第 08 行代码不是必需的，如果没有配置这些参数，将会采用默认的参数值。

### 3.2.2　发送自定义消息对象

前面提到，Kafka 生产者发送的消息必须经过序列化。实现序列化可以简单地总结为两步，第一步继承序列化 Serializer 接口；第二步实现接口方法，将指定类型序列化成 byte[]，或者将 byte[]反序列化成指定数据类型。接下来，我们来实现序列化/反序列化方式。

实现 Java 对象的序列化有很多不同的方式。这里我们介绍基于 FastJson 的序列化方式。Fastjson 是一个 Java 库，可以将 Java 对象转换为 JSON 格式，当然它也可以将 JSON 字符串转换为 Java 对象，加入以下依赖。

```
01  <dependency>
02      <groupId>com.alibaba</groupId>
03      <artifactId>fastjson</artifactId>
04      <version>1.2.68</version>
05  </dependency>
```

下面通过一个具体的案例演示如何使用 Fastjson 的一个 Java 对象进行序列化，并将其作为 Kafka 生产者的消息发送到 Kafka 消息集群中。使用下面的员工数据进行测试，其数据格式描述如表 3-1 所示。

表 3-1　员工数据格式

| 字　段 | 类　型 | 说　明 |
|---|---|---|
| empno | int | 员工号 |
| ename | String | 员工姓名 |
| job | String | 员工职位 |
| mgr | int | 员工经理的员工号 |
| hiredate | String | 入职日期 |
| sal | int | 月薪 |
| comm | int | 奖金 |
| deptno | int | 部门号 |

下面展示了 Employee.java 的完整代码，为了输出结果方便，这里还重写了 Employee 类的 toString 方法。

```
01  public class Employee {
02
03      private int empno;
04      private String ename;
05      private String job;
06      private int mgr;
07      private String hiredate;
08      private int sal;
09      private int comm;
10      private int deptno;
11
12      public Employee() {
13
14      }
15
```

```
16       @Override
17       public String toString() {
18           return "Employee [empno=" + empno + ", ename=" + ename
19                   + ", job=" + job + ", mgr=" + mgr + ", hiredate="
20                   + hiredate + ", sal=" + sal + ", comm="
21                   + comm + ", deptno=" + deptno + "]";
22       }
23
24       public int getEmpno() {
25           return empno;
26       }
27
28       public void setEmpno(int empno) {
29           this.empno = empno;
30       }
31
32       public String getEname() {
33           return ename;
34       }
35
36       public void setEname(String ename) {
37           this.ename = ename;
38       }
39
40       public String getJob() {
41           return job;
42       }
43
44       public void setJob(String job) {
45           this.job = job;
46       }
47
48       public int getMgr() {
49           return mgr;
50       }
51
52       public void setMgr(int mgr) {
53           this.mgr = mgr;
54       }
55
56       public String getHiredate() {
57           return hiredate;
58       }
59
```

```
60      public void setHiredate(String hiredate) {
61          this.hiredate = hiredate;
62      }
63
64      public int getSal() {
65          return sal;
66      }
67
68      public void setSal(int sal) {
69          this.sal = sal;
70      }
71
72      public int getComm() {
73          return comm;
74      }
75
76      public void setComm(int comm) {
77          this.comm = comm;
78      }
79
80      public int getDeptno() {
81          return deptno;
82      }
83
84      public void setDeptno(int deptno) {
85          this.deptno = deptno;
86      }
87  }
```

为了将 Employee 对象进行序列化，创建一个 EmployeeJSONSerializer 类并使用 Fastjson 将其序列化成一个 JSON 对象，完整的代码如下。

```
01  import org.apache.kafka.common.serialization.Serializer;
02
03  import com.alibaba.fastjson.JSON;
04
05  public class EmployeeJSONSerializer implements Serializer<Employee> {
06
07      @Override
08      public byte[] serialize(String topic, Employee data) {
09          return JSON.toJSONBytes(data);
10      }
11  }
```

最后，创建一个 EmployeeProducer 来将 Employee 对象发送到 Kafka 集群的 Broker 上。

```
01  import java.util.Properties;
```

```
02
03  import org.apache.kafka.clients.producer.KafkaProducer;
04  import org.apache.kafka.clients.producer.Producer;
05  import org.apache.kafka.clients.producer.ProducerConfig;
06  import org.apache.kafka.clients.producer.ProducerRecord;
07
08  public class EmployeeProducer {
09
10      public static void main(String[] args) throws Exception {
11          Properties props = new Properties();
12          props.setProperty(ProducerConfig.BOOTSTRAP_SERVERS_CONFIG,
13                  "kafka101:9092");
14          props.setProperty(ProducerConfig.KEY_SERIALIZER_CLASS_CONFIG,
15                  "org.apache.kafka.common.serialization.StringSerializer");
16          props.setProperty(ProducerConfig.VALUE_SERIALIZER_CLASS_CONFIG,
17                  "EmployeeJSONSerializer");
18
19          //创建生产者
20          Producer<String, Employee> producer =
21                  new KafkaProducer<String, Employee>(props);
22          for(int i=0;i<10;i++) {
23              Employee emp = new Employee();
24
25              emp.setEmpno(i);                    //设置员工号
26              emp.setEname("Ename" + i);          //设置员工姓名
27              emp.setJob("Job" + i);              //设置员工职位
28              emp.setMgr(1000+i);                 //设置员工的经理员工号
29              emp.setHiredate("2020-12-01");      //设置入职日期
30              emp.setSal(6000);                   //设置月薪
31              emp.setComm(2000);                  //设置奖金
32              emp.setDeptno(10);                  //设置部门号
33
34              producer.send(new ProducerRecord<String, Employee>(
35                              "mytopic1",
36                              String.valueOf(emp.getEmpno()),
37                              emp));
38
39              //每两秒发送一个 Employee 对象
40              Thread.sleep(2000);
41          }
42          producer.close();
43      }
44  }
```

我们创建 EmployeeConsumer 来消费消息，图 3.2 展示了程序运行的效果。

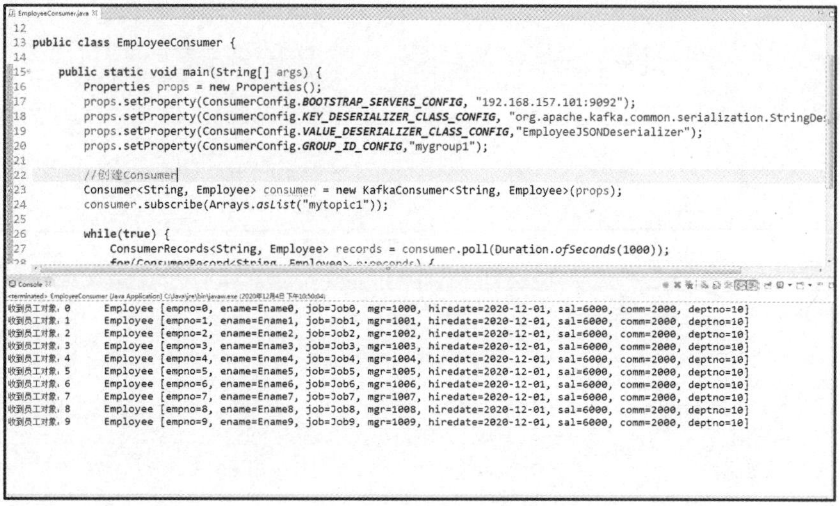

```
12
13 public class EmployeeConsumer {
14
15     public static void main(String[] args) {
16         Properties props = new Properties();
17         props.setProperty(ConsumerConfig.BOOTSTRAP_SERVERS_CONFIG, "192.168.157.101:9092");
18         props.setProperty(ConsumerConfig.KEY_DESERIALIZER_CLASS_CONFIG, "org.apache.kafka.common.serialization.StringDe
19         props.setProperty(ConsumerConfig.VALUE_DESERIALIZER_CLASS_CONFIG,"EmployeeJSONDeserializer");
20         props.setProperty(ConsumerConfig.GROUP_ID_CONFIG,"mygroup1");
21
22         //创建Consumer
23         Consumer<String, Employee> consumer = new KafkaConsumer<String, Employee>(props);
24         consumer.subscribe(Arrays.asList("mytopic1"));
25
26         while(true) {
27             ConsumerRecords<String, Employee> records = consumer.poll(Duration.ofSeconds(1000));
```

```
收到员工对象，0    Employee [empno=0, ename=Ename0, job=Job0, mgr=1000, hiredate=2020-12-01, sal=6000, comm=2000, deptno=10]
收到员工对象，1    Employee [empno=1, ename=Ename1, job=Job1, mgr=1001, hiredate=2020-12-01, sal=6000, comm=2000, deptno=10]
收到员工对象，2    Employee [empno=2, ename=Ename2, job=Job2, mgr=1002, hiredate=2020-12-01, sal=6000, comm=2000, deptno=10]
收到员工对象，3    Employee [empno=3, ename=Ename3, job=Job3, mgr=1003, hiredate=2020-12-01, sal=6000, comm=2000, deptno=10]
收到员工对象，4    Employee [empno=4, ename=Ename4, job=Job4, mgr=1004, hiredate=2020-12-01, sal=6000, comm=2000, deptno=10]
收到员工对象，5    Employee [empno=5, ename=Ename5, job=Job5, mgr=1005, hiredate=2020-12-01, sal=6000, comm=2000, deptno=10]
收到员工对象，6    Employee [empno=6, ename=Ename6, job=Job6, mgr=1006, hiredate=2020-12-01, sal=6000, comm=2000, deptno=10]
收到员工对象，7    Employee [empno=7, ename=Ename7, job=Job7, mgr=1007, hiredate=2020-12-01, sal=6000, comm=2000, deptno=10]
收到员工对象，8    Employee [empno=8, ename=Ename8, job=Job8, mgr=1008, hiredate=2020-12-01, sal=6000, comm=2000, deptno=10]
收到员工对象，9    Employee [empno=9, ename=Ename9, job=Job9, mgr=1009, hiredate=2020-12-01, sal=6000, comm=2000, deptno=10]
```

图 3.2　EmployeeConsumer 的运行效果

关于 EmployeeConsumer 的完整代码，将会在后面章节展示。

## 3.3　生产者的消息发送模式

Kafka 生产者的消息发送主要有三种模式：发后即忘（fire-and-forget）、同步模式（sync）及异步模式（async）。

发后即忘：只管向 Kafka 中发送消息而并不用关心消息是否正确到达。在大多数情况下，这种发送方式没有什么问题，不过在某些时候（比如发生不可重试异常时）会造成消息的丢失。这种发送方式的性能最高，可靠性也最差。EmployeeProducer 就是采用的这种模式。

同步模式：要实现同步的发送方式，可以利用返回的 Future 对象的阻塞等待 Kafka 的响应即可实现，直到消息发送成功。如果发生异常，就需要捕获异常并交由外层逻辑处理。改造一下之前的 EmployeeProducer 代码，使用同步模式将消息发送到 Kafka 服务器端。

```
01   Future<RecordMetadata> future = producer.send(new ProducerRecord<String,
Employee>(
02                                                     "mytopic1",
03                                                    String.valueOf(emp.getEmpno()),
04                                                     emp));
05
06  //同步模式，会阻塞在这里，直到 Kafka 服务器端返回结果
07  RecordMetadata recordMetadata = future.get();
08  if (recordMetadata != null) {
09      System.out.println("偏移量：" + recordMetadata.offset() +
```

```
10                    "; 分区:" + recordMetadata.partition());
11    }
12
13    //每两秒发送一个 Employee 对象
14    Thread.sleep(2000);
```

异步模式：为了在异步发送消息的同时能够对异常情况进行处理，生产者提供了回调支持。一般在 send()方法中指定一个 Callback 的回调函数，Kafka 在返回响应时调用该函数来实现异步的发送确认。改造一下之前的 EmployeeProducer 生产者代码，使用异步模式将消息发送到 Kafka 服务器端。

```
01    ProducerRecord record = new ProducerRecord<String,Employee>(
02                                 "mytopic1",
03                                 String.valueOf(emp.getEmpno()),
04                                 emp);
05
06    producer.send(record, new Callback() {
07        @Override
08        public void onCompletion(RecordMetadata recordMetadata, Exception e) {
09            if (e != null) {
10                e.printStackTrace();
11            }
12            if (recordMetadata != null) {
13                System.out.println("偏移量:" + recordMetadata.offset() +
14                                  ";分区:" + recordMetadata.partition());
15            }
16        }
17    });
18
19    //每两秒发送一个 Employee 对象
20    Thread.sleep(2000);
```

## 3.4　生产者的高级特性

### 3.4.1　生产者分区机制

Kafka 消息系统为什么要进行 Topic 的分区呢？我们都知道 Kafka 的主题 Topic 是由分区组成的，而将 Topic 进行分区的主要目的就是提供负载均衡和容错的能力，以及实现系统的高伸缩性和高可用性。Kafka 的消息组织方式实际上是三层结构：主题—分区—消息。主题下的每条消息只会保存在某一个分区中，而不会在多个分区中保存多份。在创建 Topic 的时候可以指定每个分区的副本数，用于支持分区中消息的容错。图 3.3 是 Kafka 官方网站上的截图，展示了 Kafka 消息模型的三层结构。

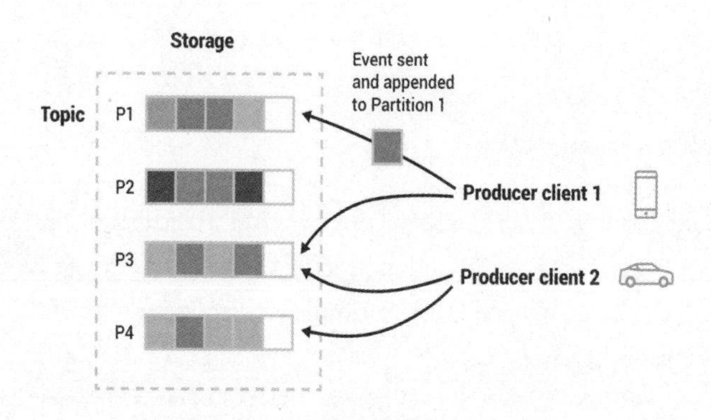


图 3.3 Kafka 消息模型的三层结构

不同的分区能够放置在 Kafka 集群中不同的节点上，而生产者和消费者在产生消息和消费消息的时候，也都是针对分区进行的，这样每个节点的机器都能独立执行各自分区的读写请求处理，并且还可以通过添加新的 Kafka 节点来增加整体系统的吞吐量。

既然 Kafka 提供了分区的机制，那么 Kafka 又为我们提供了哪些分区的策略呢？所谓的分区策略是决定生产者将消息发送到哪个分区的算法。Kafka 生产者的分区策略都实现了接口 org.apache.kafka.clients.producer.Partitioner，常见的分区策略主要有几下几种。

（1）默认分区策略（org.apache.kafka.clients.producer.internals.DefaultPartitioner）。

下面的注释摘至源码中的说明。

```
01   /**
02    * The default partitioning strategy:
03    * <ul>
04    * <li>If a partition is specified in the record, use it
05    * <li>If no partition is specified but a key is present choose a partition
based on a hash of the key
06     * <li>If no partition or key is present choose the sticky partition that
changes when the batch is full.
07    *
08    * See KIP-480 for details about sticky partitioning.
09    */
```

从注释的说明可以看出，默认的分区机制将按照以下的逻辑进行分区。

- 如果记录中指定了分区，则可以直接使用。
- 如果记录中未指定分区，但指定了 key 值，则根据 key 的 hash 值选择一个分区。这种策略的本质就是 Hash 分区。
- 如果记录中未指定分区，也未指定 key 值，则以黏性分区策略选择一个分区。

下面的 partition 方法实现了 Kafka 默认的分区机制。

```
01   /**
02    * Compute the partition for the given record.
03    *
```

```
04    * @param topic The topic name
05    * @param key The key to partition on (or null if no key)
06    * @param keyBytes serialized key to partition on (or null if no key)
07    * @param value The value to partition on or null
08    * @param valueBytes serialized value to partition on or null
09    * @param cluster The current cluster metadata
10    */
11   public int partition(String topic, Object key,
12                        byte[] keyBytes,Object value,
13                            byte[] valueBytes, Cluster cluster) {
14       if (keyBytes == null) {
15           return stickyPartitionCache.partition(topic, cluster);
16       }
17       List<PartitionInfo> partitions = cluster.partitionsForTopic(topic);
18       int numPartitions = partitions.size();
19       // hash the keyBytes to choose a partition
20       return Utils.toPositive(Utils.murmur2(keyBytes)) % numPartitions;
21   }
```

（2）轮询分区策略（org.apache.kafka.clients.producer.RoundRobinPartitioner）。

如果 key 值为 null，并且使用了默认的分区器，Kafka 会根据轮询（Random Robin）策略将消息均匀地分布到各个分区上。下面是 RoundRobinPartitioner 的核心代码。

```
01   /**
02    * Compute the partition for the given record.
03    *
04    * @param topic The topic name
05    * @param key The key to partition on (or null if no key)
06    * @param keyBytes serialized key to partition on (or null if no key)
07    * @param value The value to partition on or null
08    * @param valueBytes serialized value to partition on or null
09    * @param cluster The current cluster metadata
10    */
11   @Override
12   public int partition(String topic, Object key,
13                        byte[] keyBytes, Object value,
14                            byte[] valueBytes, Cluster cluster) {
15       List<PartitionInfo> partitions = cluster.partitionsForTopic(topic);
16       int numPartitions = partitions.size();
17       int nextValue = nextValue(topic);
18       List<PartitionInfo> availablePartitions = cluster.
availablePartitionsForTopic(topic);
19       if (!availablePartitions.isEmpty()) {
20           int part = Utils.toPositive(nextValue) % availablePartitions.size();
21           return availablePartitions.get(part).partition();
```

```
22        } else {
23            // no partitions are available, give a non-available partition
24            return Utils.toPositive(nextValue) % numPartitions;
25        }
26    }
27
28    private int nextValue(String topic) {
29        AtomicInteger counter = topicCounterMap.computeIfAbsent(topic, k -> {
30            return new AtomicInteger(0);
31        });
32        return counter.getAndIncrement();
33    }
```

（3）黏性分区策略（org.apache.kafka.clients.producer.UniformStickyPartitioner）。

黏性分区策略就像黏住这个分区一样，只要这个分区没有被填满，就会尽可能地坚持使用该分区。这种策略首先会选择单个分区发送所有无 key 的消息，一旦这个分区已填满，黏性分区策略就会随机选择另一个分区。

通过查看源码，可以得到黏性分区策略是通过 org.apache.kafka.clients.producer.internals.StickyPartitionCache 来实现的，下面展示了 StickyPartitionCache 中的核心代码。

```
01    public int nextPartition(String topic, Cluster cluster, int prevPartition)
{
02        List<PartitionInfo> partitions = cluster.partitionsForTopic(topic);
03        Integer oldPart = indexCache.get(topic);
04        Integer newPart = oldPart;
05        // Check that the current sticky partition for the topic is either not
set or that the partition that
06        // triggered the new batch matches the sticky partition that needs to be
changed.
07        if (oldPart == null || oldPart == prevPartition) {
08            List<PartitionInfo> availablePartitions = cluster.
availablePartitionsForTopic(topic);
09            if (availablePartitions.size() < 1) {
10                Integer random = Utils.toPositive(ThreadLocalRandom.current().
nextInt());
11                newPart = random % partitions.size();
12            } else if (availablePartitions.size() == 1) {
13                newPart = availablePartitions.get(0).partition();
14            } else {
15                while (newPart == null || newPart.equals(oldPart)) {
16                    Integer   random   =   Utils.toPositive(ThreadLocalRandom.
current().nextInt());
17                    newPart  =  availablePartitions.get(random  %  availablePartitions.
size()).partition();
18                }
```

```
19          }
20          // Only change the sticky partition if it is null or prevPartition
matches the current sticky partition.
21          if (oldPart == null) {
22              indexCache.putIfAbsent(topic, newPart);
23          } else {
24              indexCache.replace(topic, prevPartition, newPart);
25          }
26          return indexCache.get(topic);
27      }
28      return indexCache.get(topic);
29  }
```

（4）散列分区策略。

如果键值不为 null，并且使用了默认的分区器，Kafka 会对键进行散列，然后根据散列值把消息映射到对应的分区上。下面的代码是在 DefaultPartitioner 中使用的散列分区。

```
01  // hash the keyBytes to choose a partition
02  return Utils.toPositive(Utils.murmur2(keyBytes)) % numPartitions;
```

（5）自定义分区策略。前面提到 Kafka 生产者的分区策略都实现了接口 org.apache.kafka.clients. producer.Partitioner，用户可以根据需要对数据使用不一样的分区策略，只需要实现该接口即可。用户创建了自定义分区策略后，只需要在生产者的 Properties 中指定 ProducerConfig.PARTITIONER_CLASS_CONFIG 参数即可，代码如下所示。

```
01  props.setProperty(ProducerConfig.PARTITIONER_CLASS_CONFIG,"自定义分区策略")
```

下面我们通过一个具体的例子来实现 Kafka 生产者的自定义分区。在发送到 Kafka 系统的消息中，key 将包含员工所在的部门号，这里将根据部门号来建立分区。例如，10 号部门的员工数据将发送到 0 号分区；20 号部门的员工数据将发送到 1 号分区；30 号部门的员工数据将发送到 2 号分区；其他部门的员工数据将发送到 3 号分区。

完整的代码如下。

```
01  package demo;
02
03  import java.util.List;
04  import java.util.Map;
05
06  import org.apache.kafka.clients.producer.Partitioner;
07  import org.apache.kafka.common.Cluster;
08  import org.apache.kafka.common.PartitionInfo;
09
10  public class CustomerPartitioner implements Partitioner {
11
12      @Override
13      public void configure(Map<String, ?> configs) {
```

```
14      }
15
16      @Override
17      public int partition(String topic, Object key,
18                         byte[] keyBytes, Object value,
19                         byte[] valueBytes, Cluster cluster) {
20
21          //获取员工的部门号
22          int deptno = (Integer) key;
23                              List<PartitionInfo>    partitionInfoList    =
cluster.availablePartitionsForTopic(topic);
24          int partitionCount = partitionInfoList.size();
25
26          /*
27           * 10 号部门的员工数据将发送到 0 号分区;
28           * 20 号部门的员工数据将发送到 1 号分区;
29           * 30 号部门的员工数据将发送到 2 号分区;
30           * 其他部门的员工数据将发送到 3 号分区。
31           */
32          if(deptno == 10) {
33           return 0%partitionCount;
34          }else if(deptno == 20) {
35           return 1%partitionCount;
36          }else if(deptno == 30) {
37           return 2%partitionCount;
38          }else {
39           return 3%partitionCount;
40          }
41      }
42
43      @Override
44      public void close() {
45      }
46  }
```

自定义分区策略创建完成后，将其加入生产者的配置参数中。

```
01  props.setProperty(ProducerConfig.PARTITIONER_CLASS_CONFIG,
02             "demo.CustomerPartitioner");
```

## 3.4.2 生产者压缩机制

Kafka 发送消息的时候，可以在生产者端和 Broker 端进行消息的压缩。在一般情况下，建议采用的压缩机制是：生产者端负责压缩；Broker 端负责保持；消费者端负责解压。Kafka 采用这样的压缩机制，主要是节约 CPU 的时间去换磁盘存储的空间，以及网络 I/O 的传输量。这样的做法可以以较小的 CPU 开销带来更少的磁盘占用或更少的网络 I/O 传输。

在目前的 Kafka 版本中，支持 GZIP、Snappy 和 LZ4 三种压缩方式，表 3.2 列举了这三种压缩方式的对比。

表 3.2　三种压缩方式

| 压缩方式 | Size（Before） | Size（After） | 压缩消耗 | 解压消耗 | 最大 CPU |
|---|---|---|---|---|---|
| GZIP | 35984 | 8804 | 2179 | 389 | 26.5 |
| Snappy | 35984 | 13602 | 424 | 88 | 11 |
| LZ4 | 35984 | 16355 | 327 | 147 | 12.6 |

在了解了 Kafka 的压缩方式后，可以在生产者端的配置参数中指定相应的压缩方式，代码如下。

```
01  props.setProperty(ProducerConfig.COMPRESSION_TYPE_CONFIG,"Snappy");
```

## 3.4.3　生产者拦截器

从图 3.1 可以看出，在 KafkaProducer 的主线程中可以创建一个或多个 ProducerInterceptors（拦截器）。拦截器是从 Kafka 0.10 版本中引入的，在生产者端和消费者端均可设置，拦截器主要用于实现生产者端和消费者端的定制化控制逻辑。

对生产者而言，拦截器需要实现 org.apache.kafka.clients.producer. ProducerInterceptor 接口。下面展示了该接口 ProducerInterceptor 中的方法。

```
01  package org.apache.kafka.clients.producer;
02
03  import org.apache.kafka.common.Configurable;
04
05  public interface ProducerInterceptor<K, V> extends Configurable {
06
07      public ProducerRecord<K, V> onSend(ProducerRecord<K, V> record);
08
09          public void onAcknowledgement(RecordMetadata metadata, Exception
exception);
10
11      public void close();
12  }
```

下面对接口中的三个方法做出必要的解释：

（1）onSend。

该方法将在 KafkaProducer.send 方法的主线程中执行。KafkaProducer 确保在消息被序列化以前，调用 ProducerInterceptor.onSend 方法。用户可以在该方法中对消息进行操作，但最好不要修改消息所属的 Topic 和分区，否则会影响目标分区的计算。

（2）onAcknowledgement。

该方法会在消息被确认应答之前或消息发送失败时调用。如果生产者采用的是异步发送机制，该方法通常是在生产者回调逻辑触发之前被调用的。需要注意的是，该方法运行在生产者的 I/O 线程中，因此不要在该方法中放入很重的逻辑，否则会影响生产者的消息发送性能。

（3）close。

关闭拦截器之前，可以将一些资源清理工作放在 close 方法中。

需要注意的是，如果指定了多个连接器，生产者将按照指定顺序调用它们。如果拦截器中出现了异常，生产者会将异常的错误信息记录到错误日志中，而不是向上传递。

当创建完拦截器后，可以通过以下的代码在生产者端指定它们。

```
01   List<String> interceptors = new ArrayList<>();
02   interceptors.add("拦截器 1");
03   interceptors.add("拦截器 2");
04   props.put(ProducerConfig.INTERCEPTOR_CLASSES_CONFIG, interceptors);
```

在了解了拦截器的功能特点后，我们通过一个具体的例子来演示拦截器的使用方法。在这个例子中，将开发两个拦截器，具体的功能需求如下。

（1）拦截器 1：将当前系统的时间戳设置到 Employee 对象的 hiredate 属性上。

（2）拦截器 2：统计生产者发送成功和发送失败的消息总数。

下面是拦截器 1 的完整代码：HireDateTimeStampInterceptor。

```
01   import java.util.Map;
02
03   import org.apache.kafka.clients.producer.ProducerInterceptor;
04   import org.apache.kafka.clients.producer.ProducerRecord;
05   import org.apache.kafka.clients.producer.RecordMetadata;
06
07   public class HireDateTimeStampInterceptor
08           implements ProducerInterceptor<String,Employee> {
09
10       @Override
11       public void configure(Map<String, ?> configs) {
12
13       }
14
15       @Override
16       public ProducerRecord<String, Employee>
17           onSend(ProducerRecord<String, Employee> record) {
18
19           //获取要发送的员工对象，并将入职日期设置为当前的时间戳
20           Employee emp = record.value();
```

```
21            emp.setHiredate(String.valueOf(System.currentTimeMillis()));
22
23            return new ProducerRecord(record.topic(),
24                                       record.partition(),
25                                       record.timestamp(),
26                                       record.key(),emp);
27        }
28
29        @Override
30        public    void    onAcknowledgement(RecordMetadata    metadata,    Exception
exception) {
31        }
32
33        @Override
34        public void close() {
35        }
36    }
```

下面是拦截器 2 的完整代码：ProducerSendCounterInterceptor。

```
01    import java.util.Map;
02
03    import org.apache.kafka.clients.producer.ProducerInterceptor;
04    import org.apache.kafka.clients.producer.ProducerRecord;
05    import org.apache.kafka.clients.producer.RecordMetadata;
06
07    public class ProducerSendCounterInterceptor
08        implements ProducerInterceptor<String,Employee> {
09
10       private int errorCounter = 0;
11       private int successCounter = 0;
12
13        @Override
14        public void configure(Map<String, ?> configs) {
15        }
16
17        @Override
18        public ProducerRecord<String, Employee>
19                onSend(ProducerRecord<String, Employee> record) {
20            return record;
21        }
22
23        @Override
24        public    void    onAcknowledgement(RecordMetadata    metadata,    Exception
exception) {
25           if (exception == null) {
```

```
26              successCounter++;
27          } else {
28              errorCounter++;
29          }
30      }
31
32      @Override
33      public void close() {
34          // 输出结果
35          System.out.println("发送成功的消息总数: " + successCounter);
36          System.out.println("发送失败的消息总数: : " + errorCounter);
37      }
38  }
```

改造 EmployeeProducer 代码，将拦截器加入 Kafka 的生产者中，下面展示了需要添加的代码。

```
01  //加入拦截器
02  List<String> interceptors = new ArrayList<>();
03  interceptors.add("HireDateTimeStampInterceptor");
04  interceptors.add("ProducerSendCounterInterceptor");
05  props.put(ProducerConfig.INTERCEPTOR_CLASSES_CONFIG, interceptors);
```

代码开发完成后，就可以进行测试了。首先，启动 Kafka 集群，然后我们使用 Kafka 自带的 Consumer Console 工具来测试收到的数据。执行下面的命令启动 Consumer Console。

```
bin/kafka-console-consumer.sh --bootstrap-server kafka101:9092 --topic mytopic1
```

再启动 EmployeeProducer 程序，观察输出的结果。可以在 Consumer Console 上输出如下信息。可以看到通过拦截器已经将 Employee 对象的 hiredate 属性设置为当前系统的系统时间戳，如图 3.4 所示。

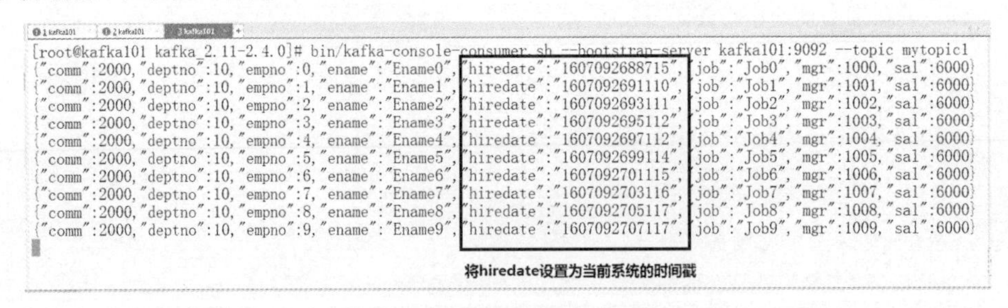


图 3.4　在 Consumer Console 上输出结果

这里我们使用 Kafka 自带的 Consumer Console 命令行工具将接收到的消息直接输出到命令行中。后面章节将介绍如何使用代码程序接收自定义的消息对象。

图 3.5 展示了 EmployeeProducer 运行后的输出结果，从输出的结果可以看到生产者发送成功的消息总数和发送失败的消息总数。

```
14 public class EmployeeProducer {
15
16     public static void main(String[] args) throws Exception {
17         Properties props = new Properties();
18         props.setProperty(ProducerConfig.BOOTSTRAP_SERVERS_CONFIG,
19                 "192.168.157.101:9092");
20         props.setProperty(ProducerConfig.KEY_SERIALIZER_CLASS_CONFIG,
21                 "org.apache.kafka.common.serialization.StringSerializer");
22         props.setProperty(ProducerConfig.VALUE_SERIALIZER_CLASS_CONFIG,
23                 "EmployeeJSONSerializer");
24
25         //加入拦截器
26         List<String> interceptors = new ArrayList<>();
27         interceptors.add("HireDateTimeStampInterceptor");
28         interceptors.add("ProducerSendCounterInterceptor");
29         props.put(ProducerConfig.INTERCEPTOR_CLASSES_CONFIG, interceptors);
30
31         //创建生产者
32         Producer<String, Employee> producer =
33                 new KafkaProducer<String, Employee>(props);
34         for(int i=0;i<10;i++) {
35             Employee emp = new Employee();
36
37             emp.setEmpno(i);//设置员工号

Console
SLF4J: Failed to load class "org.slf4j.impl.StaticLoggerBinder".
SLF4J: Defaulting to no-operation (NOP) logger implementation
SLF4J: See http://www.slf4j.org/codes.html#StaticLoggerBinder for further details.
成功发送的消息总数：10
发送失败的消息总数：0
```

通过拦截器统计得到了生产者Producer发送成功与发送失败的消息数目。

图 3.5　EmployeeProducer 执行的输出

## 3.5　生产者的参数配置

Kafka 生产者端的配置参数，除了之前介绍过的 bootstrap.servers、key.serializer 和 value.serializer 三个必须参数，还有很多可选的参数。下面列举了生产者 Producer 的配置参数及它们的含义。

### 1. acks

这个参数控制了发送消息的耐用性，用于指定分区中必须有多少个副本成功接收到消息，之后生产者才会认为这条消息写入是成功的，即生产者需要 leader 确认请求完成之前接收的应答数。通过查看 ProducerConfig 的源码可以看出 acks 参数的本质其实就是一个字符串。

```
01  public static final String ACKS_CONFIG = "acks";
```

acks 参数有三种类型的值。

（1）acks=1。

这是 acks 参数的默认值。Kafka 的生产者将消息发送到 Kafka 的 Broker 服务器端。只要 Topic 分区中的 Leader 成功写入消息，就算该消息成功发送。这时候，生产者就会收到 Kafka 服务器端 Broker 的成功确认信息，说明发送成功。

如果在生产者写入消息的过程中，Leader 分区所在的 Broker 出现了宕机，将会造成消息无法正常写入。在重新选举 Leader 的过程中，生产者 Producer 会受到一个服务器端返回的错误信息。生产者为了支持容错，避免消息的丢失，会尝试重新发送该消息。直至消息成功写入 Leader 分区。

这里需要注意的是，如果消息已经成功写入 Leader 所在的分区，但还未同步至其他 Follower 分区的时候，如果 Leader 分区所在的 Broker 出现了宕机，这时候会造成写入 Leader 分区的消息丢失。所以在这种参数值的设置下，消息是可能丢失的。

可以通过下面的代码进行设置。

```
01  props.setProperty(ProducerConfig.ACKS_CONFIG,"1");
```

（2）acks=0。

在这种参数设置下，Kafka 的生产者不需要等待任何服务器端的响应，所以这时 Kafka 集群可以达到最大的吞吐量。如果消息从生产者发送到写入 Kafka 消息系统的过程中出现异常，比如 Broker 宕机，生产者将不会得到任何反馈信息，也不会重发消息，导致消息丢失。

可以通过下面的代码进行设置。

```
01  props.setProperty(ProducerConfig.ACKS_CONFIG,"0");
```

（3）acks=-1 或 acks=all。

在这种参数设置下，Kafka 集群将达到最高的可靠性。生产者发送完消息后，需要等待 Leader 分区和所有 Follower 分区都成功写入消息后，才返回给生产者一个成功写入消息的应答响应。

可以通过下面的代码进行设置。

```
01  props.setProperty(ProducerConfig.ACKS_CONFIG,"-1");
```

最后，需要强调的是，acks 参数的值是一个字符串，不能是其他数据类型。如果像下面这样设置 acks 的参数值。

```
01  props.put(ProducerConfig.ACKS_CONFIG, 1);
```

将会抛出如下 Exception 错误信息。

```
01  Exception in thread "main" org.apache.kafka.common.config.ConfigException:
02  Invalid value 1 for configuration acks:
03  Expected value to be a string, but it was a java.lang.Integer
04      at
org.apache.kafka.common.config.ConfigDef.parseType(ConfigDef.java:665)
05      at
org.apache.kafka.common.config.ConfigDef.parseValue(ConfigDef.java:474)
06      at org.apache.kafka.common.config.ConfigDef.parse(ConfigDef.java:467)
07      at
org.apache.kafka.common.config.AbstractConfig.<init>(AbstractConfig.java:108)
08      at
org.apache.kafka.common.config.AbstractConfig.<init>(AbstractConfig.java:129)
09      at
org.apache.kafka.clients.producer.ProducerConfig.<init>(ProducerConfig.java:409)
10      at
org.apache.kafka.clients.producer.KafkaProducer.<init>(KafkaProducer.java:326)
```

```
11      at
org.apache.kafka.clients.producer.KafkaProducer.<init>(KafkaProducer.java:298)
12      at EmployeeProducer.main(EmployeeProducer.java:27)
```

### 2. buffer.memory

Kafka 生产者的 Sender 线程在将消息发送到 Kafka 服务器端之前，会把消息缓存到内存中，这个参数就决定了消息缓存的内存大小，其默认值是 32MB。如果生产者产生消息的速度大于将消息发送到服务器端的速度，那么生产者将会被阻塞，并最终导致生产者抛出一个 RecordTooLargeException 的异常，如下面 KafkaProducer 中的源码所示。

```
01  /**
02   * Validate that the record size isn't too large
03   */
04  private void ensureValidRecordSize(int size) {
05      if (size > this.maxRequestSize)
06          throw new RecordTooLargeException("The message is " + size +
07                  " bytes when serialized which is larger than the maximum
request size you have configured with the " +
08                  ProducerConfig.MAX_REQUEST_SIZE_CONFIG +
09                  " configuration.");
10      if (size > this.totalMemorySize)
11          throw new RecordTooLargeException("The message is " + size +
12                  " bytes when serialized which is larger than the total memory
buffer you have configured with the " +
13                  ProducerConfig.BUFFER_MEMORY_CONFIG +
14                  " configuration.");
15  }
```

在实际的生产环境下，应该根据实际情况进行测试最终决定 buffer.memory 参数值的大小。例如，客户端线程每秒会写入多少消息的数据量？按照默认值 32MB 的大小，是否会经常把内存缓冲写满？如果内存很快会写满，再调整 buffer.memory。经过这样的压测，你可以调试出来一个合理的内存大小。

- batch.size。

当 Kafka 客户端将多个消息发送到同一个分区的时候，生产者为了减少客户端与服务器端的请求交互，会尝试将消息批量打包在一起，进行统一发送，这样有助于提升客户端和服务器端的性能。该配置的默认批次大小（以字节为单位）是 16 384 字节。如果消息的内存大小大于该参数的配置，将不会进行批量打包的过程。

通过提升 batch.size 的大小，可以允许更多数据缓冲在分区中，那么一次请求服务器端所发送出去的数据量就更多了，这样吞吐量可能会有所提升。但是这样会造成大量内存的浪费。反过来如果减小 batch.size 的大小，则会系统地降低吞吐量。如果将 batch.size 设置为 0，则批处理机制被禁用。所以需要在这里按照生产环境的发消息速率，调节不同的

batch.size 大小，从而设置一个最合理的参数。

- compression.type。

该参数指定给到 Topic 中数据的压缩类型，其有效值的设置可以是标准的压缩方式，例如，'gzip'、'snappy'、'lz4'、'zstd'，同时该参数也可以是'uncompressed'，在这种设置下，消息数将不会被压缩。

- client.id。

当生产者向服务器端发送请求时，传递给服务器端 ID 字符串。通过这个 ID 字符串，Kafka 服务器端就可以追踪请求的资源，其本质就是将生产者及其请求的资源进行逻辑上的隔离。

- connections.max.idle.ms。

当生产者不再往服务器端发送消息时，这个参数用来决定关闭生产者连接的时间阈值，其默认值是 9min。

- linger.ms。

该参数决定消息在由生产者发送到服务器端之前，在客户端延长发送的时间。通过这样的延时发送机制，可以将多个消息组合成一个批处理进行统一发送。从本质上讲，该参数与之前提到过的 batch.size 参数类似。合理设置 batch.size 参数和 linger.ms 参数，将很好地利用 Kafka 批处理机制。把 linger.ms 设置得太小了，比如默认就是 0ms，或者设置为 5ms，那可能导致 Batch 虽然设置了 32KB，但是经常是还没凑够 32KB 的数据，5ms 之后就直接强制 Batch 将数据发送出去，这会导致你的 Batch 形同虚设，一直凑不满数据。

- max.block.ms。

该配置控制 KafkaProducer.send()和 KafkaProducer.partitionsFor()将消息阻塞多长时间。此外也可能是因为缓冲区已满或元数据不可用，导致这些方法被阻止。在用户提供的序列化程序或分区器中的锁定不会计入此超时，其默认值为 60 000ms。

- max.request.size。

Kafka 生产者能发送消息的最大值，默认值为 1MB。此设置将限制生产者的单个请求中发送的消息批次数，以避免发送过大的请求。这个参数涉及其他一些相关参数，比如服务器 Broker 端的 message.max.bytes 参数，如果 message.max.bytes 参数设置为 10，而 max.request.size 设置为 20，这时候就可以造成生产者报错。

- retries 和 retry.backoff.ms。

如果生产者出现了异常，或者消息没有成功写入 Kafka 的服务器端，生产者可以配置重试的参数值，通过生产者端的内部重试机制来执行恢复，并不是直接将异常抛出。如果重试达到设定次数，生产者才会放弃重试并抛出异常。retries 参数的默认值是 0。同时，生产者的重试还与 retry.backoff.ms 参数有关，该参数用来设定两次重试之间的时间间隔，其默认值是 100ms，从而避免无效的频繁重试。在配置 retries 参数和 retry.backoff.ms 参数之前，可以设定总重试时间要大于异常恢复时间，最好先估算一下异常恢复时间，避免生产者过早放弃重试。

- receive.buffer.bytes。

这个参数用来设置 socket 接收消息缓冲区的大小，该缓存区大小的默认值 32KB。如果将其值设置为-1，则使用操作系统的默认值。

- send.buffer.bytes。

这个参数用来设置 socket 发送消息缓冲区的大小，默认值为 128KB。与 receive.buffer.bytes 参数一样，如果将其值设置为-1，则使用操作系统的默认值。

- request.timeout.ms。

消息由生产者发出后，该参数用于决定生产者等待请求响应的最长时间，其默认值为 40s。如果响应的时间超过了该参数的设置，客户端将按照重试策略进行重试。注意，这个参数值需要比 Broker 端的参数 replica.lag.time.max.ms 值要大，这样可以减少因客户端重试引起的消息重复的概率。

- reconnect.backoff.max.ms。

该参数表示 Kafka 客户端重连的最大时间。每次连接失败，重连时间都会成指数级增加，每次增加的时间会存在 20%的随机浮动，以避免连接风暴。

- reconnect.backoff.ms。

该参数表示 Kafka 客户端每次重连时候的间隔时间。

- delivery.timeout.ms。

当生产者调用 send 方法后，该参数用于指定客户端等待发送成功或失败报告时，客户端等待时间的上限。这个时间上限包含以下几部分。

  - 一条消息在发送前的延时时间。
  - 生产者等待服务器端 Broker 确认信息的等待时间。
  - 失败时的重试时间。

- partitioner.class。

该参数表示一个实现了 org.apache.kafka.clients.producer.Partitioner 接口的类。Kafka 将使用这个类进行分区操作，其默认值是 org.apache.kafka.clients.producer.internals.DefaultPartitioner。通过实现这个接口，可以实现自定义分区。

- transaction.timeout.ms。

该参数表示生产者主动终止当前正在进行的操作之前，Kafka 等待操作状态更新的最大时间，其默认值是 1min。如果该值大于 Broker 中 max.transaction.timeout.ms 的设置，则请求失败，并报"InvalidTransactionTimeout"错误。

- transactional.id。

在事务传递过程中该参数用于表示某个事务的 ID。这样可以保证跨多个生产者会话时语义的可靠性。因为它允许客户端保证在开始任何新事务之前使用相同的 Transactional

Id 的事务来完成。

- max.in.flight.requests.per.connection。

该参数表示在消息被阻塞前，每个客户端上发送的未应答请求的最大数量，其默认值是 5。注意，如果该参数值设置大于 1，并且消息发送失败，则由于客户端的重试增加消息重新排序的风险。

- metadata.max.age.ms。

该参数表示当超过这个时间间隔时，系统就会更新元信息，其默认值 5min。Kafka 的元数据信息由 ZooKeeper 维护，包含 Topic 信息、副本信息、分区信息、Broker 信息。

- metadata.max.idle.ms。

当 Topic 处于空闲状态时，该参数用于控制生产者抓取 Topic 元信息的时间。

# 第 4 章　Kafka 的消费者

Kafka 采用消费者组的方式来消费消息，一个消费者组中可以包含多个消费者。消费者对象订阅主题并接收 Kafka 的消息，然后验证消息并保存结果。尽管一个消费者组中可以包含多个消费者，但是它们订阅的都是同一个主题的消息。那么这些消费者如何接收处理 Kafka 的消息？接收订阅的过程是怎么样的？本章将详细讨论 Kafka 消费者的执行逻辑。

## 4.1　Kafka 消费者的消费模式

当生产者将消息发送到 Kafka 集群后，会转发给消费者进行消费。消息的消费模型有两种，推送模式（push）和拉取模式（pull）。

### 4.1.1　消息的推送模式

消息的推送模式需要记录消费者的消费状态。当把一条消息推送给消费者后，需要维护消息的状态，如标记这条消息已经被消费，这种方式无法很好地保证消息被处理。如果要保证消息被处理，发送完消息后，需要将其状态设置为“已发送”。收到消费者的确认收到消息后，才将其状态更新为“已消费”，这就需要我们记录所有消息的消费状态。显然这种方式不可取。这种方式还存在一个明显的缺点，就是消息被标记为“已消费”后，其他消费者就不能再进行消费了。

### 4.1.2　消息的拉取模式

由于推送模式存在一定的缺点，因此 Kafka 采用消费拉取的模式来消费消息。由每个消费者维护自己的消费状态，并且每个消费者互相独立地顺序拉取每个分区的消息。消费者通过偏移量的信息来控制从 Kafka 中消费的消息，如图 4.1 所示。

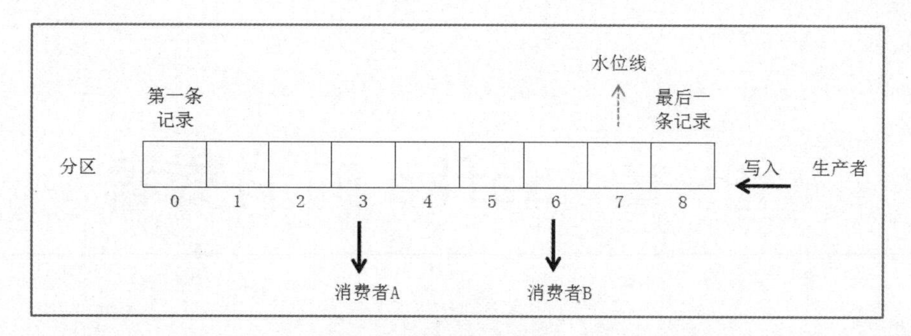


图 4.1　消息的拉取模式

由消费者通过偏移量进行消费控制的优点在于，消费者可以按照任意的顺序消费消息。例如，消费者可以通过重置偏移量信息，重新处理之前已经消费过的消息；或者直接跳转到某一个偏移量位置，并开始消费。

这里需要特别说明的是，当生产者最新写入的消息还没有达到备份数量，即新写入的消息还没有达到冗余度要求时，对消费者是不可见的。消费者只能消费到水位线（Watermark）的位置，如图 4.1 所示。

另外，如果消费者已经将消息进行了消费，Kafka 并不会立即将消息删除，而是会将所有消息进行保存，即持久化保存到 Kafka 的消息日志中。无论消息有没有被消费，用户可以通过设置保留时间来清理过期的消息数据。关于 Kafka 持久化机制与日志的清理策略，将在后续章节进行详细的介绍。

### 4.1.3　推送模式与拉取模式的区别

在了解了推送模式和拉取模式后，这两者的区别是什么？

由于消息的发送速率是由 Kafka 的 Broker 决定的，Broker 的目标是尽可能以最快的速度传递消息。所以在推送模式下，很难适应消费速率不同的消费者，从而造成消费者来不及处理消息。消费者来不及处理消息就可能造成消息的阻塞，从而降低系统的处理能力。

在拉取模式下，用户可以根据消费者的处理能力调整消息消费的速率，但在这种模式下也存在一定的缺点。如果消息的生产者没有产生消息，就可能造成消费者陷入循环中，一直等待数据到达。为了避免这种情况出现，可以在拉取过程中指定允许消费者在等待数据到达时进行阻塞，并且还可以指定消费的字节数，从而保证传输时的数据量。

### 4.1.4　消息者组

消费者是以消费者组的方式工作的，即一个消费者组由一个或多个消费者组成，它们共同消费一个 Topic 中的消息。在同一个时间点上，Topic 中的分区只能由一个组中的一个消费者进行消费，同一个分区可以被不同组中的消费者进行消费，如图 4.2 所示。

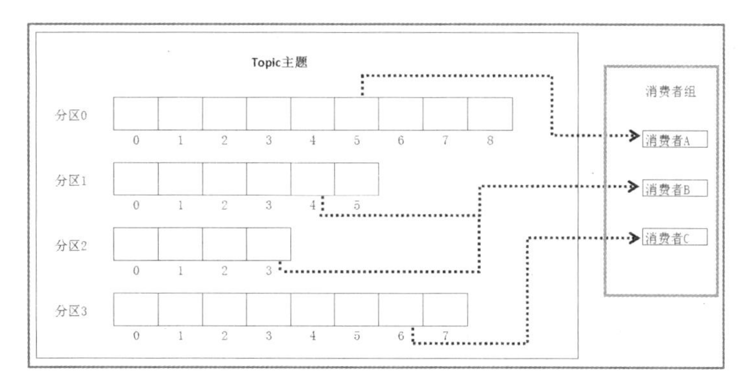


图 4.2　消费者组

图 4.2 中的消费者组由三个消费者组成，并且 Topic 由 4 个分区组成。其中，消费者 A 消费读取一个分区，消费者 B 消费读取两个分区，消费者 C 也消费读取一个分组。在这种情况下，消费者可以通过水平扩展的方式同时读取大量的消息。另外，如果一个消费者失败了，那么其他 group 成员会自动负载均衡读取之前失败的消费者读取的分区。

关于消费者组和消费者的内容，我们会在 4.3 节中进行详细介绍。

## 4.2　创建 Kafka 消费者

### 4.2.1　创建基本的消息消费者

要从 Kafka 消息集群中读取消息，需要先创建一个 KafkaConsumer 对象。创建 KafkaConsumer 对象与创建 KafkaProducer 对象非常相似。一般只需要指定以下三个必要的参数。

（1）bootstrap.servers。

该参数指定了 Kafka 集群的连接字符串，它的用途与在 KafkaProducer 中的用途是一样的。

（2）key.deserializer。

该参数与生产者中的 key.serializer 参数含义类似。消费者从 Kafka 消息集群上获取的任何消息都是字节数组的格式，因此消息的每个组成部分都要执行相应的反序列化操作才能得到原来的对象格式。该参数将消息的 key 进行反序列化，其参数值必须实现 org.apache.kafka.common.serialization.Deserializer 接口。针对绝大多数基本数据类型，Kafka 都提供了现成的反序列化器，例如，org.apache.kafka.common.serialization.StringDeserializer。该数据类型的主要作用是将接收到的字节数组转换为 UTF-8 的字符串。

当然，也可以通过实现 Deserializer 接口，自定义反序列化机制，但是需要与生产者端定义的序列化机制保持一致。

（3）value.deserializer。

该参数与 key.deserializer 类似，用来将接收到的 Kafka 消息的消息体（即 value）进行反序列化，从而得到 KafkaProducer 发送的原始数据。这里需要注意的是，key.deserializer 和 value.deserializer 可以是不同的设置。

在第 3 章的最后演示了如何创建一个 Kafka 消费者，代码如下。

```
01  Properties props = new Properties();
02  props.put("bootstrap.servers", "kafka101:9092");
03  props.put("group.id", "mygroup");
04  props.put("enable.auto.commit", "true");
05  props.put("auto.commit.interval.ms", "1000");
06  props.put("key.deserializer",
07        "org.apache.kafka.common.serialization.StringDeserializer");
08  props.put("value.deserializer",
09        "org.apache.kafka.common.serialization.StringDeserializer");
10
11  KafkaConsumer<String, String> consumer =
12        new KafkaConsumer<String, String>(props);
```

其中，第 04 行～第 05 行代码不是必须的，如果没有配置这些参数，将会采用默认的参数值。

现在可以结合第 3 章中的生产者代码来执行以下程序，程序输出的结果，如图 4.3 所示。

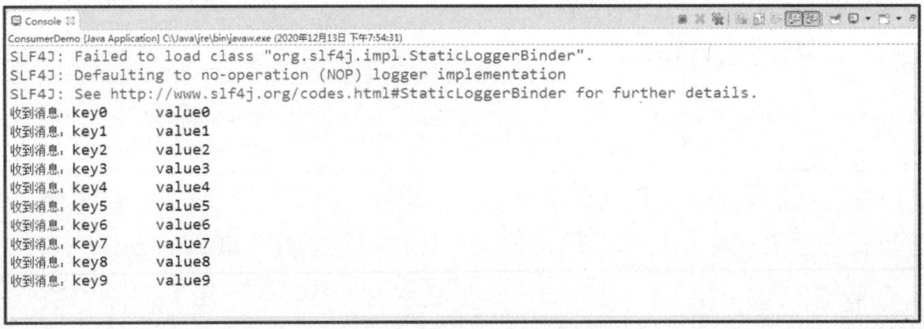

```
Console
ConsumerDemo [Java Application] C:\Java\jre\bin\javaw.exe (2020年12月13日 下午7:54:31)
SLF4J: Failed to load class "org.slf4j.impl.StaticLoggerBinder".
SLF4J: Defaulting to no-operation (NOP) logger implementation
SLF4J: See http://www.slf4j.org/codes.html#StaticLoggerBinder for further details.
收到消息：key0      value0
收到消息：key1      value1
收到消息：key2      value2
收到消息：key3      value3
收到消息：key4      value4
收到消息：key5      value5
收到消息：key6      value6
收到消息：key7      value7
收到消息：key8      value8
收到消息：key9      value9
```

图 4.3　Kafka 消费者执行的效果

## 4.2.2　接收自定义消息对象

前面章节介绍了通过生产者发送用户自定义的消息对象。我们使用了 Fastjson 的 Java 库，可以将 Java 对象转换为 JSON 格式。前面的内容只展示了 EmployeeProducer 的程序代码和最终运行的结果，下面的代码展示了 EmployeeConsumer 的完整代码。

```
01  import java.time.Duration;
02  import java.util.Arrays;
```

```
03  import java.util.Properties;
04
05  import org.apache.kafka.clients.consumer.Consumer;
06  import org.apache.kafka.clients.consumer.ConsumerConfig;
07  import org.apache.kafka.clients.consumer.ConsumerRecord;
08  import org.apache.kafka.clients.consumer.ConsumerRecords;
09  import org.apache.kafka.clients.consumer.KafkaConsumer;
10
11  public class EmployeeConsumer {
12
13      public static void main(String[] args) {
14          Properties props = new Properties();
15
16          //指定 Kafka Broker 的地址
17          props.setProperty(ConsumerConfig.BOOTSTRAP_SERVERS_CONFIG,
18                      "kafka101:9092");
19
20          //指定 Key 的反序列化方式
21          props.setProperty(ConsumerConfig.KEY_DESERIALIZER_CLASS_CONFIG,
22              "org.apache.kafka.common.serialization.StringDeserializer");
23
24          //使用我们自定义的 JSON 反序列化机制
25          props.setProperty(ConsumerConfig.VALUE_DESERIALIZER_CLASS_CONFIG,
26              "EmployeeJSONDeserializer");
27
28          //指定消费者组
29          props.setProperty(ConsumerConfig.GROUP_ID_CONFIG,
30              "mygroup1");
31
32          //创建 Consumer
33          Consumer<String, Employee> consumer =
34                      new KafkaConsumer<String, Employee>(props);
35          consumer.subscribe(Arrays.asList("mytopic1"));
36
37          while(true) {
38              ConsumerRecords<String, Employee> records =
39                      consumer.poll(Duration.ofSeconds(1000));
40
41              for(ConsumerRecord<String, Employee> r:records) {
42                  System.out.println("收到员工对象："+
43                                  r.key()+"\t"+r.value());
44              }
45          }
46      }
```

```
47  }
```

在上面创建的消费者代码中，我们使用了一个 EmployeeJSONDeserializer 的反序列化器，下面是它的完整代码。

```
01  import org.apache.kafka.common.serialization.Deserializer;
02  import com.alibaba.fastjson.JSON;
03
04  public class EmployeeJSONDeserializer
05      implements Deserializer<Employee> {
06
07      @Override
08      public Employee deserialize(String topic, byte[] data) {
09          return JSON.parseObject(data,Employee.class);
10      }
11
12  }
```

程序运行的效果图，请参看 3.2.2 节。

# 4.3 消费者与消费者组

本节将详细介绍消费者组、消费者和分区，以及它们之间的关系。

## 4.3.1 消费者和消费者组与分区的关系

图 4.4 展示了消费者组、消费者和分区之间的关系。可以看出，在 Kafka 消息系统的 Topic 有 4 个分区，即 P0、P1、P2 和 P3。消费者以消费者组为单位订阅 Topic 中的消息，而 Topic 又包含多个分区，消费者组中有多个消费者实例。那么消费者组中的每一个消费者负责处理消费哪些分区中的数据？这个分配对应关系是如何确定的？换句话说，消费者实例和分区之间的对应关系是怎样的？

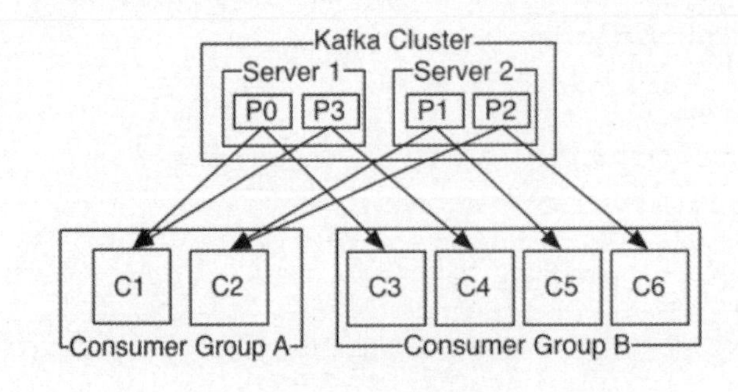


图 4.4　消费者组、消费者和分区之间的关系

前面提到，同一时刻每个分区中的一条消息只能被消费者组中的一个消费者消费。这

里的分配原则是，Topic 下的每个分区只能被消费者组中的一个消费者消费，也就是只能从属于一个消费者组成员，不会发生同一个消费者组中的两个不同的消费者负责处理消费同一分区。这里就可能存在几种不同的情况。

（1）Topic 的分区数大于消费者组中的消费者数。

如图 4.5 所示。假设有一个 Topic1 的主题，该主题有四个分区，并且有一个消费组 Group1，这个消费者组只有一个消费者 Consumer1，那么消费者 Consumer1 将会收到这四个分区的消息。

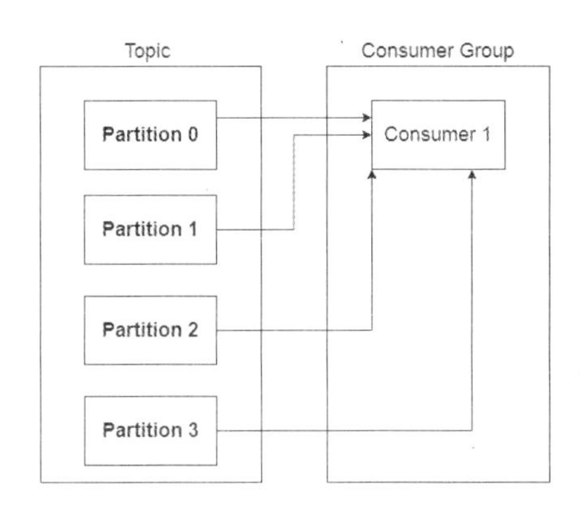


图 4.5　一个消费者和四个分区

如果增加消费者组中的消费者，例如，增加到两个消费者时，如图 4.6 所示。每个消费者将分别接收到两个分区中的消息。

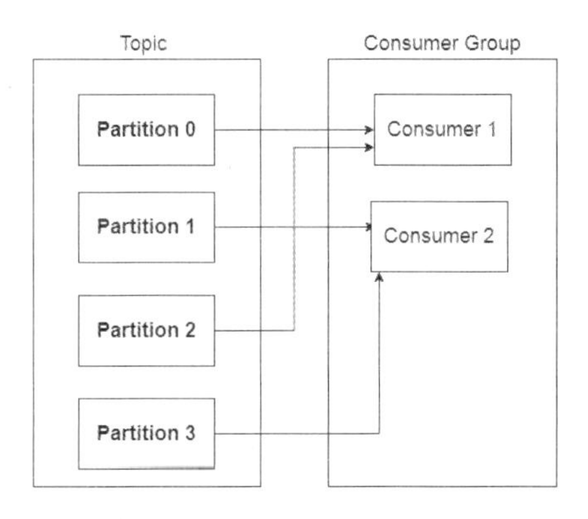


图 4.6　两个消费者和四个分区

（2）Topic 的分区数等于消费者组中的消费者数。

如果继续增加到四个消费者，那么每个消费者将分别收到一个分区的消息，如图 4.7 所示。

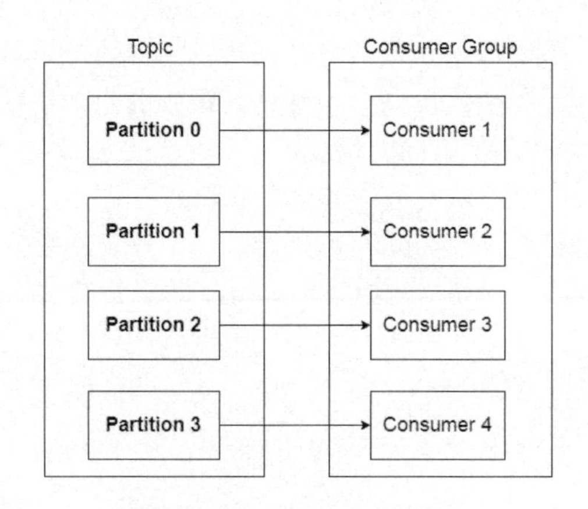


图 4.7　四个消费者和四个分区

（3）Topic 的分区数小于消费者组中的消费者数。

我们继续在这个消费组中增加消费者，如图 4.8 所示。这时消费者组中有五个消费者，有一个剩余的消费者将会空闲，它不会接收到任何消息。Kafka 不建议使用这样的情况。

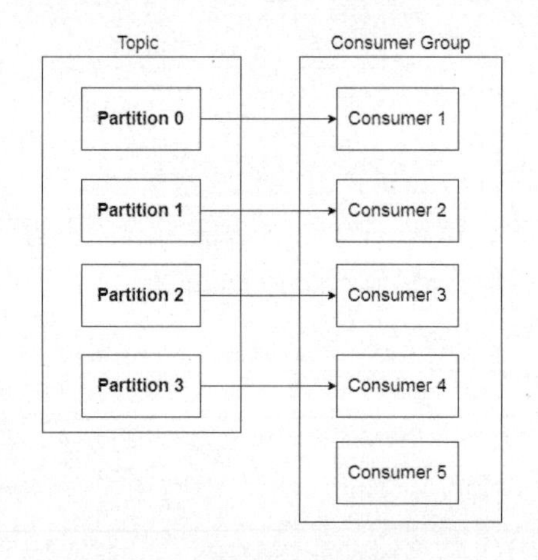


图 4.8　五个消费者和四个分区

通过在消费者组中添加消费者，可以提升系统的水平扩展能力，从而提升消费者的消费能力。Kafka 建议创建 Topic 时使用比较多的分区数，一般分区数要大于消费者的个数，这样可以在消费负载高的情况下增加消费者来提升性能。通过前面的介绍，Kafka 不建议消费者的个数大于分区数。因为多出来的消费者没有任何帮助，它一直处于空闲的状态。

这里还需要说明的是，如果存在多个消费者组，那么又会是什么样的情况呢？

由于 Kafka 支持写入一次消息，支持任意多的应用读取这个消息。应用需要包含不同的消费组，这样可以使得每个应用都能读到全量消息，如图 4.9 所示。这里，我们创建了两个消费者组，它们都可以接收到 Topic 中的全部消息，并且它们可能属于不同的应用系统。

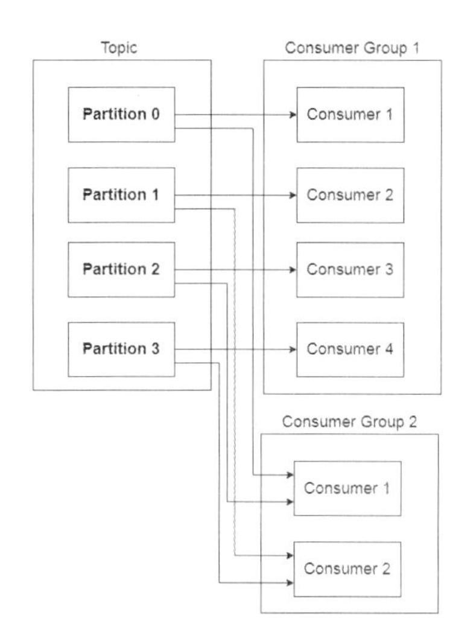


图 4.9 四个消费者和两个消费者组

总结一下，如果一个应用的消费能力不足，可以向消费者组中增加消费者；如果是不同的应用系统，为了使应用获取到全部消息，可以单独为该应用设置一个消费者组。

## 4.3.2 分区的重平衡

分区的重平衡（Rebalance）是 Kafka 一个很重要的性质，它可以保证系统的高可用和系统的水平扩展。以下几种情况会触发 Kafka 发生重平衡。

- 新的消费者加入消费组。新加入的成员会消费一个或多个分区，而这些分区中的消息在新成员加入之前被其他成员消费处理过。
- 消费者离开消费者组，例如，发生了宕机或重启。这种情况会导致之前由该消费者负责的分区分配给其他分区。

重平衡的优点是保证高可用性和扩展性，但是它也会带来一些问题，其中最主要的问题是在重平衡期间，整个消费组是不可用的，会造成所有消费者都不能消费消息；并且，重平衡会导致消费者需要重新更新状态，从而导致原来的消费者状态过期。这些都会导致系统的消费能力降低。

另外需要说明的是，在消费者组中维护一个协调者（Coordinator）用于感知消费者的心跳信息。无论是增加新的消费者，还是有消费者退出，都会由这个协调者来感知其心跳信息。如果消费者超过一定时间没有发送心跳信息，那么它的状态就会过期，协调者会认为该消费者已经宕机，然后触发重平衡。从消费者宕机到其被协调者感知到，这中间是有一段时间间隔的，这段时间内该消费者是不能进行消息消费的。

# 4.4　消费者的偏移量与提交

首先需要知道什么是消费者的偏移量。Kafka 的消费者每次拉取服务器端的消息时，总是拉取由生产者写入 Kafka 但还没有被消费者处理过的数据。因此，需要一种机制来记录哪些消息是被消费者组里的哪个消费者消费过的。与其他消息系统不同的是，Kafka 消费者每次拉取完消息后，会记录最新的偏移量地址。下次拉取消息的时候，将会从偏移量往后拉取最新的消息数据。我们把消费者更新到当前拉取分区中的位置（即偏移量）称为提交。

## 4.4.1　偏移量与重平衡

消费者需要定期提交拉取的偏移量，一方面用于记录最新消费的位置信息，以便下次的拉取操作；另一方面，当消费者退出或有新的消费者加入消费者组的时候，都会触发重平衡的操作，完成重平衡后，每个消费者可能会分配到新的分区，读取新分区中的数据。为了能够继续之前的拉取工作，消费者需要读取每个 partition 最后一次提交的偏移量，然后从偏移量指定的地方继续处理。这里就可能存在几种不同的情况。

- 情况一：如果提交的偏移量小于客户端处理的最后一个消息的偏移量，会导致两个偏移量之间的消息被重复处理，如图 4.10 所示。

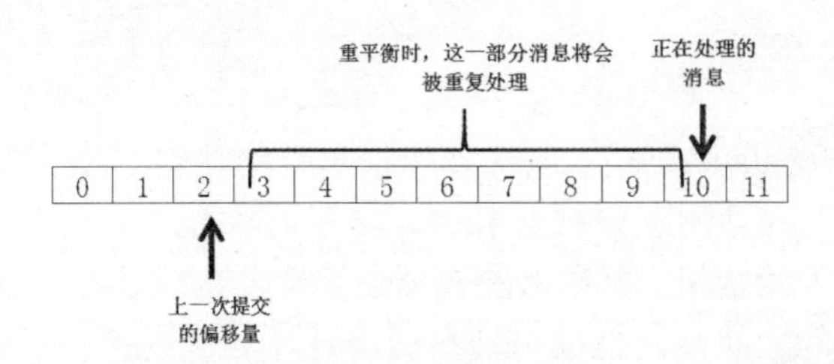


图 4.10　提交偏移量小于客户端处理的最后一个消息的偏移量

- 情况二：如果提交的偏移量大于客户端处理的最后一个消息的偏移量，会导致两个偏移量之间的消息丢失，如图 4.11 所示。

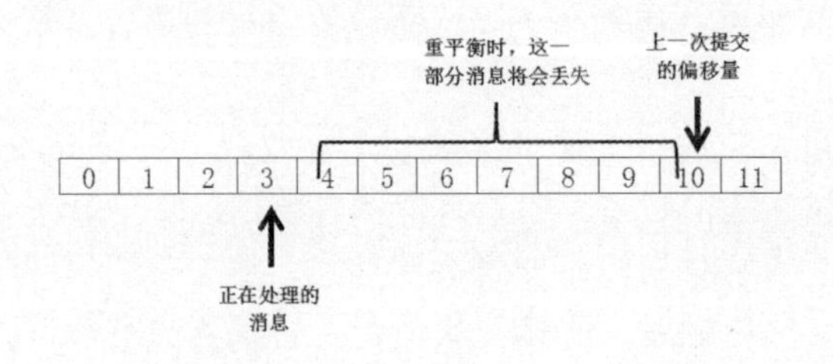


图 4.11　提交偏移量大于客户端处理的最后一个消息的偏移量

通过对以上两种情况的介绍，我们会发现如何正确提交消费者的偏移量将会对客户端产生很大的影响。因此，Kafka Consumer API 提供了很多种方式来提交偏移量。

## 4.4.2　偏移量的提交方式

（1）自动提交。

这种提交方式是一种简单的提交方式，需要把参数 enable.auto.commit 设置为 true，那么在默认情况下，每隔 5 秒消费者会自动把从 poll()方法接收到的最大偏移量提交上去，这个时间间隔可以通过参数 auto.commit.interval.ms 进行修改。当然这个自动提交是在每次进行轮询时，即调用 poll()方法时进行的。消费者会检查是否该提交偏移量了，如果已经提交，就会返回上次提交时的偏移量。

自动提交虽然方便，但是也可能存在一些问题，其中最主要的问题就是可能造成消息的重复消费。在前面的内容中介绍过，重平衡的发生会有不同的情况。其中的一种情况就是当提交的偏移量小于客户端处理的最后一个消息的偏移量，会造成消息的重复消费。下面我们来举例，按照默认的 5 秒，系统会自动提交一次。如果在最后一次提交之后的 2 秒发生了重平衡。那么重平衡完成后，消费者从最后一次提交的偏移量位置开始读取消息，这时偏移量已经落后了 2 秒，这样就会造成这 2 秒内的消息被重复消费处理。

（2）提交当前偏移量。

这种提交方式其实就是手动提交偏移量，将 enable.auto.commit 设置成 false，让应用程序决定何时提交偏移量，即使用 commitSync()方法提交偏移量。这种方式非常简单，也很可靠。它可以减少在平衡时重复处理的消息数量，并同时消除丢失消息的可能性。需要注意的是，commitSync()方法将提交由 poll()方法返回的最新偏移量，所以在处理完所有记录后要确保调用了 commitSync()方法，否则还会有丢失消息的风险。

这种提交方式的本质是同步提交偏移量，因此在提交的过程中，其应用程序会被阻塞。代码程序如下所示。

```
01  KafkaConsumer<String, String> consumer =
02                  new KafkaConsumer<String, String>(props);
03
04  consumer.subscribe(Arrays.asList("mytopic1"));
05  while (true) {
06   ConsumerRecords<String, String> records =
07                  consumer.poll(Duration.ofMillis(100));
08
09   for (ConsumerRecord<String, String> record : records) {
10       System.out.println("收到消息: "+ record.key() + "\t"
11                                 + record.value());
12
13       //手动提交当前偏移量
```

```
14        consumer.commitSync();
15    }
16  }
```

在这段程序代码中，我们把消息的内容打印就算处理完成了，然而在实际的场景下，这应该取决于业务逻辑。处理完当前批次的消息后，在进入下一次轮询之前，调用commitSync()方法提交当前批次最新的偏移量。只要没有发生不可恢复的错误，commitSync()方法会一直尝试提交，直至提交成功。

（3）异步提交。

上面提到的同步提交方式会造成应用程序一直阻塞，这样会限制应用程序的吞吐量。其中的一种解决办法就是降低提交频率；另一种方式可以使用异步提交 API。消费者只需要发送提交偏移量的请求，而不需要等待服务器端的响应。

程序代码如下所示。

```
01  KafkaConsumer<String, String> consumer =
02                  new KafkaConsumer<String, String>(props);
03
04  consumer.subscribe(Arrays.asList("mytopic1"));
05  while (true) {
06   ConsumerRecords<String, String> records =
07                  consumer.poll(Duration.ofMillis(100));
08
09   for (ConsumerRecord<String, String> record : records) {
10       System.out.println("收到消息: "+ record.key() + "\t"
11                                   + record.value());
12
13       //异步提交当前偏移量
14       consumer.commitAsync();
15   }
16  }
```

与同步提交不同的是，在成功提交偏移量之前，异步提交不会进行重试，只是根据服务器端的响应做出相应的动作。如果在得到服务器端返回响应之前，有另一个较大的偏移量信息被成功提交，就可能造成消息的重复消费。假设我们发出一个异步请求用于提交偏移量 1000，但服务器端并没有收到这样的请求，例如，网络出现了问题，服务器端也不会做出任何响应。与此同时，消费者处理了另一个批次的消息，偏移量是 2500，并且成功进行提交。由于之前偏移量为 1000 的提交可能晚于偏移量为 2500 的提交，这时如果偏移量为 1000 的提交成功了，就可能造成偏移量从 1000 到 2500 之间的消息被重复消费。

异步提交方式也支持回调，代码如下所示。

```
01  //异步提交当前偏移量
02  consumer.commitAsync(new OffsetCommitCallback() {
03
```

```
04  @Override
05  public void onComplete(Map<TopicPartition, OffsetAndMetadata> offsets,
06                          Exception exception) {
07      //回调方法
08      if(exception != null) {
09          System.out.println(exception.getMessage());
10      }
11   }
12  });
```

（4）组合同步提交和异步提交。

既然消费者执行同步提交偏移量和异步提交偏移量的两种方式，我们就可以组合使用 commitSync()方法和 commitAsync()方法来提交偏移量信息。这样针对偶尔出现的提交失败，不必提交偏移量的重试也不会有太大问题。代码如下所示。

```
01  try {
02      while (true) {
03          ConsumerRecords<String,          String>          records          =
consumer.poll(Duration.ofMillis(100));
04
05          for (ConsumerRecord<String, String> record : records) {
06              System.out.println(" 收 到 消 息 ： "  +  record.key()  +  "\t"  +
record.value());
07
08              //异步提交当前偏移量
09              consumer.commitAsync();
10          }
11      }
12  }catch(Exception ex) {
13      ex.printStackTrace();
14  }finally {
15      try {
16          //同步提交当前偏移量
17          consumer.commitSync();
18      }finally {
19          consumer.close();
20      }
21  }
```

（5）提交指定的偏移量。

在使用同步提交偏移量和异步提交偏移量时，可以在调用 commitSync()方法和 commitAsync()方法时，传入希望提交的 partition 和 offset 的 map，即提交特定的偏移量，代码如下所示。

```
01      KafkaConsumer<String,  String>  consumer  =  new  KafkaConsumer<String,
String>(props);
```

```
02   consumer.subscribe(Arrays.asList("mytopic1"));
03
04   //用于跟踪偏移量的 map
05   Map<TopicPartition, OffsetAndMetadata> currentOffsets = new HashMap<>();
06   //计数器，每 200 条消息提交一次
07   int count = 0;
08
09   while (true) {
10       ConsumerRecords<String,              String>          records          =
consumer.poll(Duration.ofMillis(100));
11
12       for (ConsumerRecord<String, String> record : records) {
13           System.out.println("收到消息:" + record.key() + "\t" + record.value());
14
15           //记录当前消息的偏移量
16           currentOffsets.put(new                    TopicPartition(record.topic(),
record.partition()),
17                              new OffsetAndMetadata(record.offset() + 1, null));
18           //消息数量加 1
19           count++;
20
21           if (count % 200 == 0) {
22               //每处理 200 条消息就提交一次偏移量
23               consumer.commitAsync(currentOffsets, null);
24           }
25       }
26   }
```

## 4.5 消费者的高级特性

### 4.5.1 消费者的分区策略

我们知道 Kafka 的 Topic 是由分区组成的，并且还可以配置分区的冗余度。一个分区在多个 Broker 中选举出一个 Leader，消费者只访问这个 Leader 的分区副本。这里重点介绍一个消费者如何选定一个 Topic 多个分区中的一个分区和 Kafka 消费者支持的分区策略。

通过前面内容的介绍可知，一条消息只能被消费者组中的一个消费者消费。消费者组订阅 Topic，意味着该 Topic 下的所有分区都会被消费者组中的消费者消费，如果按照从属关系来说，Topic 下的每个分区只属于消费者组中的一个消费者，不可能出现组中的两个消费者负责同一个分区。

Kafka 通过配置消费者分区分配策略来决定分区中的消息被哪一个消费者消费。消费者分区的分配策略都应该实现 org.apache.kafka.clients.consumer.internals.

AbstractPartitionAssignor 接口。通过实现这个接口，用户可以自定义分区分配策略。Kafka 提供了 3 种实现的方式，可以通过参数 partition.assignment.strategy 进行指定。

（1）RangeAssignor。

这是默认的分区分配策略。这种分配策略是根据 Kafka Consumer 端的总数和 Topic 中的分区总数来获取一个范围的，然后将分区按照范围进行平均分配，以保证分区尽可能均匀地分配给所有消费者。下面给出了这种分配方式的源代码。

```
01  @Override
02      public  Map<String,  List<TopicPartition>>  assign(Map<String,  Integer>
partitionsPerTopic,
03                                          Map<String, Subscription> subscriptions)
{
04       //用于记录 Topic 与 Consumer 之间的关系映射

05       Map<String,  List<MemberInfo>>  consumersPerTopic  =  consumersPerTopic
(subscriptions);
06
07       Map<String, List<TopicPartition>> assignment = new HashMap<>();
08       for (String memberId : subscriptions.keySet())
09           assignment.put(memberId, new ArrayList<>());
10
11       for (Map.Entry<String, List<MemberInfo>> topicEntry : consumersPerTopic.
entrySet()) {
12           //获取 Topic 信息
13           String topic = topicEntry.getKey();
14
15           //获取消费者列表
16           List<MemberInfo> consumersForTopic = topicEntry.getValue();
17
18           //Topic 下有多少个分区
19           Integer numPartitionsForTopic = partitionsPerTopic.get(topic);
20           if (numPartitionsForTopic == null)
21               continue;
22
23           //将消费者按字典序列排序
24           Collections.sort(consumersForTopic);
25
26           //分区数量除以消费者数量
27       int numPartitionsPerConsumer = numPartitionsForTopic/consumersForTopic. size();
28
29           //取模运算，余数就是额外的分区
30           int consumersWithExtraPartition = numPartitionsForTopic %
consumersForTopic.size();
31
```

```
32            List<TopicPartition> partitions = AbstractPartitionAssignor.partitions
(topic, numPartitionsForTopic);
33            for (int i = 0, n = consumersForTopic.size(); i < n; i++) {
34                int start = numPartitionsPerConsumer * i + Math.min(i,
consumersWithExtraPartition);
35                int length = numPartitionsPerConsumer + (i + 1 >
consumersWithExtraPartition ? 0 : 1);
36
37            //分配分区
38
    assignment.get(consumersForTopic.get(i).memberId).addAll(partitions.subList
(start, start + length));
39            }
40        }
41        return assignment;
42    }
```

（2）RoundRobinAssignor。

这种分区分配策略对应的 partition.assignment.strategy 参数值为：org.apache.kafka.clients.consumer.RoundRobinAssignor。这种方式将 Consumer Group 中的所有消费者及其订阅 Topic 的分区按照字典序列排序，然后通过轮询的方式逐个将分区分配给每个消费者。下面给出了这种分配方式的源代码。

```
01   @Override
02      public  Map<String,  List<TopicPartition>>  assign(Map<String,  Integer>
partitionsPerTopic,
03                                       Map<String, Subscription> subscriptions)
{
04       Map<String, List<TopicPartition>> assignment = new HashMap<>();
05       List<MemberInfo> memberInfoList = new ArrayList<>();
06
07       //获取消费者订阅的所有 Topic 信息
08       for (Map.Entry<String, Subscription> memberSubscription : subscriptions.
entrySet()) {
09           assignment.put(memberSubscription.getKey(), new ArrayList<>());
10           memberInfoList.add(new MemberInfo(memberSubscription.getKey(),
11
memberSubscription.getValue().groupInstanceId()));
12       }
13
14       //消费者及其订阅 Topic 的分区按照字典序列排序
15       CircularIterator<MemberInfo>   assigner   =   new   CircularIterator<>
(Utils.sorted(memberInfoList));
16
17       //使用轮询的方式逐个将分区分配给每个消费者
```

```
18        for (TopicPartition partition : allPartitionsSorted(partitionsPerTopic,
subscriptions)) {
19            final String topic = partition.topic();
20            while
(!subscriptions.get(assigner.peek().memberId).topics().contains(topic))
21                assigner.next();
22            assignment.get(assigner.next().memberId).add(partition);
23        }
24        return assignment;
25    }
```

（3）StickyAssignor。

这种分区分配策略采用黏性分配策略，该策略从 Kafka 0.11 版本引入。所谓黏性分配策略，既要保证分区的分配要尽可能均匀，又要保证每次分区的分配尽可能与上次分配的保持相同，就像进行粘贴一样。如果这两点发生冲突，优先考虑第一点，即分区的分配要尽可能均匀。

## 4.5.2　重平衡监听器

前面介绍到当有新的消费者成员加入消费者组或有消费者退出，都可能触发重平衡的操作。但是在重平衡期间，消费者组内的消费者是无法读取消息的，在重平衡这一段时间内，消费者组不可用。如果在这段时间内，我们想要执行一些应用程序代码，在调用 subscribe()方法时传入一个 org.apache.kafka.clients.consumer.ConsumerRebalanceListener 接口的实例就可以了，而这个实例就是重平衡监听器。

下面给出了一个重平衡监听器的实现。当发生重平衡操作的时候，该监听器会将当前消费者的偏移量进行提交，从而保证重平衡完成后，消费者能够从上次完成的偏移量地址进行消费。

```
01  import java.util.Collection;
02  import java.util.Map;
03
04  import org.apache.kafka.clients.consumer.ConsumerRebalanceListener;
05  import org.apache.kafka.clients.consumer.KafkaConsumer;
06  import org.apache.kafka.clients.consumer.OffsetAndMetadata;
07  import org.apache.kafka.common.TopicPartition;
08
09  public class ProcessConumserRebalance implements ConsumerRebalanceListener
{
10
11      //消费者当前的偏移量
12      private Map<TopicPartition, OffsetAndMetadata> currentOffsets;
13
```

```
14      private KafkaConsumer<String, String> consumer;
15
16      public ProcessConumserRebalance
17              (Map<TopicPartition, OffsetAndMetadata> currentOffsets,
18               KafkaConsumer<String, String> consumer) {
19          this.currentOffsets = currentOffsets;
20          this.consumer = consumer;
21      }
22
23      /*
24       * 这个方法会在重平衡开始之前和消费者停止读取消息之后被调用
25       * 可以在此处将位移提交。这里的 partitions 表示在平衡之前的分区
26       */
27      @Override
28      public void onPartitionsRevoked(Collection<TopicPartition> partitions) {
29          System.out.println("重平衡发生。提交当前的偏移量" + currentOffsets);
30          consumer.commitSync(currentOffsets);
31      }
32
33      /*
34       *  这个方法在重新分配分区之后和开始消费消息之前被调用
35       * partitions 表示在平衡后分配的分区
36       */
37      @Override
38      public void onPartitionsAssigned(Collection<TopicPartition> partitions)
{
39      }
40  }
```

在这段代码中，我们定义了一个 currentOffsets 变量，将其用于保存消费者当前消费的偏移量。这里也可以不定义该变量，直接使用下面的方式进行提交。

```
01  /*
02   * 这个方法会在重平衡开始之前和消费者停止读取消息之后被调用
03   * 可以在此处将位移提交。这里的 partitions 表示在平衡之前的分区
04   */
05  @Override
06  public void onPartitionsRevoked(Collection<TopicPartition> partitions) {
07      System.out.println("重平衡时，发生分区的丢失。提交当前的偏移量");
08      consumer.commitSync();
09  }
```

当监听器创建完成后，就可以在消费者订阅 Topic 时使用监听器了，代码如下所示。

```
01  consumer.subscribe(Arrays.asList("mytopic1"),
02                   new ProcessConumserRebalance(currentOffsets,consumer));
```

### 4.5.3　消费者的拦截器

消费者的拦截器允许用户在消费者消费消息之前或消费者提交偏移量之后，执行特定的业务逻辑。消费者拦截器都是实现了 org.apache.kafka.clients.consumer.ConsumerInterceptor 接口的实例。可以通过消费者拦截器配置参数 interceptor.classes 进行配置。下面给出了该接口的源代码程序。

```
01  package org.apache.kafka.clients.consumer;
02
03  import org.apache.kafka.common.Configurable;
04  import org.apache.kafka.common.TopicPartition;
05  import java.util.Map;
06
07     public  interface  ConsumerInterceptor<K,  V>  extends  Configurable,
AutoCloseable {
08
09     public ConsumerRecords<K, V> onConsume(ConsumerRecords<K, V> records);
10
11     public void onCommit(Map<TopicPartition, OffsetAndMetadata> offsets);
12
13     public void close();
14  }
```

该接口有两个主要的方法：onConsume()和 onCommit()。

onConsume()：该方法在消息返回给 Consumer 之前被调用。

onCommit()：该方法在 Consumer 提交偏移量信息后被调用。

下面的代码实现了一个自定义的消费者拦截器。在 onConsume()方法中，我们执行了一个简单的过滤操作，该过滤操作只将消息中包含字符串“***DemoTest***”的消息返回给消费者；而在 onCommit()方法中，只是简单地将 Topic 分区中的偏移量信息打印出来。

```
01  import java.util.ArrayList;
02  import java.util.HashMap;
03  import java.util.List;
04  import java.util.Map;
05  import java.util.Set;
06
07  import org.apache.kafka.clients.consumer.ConsumerInterceptor;
08  import org.apache.kafka.clients.consumer.ConsumerRecord;
09  import org.apache.kafka.clients.consumer.ConsumerRecords;
10  import org.apache.kafka.clients.consumer.OffsetAndMetadata;
11  import org.apache.kafka.common.TopicPartition;
12
13   public class MyConsumerInterceptor implements ConsumerInterceptor<String,
String> {
```

```
14
15      @Override
16      public void configure(Map<String, ?> configs) {
17
18      }
19
20      @Override
21      public ConsumerRecords<String, String>
22              onConsume(ConsumerRecords<String, String> records) {
23
24          List<ConsumerRecord<String, String>> filterRecords = new ArrayList<>();
25
26          Map<TopicPartition, List<ConsumerRecord<String, String>>>
27                  newRecords = new HashMap<>();
28
29          Set<TopicPartition> partitions = records.partitions();
30
31          for (TopicPartition tp : partitions) {
32              List<ConsumerRecord<String, String>> consumerRecords
33                                              = records.records(tp);
34
35              for (ConsumerRecord<String, String> record : consumerRecords) {
36                  if (record.value().contains("***DemoTest***")) {
37                      filterRecords.add(record);
38                  }
39              }
40
41              if (filterRecords.size() > 0) {
42                  newRecords.put(tp, filterRecords);
43              }
44          }
45          return new ConsumerRecords<>(newRecords);
46      }
47
48      @Override
49      public void onCommit(Map<TopicPartition, OffsetAndMetadata> offsets) {
50          Set<TopicPartition> keySet = offsets.keySet();
51
52          for(TopicPartition tp:keySet) {
53              OffsetAndMetadata value = offsets.get(tp);
54          System.out.println("Topic Partition = " + tp + "\t offset = " + value);
55          }
56      }
57
```

```
58      @Override
59      public void close() {
60      }
61  }
```

消费者拦截器创建后，就可以在创建 KafkaConsumer 时，通过参数 interceptor.classes 进行指定了，代码如下所示。

```
01  Properties props = new Properties();
02
03  props.put("bootstrap.servers", "kafka101:9092");
04  props.put("group.id", "mygroup");
05  props.put("key.deserializer",
06              "org.apache.kafka.common.serialization.StringDeserializer");
07  props.put("value.deserializer",
08              "org.apache.kafka.common.serialization.StringDeserializer");
09
10  props.put(ConsumerConfig.INTERCEPTOR_CLASSES_CONFIG,
11              "MyConsumerInterceptor");
12
13      KafkaConsumer<String, String> consumer = new KafkaConsumer<String,
String>(props);
14  consumer.subscribe(Arrays.asList("mytopic1"));
```

## 4.5.4　消费者的优雅退出

在一般情况下，在一个主线程中循环 poll 消息并进行处理。当需要退出 poll 循环时，可以使用另一个线程调用 consumer.wakeup()方法，调用此方法会使 poll()方法抛出 WakeupException。主线程在捕获 WakeUpException 后，需要调用 consumer.close()方法将消费者关闭。另外，在调用 consumer.wakeup()方法时，主线程正在处理消息，在下一次主线程调用 poll 时会抛出异常。

下面给出了调用 consumer.wakeup()方法的方式。

```
01  //调用 consumer.wakeup()方法通知主线程退出
02  Runtime.getRuntime().addShutdownHook(new Thread() {
03          public void run() {
04              System.out.println("Starting exit...");
05
06              //调用消费者的 wakeup()方法通知主线程退出
07              consumer.wakeup();
08
09              try {
10                  //等待主线程退出
11                  mainThread.join();
12              } catch (InterruptedException e) {
```

```
13                  e.printStackTrace();
14              }
15          }
16  });
```

下面的代码展示了主线程捕获异常进行处理的方式。

```
01      KafkaConsumer<String,  String>  consumer  =  new  KafkaConsumer<String,
String>(props);
02  consumer.subscribe(Arrays.asList("mytopic1"));
03
04  try {
05      while (true) {
06          ConsumerRecords<String,          String>          records          =
consumer.poll(Duration.ofMillis(100));
07
08          for (ConsumerRecord<String, String> record : records) {
09          System.out.println("收到消息: " + record.key() + "\t" + record.value());
10          }
11      }
12  } catch (WakeupException e) {
13      //这里忽略产生的异常
14  } finally {
15      consumer.close();
16      System.out.println("消费者已经被关闭！！！");
17  }
```

## 4.6 消费者的参数配置

Kafka 消费者端的配置参数，除了 bootstrap.servers、key.deserializer、value.deserializer 三个必需参数以外，还有很多可选的参数。下面列举了消费者的配置参数及它们的含义。

- bootstrap.servers。

该参数表示 Kafka Broker 集群的地址信息，其格式为 ip1:port、ip2:port 等，不需要设定全部的集群地址，设置两个或两个以上即可。

- group.id。

该参数表示消费者组名称，如果 group.id 相同则表示属于一个消费者组中的成员。如果没有指定该参数，会报出异常。

- fetch.min.bytes。

该参数用来配置 Kafka 消费者在一次拉取请求中能从 Kafka 中拉取的最小数据量，即调用 poll()方法时，每次拉取的数据量，其默认值为 1 字节。

消费者在拉取数据时，如果 Kafka 服务器端返回给消费者的数据量小于这个参数值的设定，那么消费者就需要进行等待，直到数据量满足这个参数的配置大小。因此在实际运行环境中，可以适当调大这个参数的值以提高一定的吞吐量。另外，增大这个参数值也会造成额外的延迟，因此增大该参数不适合敏感的应用。

- fetch.max.bytes。

该参数与 fetch.min.bytes 参数对应，它用来配置 Kafka 消费者在一次拉取请求中从 Kafka 服务器端中拉取的最大数据量，其默认值为 52 428 800 字节，也就是 50MB。

该参数并不是绝对的最大值。试想一下，如果该参数设置的值比任何一条由生产者写入 Kafka 服务器端中的消息字节数小，那么会不会造成无法消费呢？如果在第一个非空分区中拉取的第一条消息字节数大于该值，那么该消息仍然返回，以确保消费者继续工作。Kafka 消息系统中，能够接收的最大消息的字节数是通过服务器端参数 message.max.bytes 进行设置的。

- fetch.max.wait.ms。

该参数也和 fetch.min.bytes 参数有关。前面提到，如果 Kafka 服务器端返回给消费者的数据量小于 fetch.min.bytes 参数值的设定，消费者就需要等待，直到数据量满足这个参数的配置大小。然而有可能会一直等待而无法将消息发送给消费者，显然这是不合理的。fetch.max.wait.ms 参数用于指定 Kafka 的等待时间，默认值为 500ms。当 Kafka 满足不了 fetch.min.bytes 参数值的设定时，Kafka 集群也会根据 fetch.max.wait.ms 参数值的设定，默认等待 5s，然后将消息数据返回给消费者。综合来看，fetch.min.bytes 和 fetch.max.wait.ms 都有可能造成消息的延迟处理。如果业务应用对延迟敏感，那么可以适当调小这些参数。

- max.poll.records。

该参数用来配置 Kafka 消费者在一次拉取请求中拉取的最大消息数，其默认值为 500 条。如果消息数都比较小，则可以适当调大这个参数值来提升消费速度。

- max.partition.fetch.bytes。

该参数用来配置从每个分区里返回给消费者的最大数据量，其默认值为 1 048 576 字节，即 1MB。这个参数与 fetch.max.bytes 参数相似，只不过 max.partition.fetch.bytes 用来限制一次拉取中每个分区消息的字节数，而 fetch.max.bytes 用来限制一次拉取中整体消息的字节数。同样，如果这个参数设定的值比消息字节数小，那么也不会造成无法消费。

- connections.max.idle.ms。

该参数用来指定在多长时间之后，关闭闲置的 Kafka 消费者连接，默认值是 540 000ms，即 9min。

- send.buffer.bytes。

该参数用来设置发送消息缓冲区（SO_SNDBUF）的大小，其默认值为 131 072 字节，即 128KB。与 receive.buffer.bytes 参数一样，如果设置为-1，则使用操作系统的默认值。

- request.timeout.ms。

该参数用来配置 Kafka 消费者等待请求响应的最长时间，其默认值为 40s。

- receive.buffer.bytes。

该参数用来设置接收消息缓冲区（SO_RECBUF）的大小，其默认值为 65 536 字节，即 64KB。如果将该参数设置为-1，则使用操作系统的默认值。

- metadata.max.age.ms。

该参数用来配置元数据的过期时间，其默认值为 300 000ms，即 5min。如果元数据在此参数限定的时间范围内没有进行更新，即使没有任何分区变化或有新的 Kafka Broker 加入，也会被强制更新。

- reconnect.backoff.ms。

该参数用来配置 Kafka 消费者每次尝试重新连接指定主机之前应该等待的时间，避免频繁地连接主机，其默认值为 50s。

- auto.offset.reset。

该参数值为字符串类型，其有效值为以下三个。

- earliest：当各分区下有已提交的偏移量时，从提交的偏移量开始消费；无提交的偏移量时，从头开始消费。
- latest：当各分区下有已提交的偏移量时，从提交的偏移量开始消费；无提交的偏移量时，消费新产生的该分区下的数据。
- none：Topic 各分区都存在已提交的偏移量时，从偏移量后开始消费；只要有一个分区不存在已提交的偏移量，则抛出异常。

注意，除了以上三个有效值以外，设置其他任何值都会抛出错误。

- enable.auto.commit。

该参数值为 boolean 类型，配置是否开启自动提交消费位移的功能，默认开启。

- auto.commit.interval.ms。

该参数只有当 enable.auto.commit 参数设置为 true 时才生效，表示开启自动提交偏移量功能时自动提交消费位移的时间间隔，其默认值为 5s。

- partition.assignment.strategy。

该参数表示消费者的分区分配策略，支持轮询策略设置和范围策略设置。

```
01  轮询策略设置
02
partition.assignment.strategy=org.apache.kafka.clients.consumer.RoundRobinAssig
nor
03
04  范围策略设置
05
partition.assignment.strategy=org.apache.kafka.clients.consumer.RangeAssignor
```

- interceptor.class。

该参数用来配置消费者客户端的拦截器，该拦截器必须实现 org.apache.kafka.

clients.consumer.ConsumerInterceptor 接口。使用消费者拦截器可以允许用户截取消费者接收到的消息，从而可以进一步改变消息。在默认情况下，没有拦截器的设置。

- exclude.internal.topics

该参数用来指定 Kafka 中的内部主题是否可以向消费者公开，其默认值为 true。在 Kafka 消息系统中有两个内部的主题： __consumer_offsets 和__transaction_state。

# 第 5 章　Kafka 的服务器端

Kafka 是一种高吞吐量的分布式发布订阅消息系统，它可以处理消费者在网站中的所有动作流数据。为了实现这样的目标，Kafka 从设计原理方面进行了详尽的考虑，主要表现在以下几个方面。

- 高吞吐量，支持大量数据的事件流。
- 支持消息数据的可靠传送，能够处理积压的大量数据。
- 支持低延迟的消息传递。
- 支持系统的自动容错。
- 通过 Topic 的分区，支持消息的分布式处理。

本章将从主题与分区、消息的持久性、消息传递保障、副本和 Leader 的选举、Kafka 的日志清理、Kafka 配额的管理、Kafka 与 ZooKeeper 及 Kafka 服务器端参数设置几个方面来详细讨论 Kafka 的服务器端的核心设计原理。

## 5.1　主题与分区

在前面章节中，我们介绍了主题和分区的基本概念。本节将详细讨论主题与分区。

### 5.1.1　主题和分区的关系

在 Kafka 消息系统中要实现消息的发送与订阅，必须首先创建 Topic。因为 Topic 是 Kafka 进行消息归类的基本单元。Topic 接收消息生产者发布的消息，并将消息转发给消费者。换句话说，消息的消费者负责订阅消息进行处理消费。

Topic 其实是一个逻辑概念，由分区组成。分区则是一个物理概念，一个 Topic 可以包含多个分区，而一个分区只能属于一个 Topic，正因为 Kafka 采用这样的数据模型，从而实现了消息的分布式管理。

消息被发送到 Topic 的时候，实际上是发送给了 Topic 中的某一个分区，并且被添加到分区的最后，通过一个偏移量来指定消息的位置。同一个 Topic 的不同分区中的消息数据是不同的。

同时，在创建 Topic 时，还可以指定分区的副本数，通过增加副本数量可以提升容错能力，Kafka 通过多副本机制实现了故障的自动转移。

在同一个分区的多个副本中，存在一个 Leader 副本和多个 Follower 副本，它们保存的是相同的消息。其中 Leader 副本负责处理读写请求；Follower 副本只负责与 Leader 副本的消息同步。不同的 Follower 副本可能处于不同的节点上，当 Leader 副本出现了故障，Kafka 会从 Follower 副本中选举一个新的 Leader 副本。

图 5.1 说明了 Topic、分区及副本之间的关系。

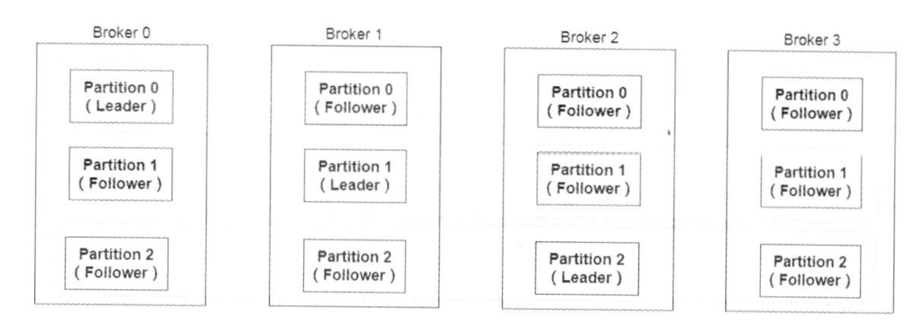


图 5.1　Topic、分区及副本之间的关系

在图 5.1 中，我们创建的 Topic 一共包含 3 个分区，每个分区的冗余度（即副本数）是 4 个。这里需要说明的是，在图 5.1 中的灰色方框表示 Leader 副本与 Follower 副本已经完成了消息数据的同步。

Kafka 是如何从多个 Follower 副本中进行选举的呢？关于这个问题，我们会在后面章节中进行详细介绍。

## 5.1.2　主题的管理

对主题进行管理与操作，主要会用到 kafka-topics.sh 和 kafka-configs.sh 这两个脚本。表 5.1 列出了 kafka-topics.sh 脚本命令中的参数含义。

表 5.1　kafka-topics.sh 脚本参数

| 参 数 选 项 | 描 述 信 息 |
| --- | --- |
| --alter | 更改主题的分区数、副本分配和/或配置 |
| --at-min-isr-partitions | 如果在描述主题时设置，则仅显示 isr 计数等于配置的最小值的分区。不支持--zookeeper 选项 |
| --bootstrap-server | 必须要连接的 Kafka 服务器。如果提供此参数，则不需要通过 ZooKeeper 连接 |
| --command-config | 包含要传递给管理客户端配置的属性文件。它只与--bootstrap-server 服务器选项一起使用，用于描述和更改 Broker 配置 |

续表

| 参数选项 | 描述信息 |
| --- | --- |
| --config | --config 参数用于重写 Topic 的配置参数。下面列出了可以修改的参数列表。本章最后会介绍这些参数的详细含义。<br>cleanup.policy<br>compression.type<br>delete.retention.ms<br>file.delete.delay.ms<br>flush.messages<br>flush.ms<br>follower.replication.throttled.replicas<br>index.interval.bytes<br>leader.replication.throttled.replicas<br>max.compaction.lag.ms<br>max.message.bytes<br>message.downconversion.enable<br>message.format.version<br>message.timestamp.difference.max.ms<br>message.timestamp.type<br>min.cleanable.dirty.ratio<br>min.compaction.lag.ms<br>min.insync.replicas<br>preallocate<br>retention.bytes<br>retention.ms<br>segment.bytes<br>segment.index.bytes<br>segment.jitter.ms<br>segment.ms<br>unclean.leader.election.enable |
| --create | 创建一个 Topic |
| --delete | 删除一个 Topic |
| --delete-config | 从 Topic 中删除某一个配置参数。参数列表参考--config |
| --describe | 列出 Topic 的详细信息 |
| --disable-rack-aware | 禁用机架感知副本分配的机制 |
| --exclude-internal | 运行 list 或 describe 命令时排除内部主题。在默认情况下，列出内部主题 |

续表

| 参 数 选 项 | 描 述 信 息 |
|---|---|
| --force | 取消控制台提示 |
| --help | 显示帮助信息 |
| --if-exists | 如果在更改、删除或描述主题时设置，则仅在主题存在时执行操作。不支持--bootstrap-server 选项 |
| --if-not-exists | 如果在更改、删除或描述主题时设置，则仅在主题不存在时执行操作。不支持--bootstrap-server 选项 |
| --list | 列出所有可用的 Topic 信息 |
| --partitions | 正在创建或更改的主题的分区数 |
| --replica-assignment | 将 Topic 中的分区手动进行分配的分区-Broker 的对应列表 |
| --replication-factor | Topic 中每个分区的副本数 |
| --topic | Topic 的名称 |
| --topics-with-overrides | 如果在描述主题时设置，则仅显示已重写配置的主题 |
| --unavailable-partitions | 如果在描述主题时设置，则仅显示 Leader 不可用的分区 |
| --under-min-isr-partitions | 如果在描述主题时设置，则仅显示 isr 计数小于配置的分区最小值。不支持--zookeeper 选项 |
| --under-replicated-partitions | 如果在描述主题时设置，仅显示已经完成同步的分区 |
| --version | 显示 Kafka 的版本信息 |
| --zookeeper | 已过期。ZooKeeper 的地址信息 |

### 1. 创建主题

使用下面的命令来创建一个新的 Topic 主题。

```
bin/kafka-topics.sh --create --zookeeper kafka101:2181 \
--replication-factor 2 --partitions 3 --topic mytopic1
```

其中，

- --zookeeper：用于指定 ZooKeeper 的地址，如果是多个 ZooKeeper 地址可以使用逗号分隔。
- --replication-factor：用于指定分区的副本数。这里设置的副本数为 2，表示同一个分区有两个副本。
- --partitions：用于指定该 Topic 包含的分区数。这里设置的分区数为 3，表示该 Topic 由 3 个分区组成。
- --topic：用于指定 Topic 的名称。

### 2. 查看主题

使用--list 参数列出所有可用的 Topic 信息。

```
01        [root@kafka101   kafka_2.11-2.4.0]#   bin/kafka-topics.sh   --zookeeper
kafka101:2181 -list
02  __consumer_offsets
03  mytopic1
04  [root@kafka101 kafka_2.11-2.4.0]#
```

使用--describe 参数查看某个 Topic 的详细信息。

```
01        [root@kafka101   kafka_2.11-2.4.0]#   bin/kafka-topics.sh   --zookeeper
kafka101:2181 \
02  > --describe --topic mytopic1
03  Topic: mytopic1 PartitionCount: 3    ReplicationFactor: 2 Configs:
04      Topic: mytopic1  Partition: 0 Leader: 0    Replicas: 0,1    Isr: 0,1
05      Topic: mytopic1  Partition: 1 Leader: 1    Replicas: 1,0    Isr: 1,0
06      Topic: mytopic1  Partition: 2 Leader: 0    Replicas: 0,1    Isr: 0,1
07  [root@kafka101 kafka_2.11-2.4.0]#
```

从上面的信息中可以看出，Topic mytopic1 包含 3 个分区，每个分区有两个副本。其中，分区 0 的 Leader 是 0 号副本；分区 1 的 Leader 是 1 号副本；分区 2 的 Leader 是 0 号副本；ISR 列表中的数据表示已经完成数据同步的副本号。

**3. 修改主题**

我们现在给 Topic mytopic1 增加一个分区。

```
01        [root@kafka101   kafka_2.11-2.4.0]#   bin/kafka-topics.sh   --zookeeper
kafka101:2181 \
02  > --alter --topic mytopic1 --partitions 4
03   WARNING: If partitions are increased for a topic that has a key, the
partition
04  logic or ordering of the messages will be affected
05  Adding partitions succeeded!
06  [root@kafka101 kafka_2.11-2.4.0]#
```

从上面的警告信息可以看出，当主题中的消息包含 key 时，根据 key 进行分区就会受到影响。

现在重新使用--describe 来查看该 Topic 的详细信息。

```
01        [root@kafka101   kafka_2.11-2.4.0]#   bin/kafka-topics.sh   --zookeeper
kafka101:2181 \
02  > --describe --topic mytopic1
03  Topic: mytopic1 PartitionCount: 4    ReplicationFactor: 2 Configs:
04      Topic: mytopic1  Partition: 0 Leader: 0    Replicas: 0,1    Isr: 0,1
05      Topic: mytopic1  Partition: 1 Leader: 1    Replicas: 1,0    Isr: 1,0
06      Topic: mytopic1  Partition: 2 Leader: 0    Replicas: 0,1    Isr: 0,1
07      Topic: mytopic1  Partition: 3 Leader: 1    Replicas: 1,0    Isr: 1,0
08  [root@kafka101 kafka_2.11-2.4.0]#
```

可以看出 Topic mytopic1 中新增加了一个 Partition 3 的分区。既然可以增加 Topic 的分区数，那么可以减少吗？可以来测试一下。

```
01        [root@kafka101   kafka_2.11-2.4.0]#   bin/kafka-topics.sh   --zookeeper
kafka101:2181 \
02  > --alter --topic mytopic1 --partitions 3
03   WARNING: If partitions are increased for a topic that has a key, the
partition
04  logic or ordering of the messages will be affected
05  Error while executing topic command : The number of partitions for a topic
can
06  only be increased. Topic mytopic1 currently has 4 partitions, 3 would not
be an increase.
07                             2020-12-15            08:787]             ERROR
org.apache.kafka.common.errors.InvalidPartitionsException:
08  The number of partitions for a topic can only be increased. Topic mytopic1
currently has
09  4 partitions, 3 would not be an increase. (kafka.admin.TopicCommand$)
10  [root@kafka101 kafka_2.11-2.4.0]#
```

可以看到出现了如下的错误，说明在目前版本的 Kafka 中，只支持分区数的增加而不支持分区数的减少。

```
Error while executing topic command : The number of partitions for a topic can
only be increased. Topic mytopic1 currently has 4 partitions, 3 would not be an
increase.
```

### 4. 配置管理

使用 kafka-configs.sh 脚本对配置进行管理和操作。这里需要注意的是，这种管理和操作只能在 Kafka 服务器端处于运行状态下进行。通过这种方式，可以达到动态修改变更参数的目的。kafka-configs.sh 脚本包含变更配置和查看配置这两种类型的指令。

在使用 kafka-configs.sh 脚本时需要注意的是，由于它使用 entity-type 参数来指定操作配置的类型，并且使用 entity-name 参数来指定操作配置的名称，这里需要单独对这两个参数进行说明，如表 5-2 所示。

表 5.2 entity-name 和 entity-type 的参数值说明

| 参 数 选 项 | 描 述 信 息 |
|---|---|
| --cntity-namc <String> | 参数的名称，可以是以下对应的名称：<br>topic name<br>clicnt id<br>user principal name<br>broker id |

续表

| 参数选项 | 描述信息 |
| --- | --- |
| --entity-type <String> | 参数配置的类型，可以是以下的值：<br>topics<br>clients<br>users<br>brokers<br>broker-loggers |

下面我们通过具体的操作来演示如何使用 kafka-configs.sh 脚本。

使用下面的命令查看 mytopic1 的配置信息。

```
01      [root@kafka101  kafka_2.11-2.4.0]#  bin/kafka-configs.sh  --zookeeper
kafka101:2181 \
02  > --describe --entity-type topics --entity-name mytopic1
03  Configs for topic 'mytopic1' are
04  [root@kafka101 kafka_2.11-2.4.0]#
```

这里没有显示出任何配置信息，其原因是采用的都是 Kafka 默认的配置信息，所有没有任务的输出信息。

下面可以尝试修改某个参数值的设定。

```
01      [root@kafka101  kafka_2.11-2.4.0]#  bin/kafka-configs.sh  --zookeeper
kafka101:2181 \
02  > --alter \
03  > --entity-type topics --entity-name mytopic1 \
04  > --add-config cleanup.policy=compact,max.message.bytes=10000
05  Completed Updating config for entity: topic 'mytopic1'.
06  [root@kafka101 kafka_2.11-2.4.0]#
```

再重新使用--describe 查看 Topic 的详细信息。

```
01      [root@kafka101  kafka_2.11-2.4.0]#  bin/kafka-configs.sh  --zookeeper
kafka101:2181 \
02  > --describe --entity-type topics --entity-name mytopic1
03              Configs      for      topic      'mytopic1'      are
max.message.bytes=10000,cleanup.policy=compact
04  [root@kafka101 kafka_2.11-2.4.0]#
```

这里就可以看到刚才我们配置的参数信息。

### 5. 删除主题

当某个 Topic 不再使用，可以直接将其删除。

```
01      [root@kafka101  kafka_2.11-2.4.0]#  bin/kafka-topics.sh  --zookeeper
kafka101:2181 \
02  --delete --topic mytopic1
03
04  Topic mytopic1 is marked for deletion.
```

```
05   Note: This will have no impact if delete.topic.enable is not set to true.
06   [root@kafka101 kafka_2.11-2.4.0]#
```

这里需要说明的是，通过上面的方式删除 Topic 可能不能保证一个 Topic 和数据被彻底删除。如果想要保证一个 Topic 被彻底删除，需要按照以下的步骤执行。

（1）如果需要删除的 Topic 此时正在被程序生产者和消费者使用，则先停止生产者和消费者的使用。

（2）设置 delete.topic.enable=true。

（3）bin/kafka-topics.sh --delete --zookeeper [zookeeper server:port] --topic [topic name]。

（4）删除 Topic 对应的 Kafka 存储目录，即在 server.properties 文件中配置 log.dirs 目录下相关的 Topic 的数据目录。如果 Kafka 集群中有多个 Broker，并且 Topic 也有多个分区和副本，则需要删除所有对应的目录。

（5）使用 ZooKeeper 客户端命令 zkCli.sh 登录 ZooKeeper，删除/brokers/topics 节点下对应的 Topic 目录，即执行命令：rmr /brokers/topics/Topic Name。

成功删除 Topic 后，再次使用--list 查看 Kafka 集群中的可用 Topic，发现 mytopic1 已经被成功删除了。

```
01       [root@kafka101   kafka_2.11-2.4.0]#   bin/kafka-topics.sh   --zookeeper
kafka101:2181 -list
02   __consumer_offsets
03   [root@kafka101 kafka_2.11-2.4.0]#
```

## 5.1.3　使用 KafkaAdminClient

在一般情况下，我们都使用 kafka-topics.sh 和 kafka-configs.sh 这两个脚本来管理查看 Kafka。但是有些时候需要将某些管理查看的功能集成到外部系统中。为了支持这样的功能，Kafka 也提供了相应的 API 来直接操作 Kafka。在目前的 Kafka 版本中，Kafka 提供了一个抽象类 AdminClient 来封装对 Kafka 的操作，这个类属于 kafka-client 包。它有一个默认的实现 org.apache.kafka.clients.admin.KafkaAdminClient。本节将重点介绍 KafkaAdminClient 的使用方法。

表 5.3 列出了 KafkaAdminClient 常用的 API。

表 5.3　KafkaAdminClient 的常用 API

| API | 说　明 |
| --- | --- |
| createTopics | 创建 Topic |
| deleteTopics | 删除 Topic |
| describeTopics | 查看 Topic 的详细信息 |
| describeCluster | 查询集群信息 |

续表

| API | 说　明 |
|---|---|
| describeConfigs | 查询配置信息 |
| alterConfigs | 修改配置信息 |
| alterReplicaLogDirs | 修改副本的日志目录 |
| describeLogDirs | 查询节点的日志目录信息 |
| describeReplicaLogDirs | 查询副本的日志目录信息 |
| createPartitions | 增加分区 |

下面的代码使用 KafkaAdminClient.createTopics 创建了一个新的 Topic。

```
01  import java.util.ArrayList;
02  import java.util.Collection;
03  import java.util.Properties;
04  import org.apache.kafka.clients.CommonClientConfigs;
05  import org.apache.kafka.clients.admin.AdminClient;
06  import org.apache.kafka.clients.admin.KafkaAdminClient;
07  import org.apache.kafka.clients.admin.NewTopic;
08  import org.junit.After;
09  import org.junit.Before;
10  import org.junit.Test;
11
12  public class KafkaAdminClientDemo {
13
14      private String TOPIC_NAME = "mytopic1";
15      private String URL = "kafka101:9092";
16      private AdminClient client;
17
18      @Before
19      public void testBefore() {
20          Properties props = new Properties();
21          props.put(CommonClientConfigs.BOOTSTRAP_SERVERS_CONFIG, URL);
22          client = KafkaAdminClient.create(props);
23      }
24
25      @Test
26      public void testCreateTopic() {
27           //使用 KafkaAdminClient 在 Kafka 上创建一个 Topic
28          NewTopic newTopic = new NewTopic(TOPIC_NAME,4, (short)2);
29          Collection<NewTopic> newTopicList = new ArrayList<>();
30          newTopicList.add(newTopic);
31          client.createTopics(newTopicList);
32      }
33
```

```
34
35      @After
36      public void testAfter() {
37          client.close();
38      }
39  }
```

其中，

```
NewTopic newTopic = new NewTopic(TOPIC_NAME,4, (short)2);
```

表示该 Topic 由 4 个分区组成，并且副本的冗余度为 2。

## 5.2　消息的持久性

当生产者把消息发送给 Kafka 服务器端后，Kafka Broker 会将消息存储在文件系统中。但是传统磁盘的读写操作是很缓慢的，那么 Kafka 基于文件系统的架构是否能够解决这一问题呢？本节将重点讨论 Kafka 消息的持久性，并且探讨 Kafka 如何解决传统磁盘 IO 的性能问题。

### 5.2.1　Kafka 消息持久性概述

消息由 Kafka Broker 接收后会进行持久化的操作，以便在消费者不可用的时候保存消息。Apache Kafka 的底层依然是基于 Java 实现的，而 Java 的磁盘 IO 操作又存在以下两点问题。

- 存储缓存对象严重影响性能。
- 堆内存数据的增加导致 Java 垃圾回收的速度越来越慢。

虽然传统的磁盘读写操作会很慢，但是磁盘线性写入的性能远远大于随机写入的性能。因为底层的操作系统对磁盘的线性写入进行了大量优化，在某些情况下，磁盘线性甚至比随机的内存读写更快。所以 Kafka 在进行消息持久化操作时，写日志文件采用的就是磁盘的线性写入方式，从而解决了传统磁盘写操作慢的问题。这里的持久化需要从读和写两个方面来考虑。

- 写操作：将数据顺序追加到日志文件中。
- 读操作：根据偏移量从日志文件中读取。

Kafka 具有以下优点。

- 实现了读写分离，数据大小不会对性能产生影响。
- 硬盘空间相对于内存空间容量限制更小。
- 访问磁盘采用线性的方式，速度更快、更稳定。

## 5.2.2 Kafka 的持久化原理解析

一个 Topic 被分成多个 Partition，每个 Partition 在存储层面是一个 append-only 日志文件，属于一个 Partition 的消息都会被直接追加到日志文件的尾部，每条消息在文件中的位置称为偏移量。Kafka 的写操作如图 5.2 所示。

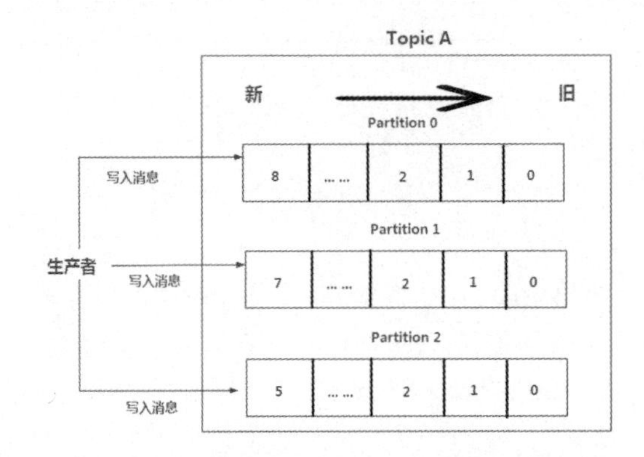


图 5.2　Kafka 的写操作

之前创建的 mytopic1 具有三个分区。可以到对应的日志目录下查看，如图 5.3 所示。

```
[root@kafka101 broker0]# pwd
/root/training/kafka_2.11-2.4.0/logs/broker0
[root@kafka101 broker0]# ls mytopic1-*
mytopic1-0:
00000000000000000000.index  00000000000000000000.log  00000000000000000000.timeindex  leader-epoch-checkpoint

mytopic1-1:
00000000000000000000.index  00000000000000000000.log  00000000000000000000.timeindex  leader-epoch-checkpoint

mytopic1-2:
00000000000000000000.index  00000000000000000000.log  00000000000000000000.timeindex  leader-epoch-checkpoint

mytopic1-3:
00000000000000000000.index  00000000000000000000.log  00000000000000000000.timeindex  leader-epoch-checkpoint
[root@kafka101 broker0]#
```

图 5.3　Kafka 的日志目录

Kafka 日志分为 index 与 log，这两个文件总数成对出现。但是它们存储的信息是不一样的。

- index 文件存储元数据，即索引文件。
- log 文件存储消息，即数据文件。

索引文件元数据指向对应数据文件中消息的偏移量。例如，1,128 指的是数据文件的第 1 条数据，偏移地址为 128；而物理地址（在索引文件中指定）+ 偏移量可以定位消息。

可以使用 Kafka 自带的工具，即 kafka.tools.DumpLogSegments 查看 log 日志文件中的数据信息。先查看一下这个工具的帮助信息。

```
01  bin/kafka-run-class.sh kafka.tools.DumpLogSegments --help
```

```
02  This tool helps to parse a log file and dump its contents to the console, useful
03  for debugging a seemingly corrupt log segment.
04  Option                              Description
05  ------                              -----------
06  --deep-iteration                   if set, uses deep instead of shallow iteration.
07                                    Automatically set if print-data-log is enabled.
08   --files <String: file1, file2, ...>          REQUIRED: The comma separated
list of data and
09                                    index log files to be dumped.
10  --help                              Print usage information.
11  --index-sanity-check                if set, just checks the index sanity without
12                                    printing its content. This is the same check
13                                      that is executed on broker startup to determine
14                                        if an index needs rebuilding or not.
15  --key-decoder-class [String]       if set, used to deserialize the keys. This class
16                                  should implement kafka.serializer.Decoder trait.
17                                Custom jar should be available in kafka/libs directory.
18                                      (default: kafka.serializer.StringDecoder)
19   --max-message-size <Integer: size>    Size of largest message. (default:
5242880)
20  --offsets-decoder                    if set, log data will be parsed as offset
data
21                                     from the __consumer_offsets topic.
22  --print-data-log                     if set, printing the messages content when
23                                     dumping data logs. Automatically set if
24                                       any decoder option is specified.
25  --transaction-log-decoder             if set, log data will be parsed as
26                                     transaction metadata from the
27                                     __transaction_state topic.
28  --value-decoder-class [String]        if set, used to deserialize the messages.
29                                     This class should implement kafka.
30                                     serializer.Decoder trait. Custom jar
31                                     should be available in kafka/libs
32                                     directory. (default: kafka.serializer.
33                                     StringDecoder)
34  --verify-index-only                   if set, just verify the index log without
35                                     printing its content.
36  --version                           Display Kafka version.
37  [root@kafka101 kafka_2.11-2.4.0]#
```

通过使用--files 参数来指定要导出的日志文件。如果需要导出多个日志文件的数据，可以使用逗号进行分隔。下面我们举例说明。

```
bin/kafka-run-class.sh kafka.tools.DumpLogSegments --files \
logs/broker0/mytopic1-0/00000000000000000000.log --print-data-log
```

其中，参数--print-data-log 会在导出日志消息数据的时候，将其打印在屏幕上。

```
Dumping logs/broker0/mytopic1-0/00000000000000000000.log
Starting offset: 0
baseOffset: 0 lastOffset: 0 count: 1 baseSequence: -1 lastSequence: -1 p
| offset: 0 CreateTime: 1589796084450 keysize: -1 valuesize: 5 sequence:
baseOffset: 1 lastOffset: 1 count: 1 baseSequence: -1 lastSequence: -1 p
| offset: 1 CreateTime: 1589955515316 keysize: -1 valuesize: 11 sequence
baseOffset: 2 lastOffset: 2 count: 1 baseSequence: -1 lastSequence: -1 p
| offset: 2 CreateTime: 1589955676344 keysize: 7 valuesize: 4 sequence:
baseOffset: 3 lastOffset: 3 count: 1 baseSequence: -1 lastSequence: -1 p
| offset: 3 CreateTime: 1589955680346 keysize: 7 valuesize: 4 sequence:
baseOffset: 4 lastOffset: 4 count: 1 baseSequence: -1 lastSequence: -1 p
```

图 5.4　Kafka 的日志目录

## 5.2.3　持久化的读写流程

图 5.5 是 Kafka 官方提供的关于 Kafka 日志持久化读写的说明。

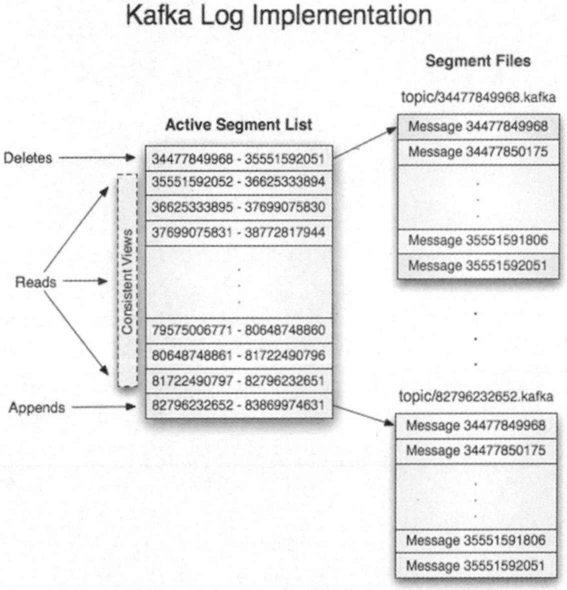

图 5.5　Kafka 日志的读写

（1）写流程。

日志文件允许串行追加，并且总是附加到最后一个日志文件的后面。每个日志文件的大小可以由参数 log.segment.bytes 来指定，其默认值是 1GB。当日志文件的大小达到或超过这个值的时候，就会产生一个新的日志文件。每个生成的日志文件称为一个 Segment

File。日志文件有两个配置参数：M 和 S。M 强制操作系统将日志文件刷新到磁盘之前写入的消息数；S 强制操作系统将日志文件刷新到磁盘之前的时间，单位是秒。在系统崩溃的情况下，最多会丢失 M 条消息或 S 秒的数据。

（2）读流程。

Kafka 在读取持久化数据的时候，通过给出的消息偏移量和数据块大小来读取数据，并且将读取的数据放入一个缓存区中。如果读取的数据非常大，可以多次重试读取，每次都将缓冲区加倍，直到成功读取消息为止。在读取数据的时候，需要先确定存储数据的日志文件，即 Segment File，再得到数据在 Segment File 中的偏移量，然后从此位置开始读取。

### 5.2.4　为什么要建立分段和索引

Kafka 为了解决在读取数据时快速定位数据，采用了将数据文件分段并建立索引的形式。每段放在一个单独的数据文件里面，数据文件以该段中最小的 offset 命名。这样在查找指定 offset 的消息数据时，用二分查找法就可以定位到该数据消息在哪个段中，即快速定位数据文件。

为了在一个段内快速定位数据，进一步提高查找的效率，Kafka 为每个分段后的数据文件建立了索引文件，文件名与数据文件名是一样的，只是文件扩展名为.index。索引文件中的每条索引表示数据文件中一条消息数据的位置信息。这里需要注意的是，在索引文件中并没有为每条消息数据建立索引，而是每隔一定字节的数据建立一条索引。这样避免索引文件占用过多的空间，从而可以将索引文件保留在内存中。

另外，在图 5.3 中还有一个.timeindex 索引文件，这个文件的格式与.index 文件格式一样，它主要用于消息的定期删除。在这个.timeindex 索引文件中记录的是消息数据发布时间与 offset 的对应关系，其实也是一种稀疏索引。

## 5.3　消息的传输保障

前面我们介绍过 Kafka 分区的副本机制，正是由于这样的副本机制保障了 Kafka 消息的高可用性。在目前版本的 Kafka 中，提供分区级别的副本策略，可以通过调节副本的相关参数达到高可用容错的目的。那么消息的生产者在向 Kafka 服务器端发送消息的时候，如何在保证消息的可靠传输的同时又能保证消费者的消费呢？本节将来讨论 Kafka 消息的传输保障。

### 5.3.1　生产者的 ack 机制

当生产者向服务器端分区的 Leader 副本发送数据时，可以通过 acks 参数来设置生产者数据可靠性的级别。ask 参数有下面三个取值。

（1）1（默认）。

这种参数设置意味着生产者在服务器端分区的 Leader 副本已成功接收到数据并得到确认后发送下一条数据。换句话说，只要 Leader 副本的分区成功接收到数据，就算生产者成功发送了数据，在这种方式下也存在数据丢失的可能性。如果分区 Leader 副本所在的节点出现了宕机，则会丢失数据。因为这时分区的 Follower 副本还没有完成与 Leader 副本的数据同步。

（2）0。

这种参数设置意味着生产者不用等待来自服务器端 Broker 的确认，继续发送下一条消息。在这种情况下数据传输效率最高，但是数据可靠性是最低的。因为生产者发送数据后，并不知道服务器端是否成功接收到数据消息。所以，一般不建议在生产环境中使用这种参数设置。

（3）all。

这种参数设置意味着生产者发送完消息数据后，需要等待服务器端 Topic 分区的所有副本都完成与 Leader 副本的数据同步后，才算数据消息成功发送。很明显，在这种方式下，Kafka 的性能最差，但是其可靠性最高。

下面的代码使用了“all”的参数设置获取最高的消息传输可靠性。

```
01  Properties props = new Properties();
02  props.put(ProducerConfig.BOOTSTRAP_SERVERS_CONFIG, "kafka101:9092");
03  props.put(ProducerConfig.KEY_SERIALIZER_CLASS_CONFIG,
04          "org.apache.kafka.common.serialization.StringSerializer");
05  props.put(ProducerConfig.VALUE_SERIALIZER_CLASS_CONFIG,
06        "org.apache.kafka.common.serialization.StringSerializer");
07
08  props.put(ProducerConfig.ACKS_CONFIG, "all");
09
10  Producer<String, String> producer = new KafkaProducer<String, String>(props);
```

## 5.3.2 消费者与高水位线

生产者将消息成功发送到服务器端，并将日志数据添加到分区的最后。消费者在消费消息的时候，能够读取到分区中的哪些消息数据呢？要讨论这个问题，就需要知道什么是高水位线。首先，介绍一个名词 LEO，它是 Log End Offset 的缩写，表示 Topic 分区的每个副本日志中最后一条消息的位置。高水位线（High Watermark）等于 Topic 分区中每个副本对应的最小的 LEO 值。Kafka 的高水位线如图 5.6 所示。

**分区Partition**

| Leader | Follower1 | Follower2 |
| --- | --- | --- |
| 6 | | |
| 5 | | 5 |
| 4 | | 4 |
| 3 | 3 | 3 |
| 2 | 2 | 2 |
| 1 | 1 | 1 |
| 0 | 0 | 0 |

**高水位线**

图 5.6　Kafka 的高水位线

在图 5.6 中的分区有 3 个副本。通过前面的介绍我们知道 Leader 副本负责分区的读写操作，可以看到 Leader 副本目前的 LEO 是 6。两个 Follower 副本从 Leader 副本上同步数据，Follower1 的 LEO 是 3，Follower2 的 LEO 是 5。我们取所有副本中最小的 LEO 值 3，就是该分区的高水位线。

高水位线与消费者有什么关系呢？由于每个分区的副本都是自己的 LEO，消费者在消费消息的时候，最多只能消费到高水位线所在的位置。通过这样的机制，当生产者写入新的消息后，消费者是不能够立即消费的。Leader 副本会等待该消息被所有副本都同步后，再去更新高水位线的位置，这样消费者才能消费生产者新写入的消息。这样就保证了如果 Leader 副本所在的 Broker 出现了宕机的情况，Kafka 选举出新的 Leader 副本后，该消息仍然可以在重新选举的 Leader 副本中获取。

## 5.4　副本和 Leader 副本的选举

前面提到如果分区存在多个副本，其中一个是 Leader 副本，其他都是 Follower 副本。如果某个分区的 Leader 副本宕机了，那么 Kafka 会自动从其他 Follower 副本中选举一个新的 Leader 副本。之后所有读写就会转移到这个新的 Leader 副本上。那么 Kafka 如何进行选举呢？要知道 Kafka 不是采用常见的多数选举的方式进行 Leader 副本的选举，而是在 ZooKeeper 上针对每个 Topic 维护一个 ISR（in-sync replica，已同步的副本）列表集合。如果不在 ISR 列表中的副本表示还没有完成与 Leader 副本的同步，也就没有被选举的资格。换句话说，只有这个 ISR 列表中的副本才有资格成为 Leader 副本。

在进行 Leader 副本选举时，首先选举 ISR 中的第一个，如果第一个选举不成功，接着选举第二个，依次类推。因为 ISR 中的是同步副本，消息最完整且各个节点都是一样的。Kafka 的选举机制相对比较简单，就是采用 ISR 列表的顺序选举的。假设某个 Topic 的分区有 $N$ 个副本，Kafka 可以容忍 $N$−1 个 Leader 副本宕机或不可用。还有一点需要说明的是，如果 ISR 列表中的副本都不可用，Kafka 则会从不在 ISR 列表的副本中选举一个 Leader 副本，这时候就可能导致数据不一致的问题。

为了方便说明这个机制，我们来创建一个新的 Topic：mytopic2，它有两个分区，并且每个分区的副本数为 3。执行下面的命令。

```
bin/kafka-topics.sh --create --zookeeper kafka101:2181 \
--replication-factor 3 --partitions 2 --topic mytopic2
```

下面通过--describe 指令查看 Topic 的详细信息，执行下面的命令。

```
[root@kafka101 kafka_2.11-2.4.0]# bin/kafka-topics.sh --zookeeper kafka101:2181
\
> --describe --topic mytopic2
Topic: mytopic2  PartitionCount: 2    ReplicationFactor: 3 Configs:
    Topic: mytopic2  Partition: 0 Leader: 0    Replicas: 0,1,2  Isr: 0,1,2
    Topic: mytopic2  Partition: 1 Leader: 1    Replicas: 1,2,0  Isr: 1,2,0
[root@kafka101 kafka_2.11-2.4.0]#
```

我们以 Partition 1 为例。当前的 Leader 是 1 号副本，并且在 ISR 列表中有（1,2,0）号副本。如果 1 号副本的 Leader 宕机，则会顺序选择 2 号副本作为新的 Leader；如果 2 号副本出现了故障，则会选举 0 号副本作为 Leader。

我们来做一个简单的测试。使用 kill -9 命令杀掉 1 号 Broker，即 1 号副本出现了宕机。再使用--describe 指令来查看 Topic 的详细信息。

如图 5.7 所示，以 Partition 1 为例，当 1 号副本出现了宕机，Kafka 自动按照 ISR 列表中的顺序选举 2 号副本作为 Partition 1 的 Leader。

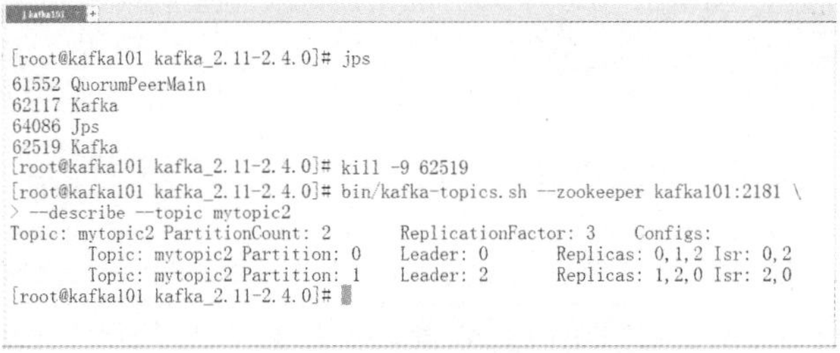

```
[root@kafka101 kafka_2.11-2.4.0]# jps
61552 QuorumPeerMain
62117 Kafka
64086 Jps
62519 Kafka
[root@kafka101 kafka_2.11-2.4.0]# kill -9 62519
[root@kafka101 kafka_2.11-2.4.0]# bin/kafka-topics.sh --zookeeper kafka101:2181 \
> --describe --topic mytopic2
Topic: mytopic2 PartitionCount: 2       ReplicationFactor: 3    Configs:
        Topic: mytopic2 Partition: 0    Leader: 0       Replicas: 0,1,2 Isr: 0,2
        Topic: mytopic2 Partition: 1    Leader: 2       Replicas: 1,2,0 Isr: 2,0
[root@kafka101 kafka_2.11-2.4.0]#
```

图 5.7　Leader 的选举

## 5.5　Kafka 配额的管理

配额（Quota）是对 Kafka 某种资源的限制。Apache Kafka 配额所能管理和配置的对象有 3 种。

- 用户级别：user。
- 客户端级别：clientid。
- 用户级别+客户端级别：user + clientid。

这三种配额的管理都是对接入 Kafka 的生产者和消费者身份的认定方式。其中，客户端级别是每个接入 Kafka 集群的生产者或消费者的一个身份标志；用户级别只有在开启身份认证的 Kafka 集群中才有。如果 Kafka 集群没有开启身份认证，则只能使用客户端级别的方式进行限流。

可以使用 kafka-configs.sh 脚本命令来为 Kafka 设定配额。Kafka 配额可配置的选项，以及它们的含义如下。

（1）producer_byte_rate。

消息的生产者单位时间内可以发布到 Kafka 消息集群中的单台 Broker 的字节数（单位：秒）。

（2）consumer_byte_rate。

消息的消费者在单位时间内可以从 Kafka 消息集群中单台 Broker 拉取的字节数（单位：秒）。

下面通过具体的命令来展示如何配置 Kafka 的配额。

客户端级别的配置，例如，client-id 为“clientA”。

```
bin/kafka-configs.sh  --zookeeper kafka101:2181 --alter --add-config \
'producer_byte_rate=1024,consumer_byte_rate=2048' \
-entity-type clients --entity-name clientA
```

使用--describe 指令查看上面的配置，如图 5.8 所示。

```
bin/kafka-configs.sh --zookeeper kafka101:2181 \
--describe --entity-type clients --entity-name clientA
```

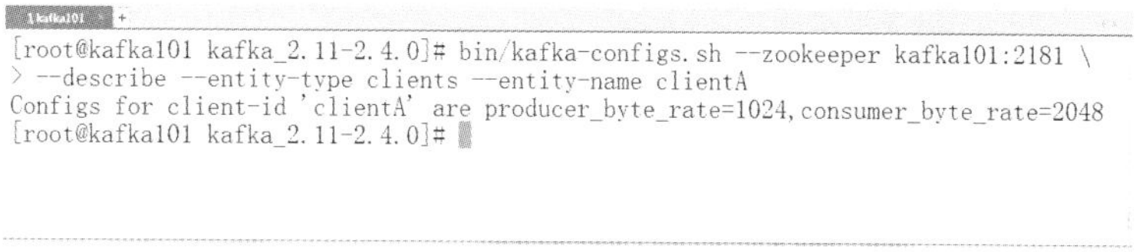

```
[root@kafka101 kafka_2.11-2.4.0]# bin/kafka-configs.sh --zookeeper kafka101:2181 \
> --describe --entity-type clients --entity-name clientA
Configs for client-id 'clientA' are producer_byte_rate=1024,consumer_byte_rate=2048
[root@kafka101 kafka_2.11-2.4.0]#
```

图 5.8　客户端级别的配额管理

用户级别的配置，例如，user 为“user1”。

```
bin/kafka-configs.sh  --zookeeper kafka101:2181 --alter --add-config \
'producer_byte_rate=1024,consumer_byte_rate=2048' \
--entity-type users --entity-name user1
```

使用--describe 指令查看上面的配置，如图 5.9 所示。

```
bin/kafka-configs.sh --zookeeper kafka101:2181 \
--describe --entity-type users --entity-name user1
```

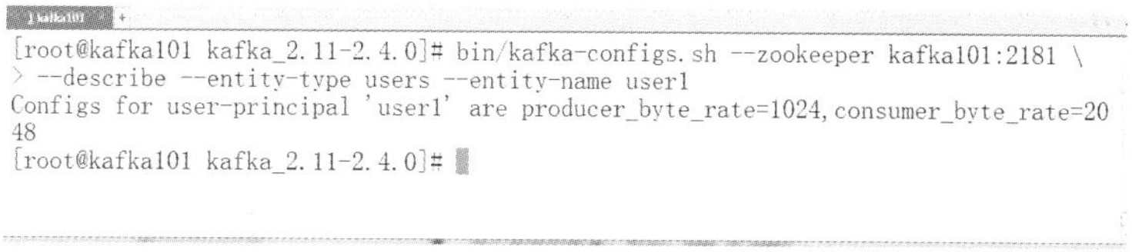

```
[root@kafka101 kafka_2.11-2.4.0]# bin/kafka-configs.sh --zookeeper kafka101:2181 \
> --describe --entity-type users --entity-name user1
Configs for user-principal 'user1' are producer_byte_rate=1024,consumer_byte_rate=20
48
[root@kafka101 kafka_2.11-2.4.0]#
```

图 5.9　用户级别的配额管理

结合客户端级别和用户级别的配置，例如，user 为“user1”，clientid 为“clientA”。

```
bin/kafka-configs.sh  --zookeeper kafka101:2181 --alter --add-config \
'producer_byte_rate=1024,consumer_byte_rate=2048'  \
--entity-type users --entity-name user1 --entity-type clients \
--entity-name clientA
```

## 5.6 Kafka 的日志删除与压缩

Apache Kafka 是一个基于日志的消息处理系统。一个 Topic 可以有若干个 Partition 的分区，而分区是数据管理的基本单元。一个分区的数据文件可以存储在若干个独立磁盘的目录中。每个 Partition 的日志文件存储时又会被分成多个 Segment。Segment 是日志清理的基本单元，需要注意的是，当前正在使用的 Segment 是不会被清理的。以 Segment 为单位，每一个 Partition 分区的日志都会被分为两部分：一部分是已清理的部分；另一部分是未清理的部分。同时未清理的部分又分为可以清理的部分和不可清理的部分。

Kafka 通过日志的压缩提供保留较为细粒度的日志记录，这种日志的压缩方式有别于基于粗粒度的基于时间的保留。

### 5.6.1 日志的删除

Kafka 中的每一条数据都包含 Key 和 Value，并且 Kafka 通过日志将数据存储在磁盘中。在一般情况下，日志并不会永久保留，否则会占用大量的磁盘空间。在数据达到一定的范围或超过一定的时间后，最早写入的数据将会被删除，这就是日志删除（Log Deletion）策略。这种日志删除策略按照一定的保留策略来直接删除不符合条件的日志分段。

### 5.6.2 日志的压缩

日志压缩（Log Compaction）是另一种清理的方式。它在默认的删除规则之外提供了另一种删除过时数据的方式，就是当 Key 相同，而数据不同时，只保留最后一条数据，前面的数据在合适的情况下被删除。换句话说，这种日志清理的策略是针对每个消息的 key 进行整合的，对于有相同 Key 的不同 Value 值，只保留最后一个版本。

### 5.6.3 清理的实现细节

通过服务器端的参数 log.cleanup.policy 来设置 Kafka 的日志清理策略，此参数默认值为“delete”，即采用日志删除的清理策略。如果要采用日志压缩的清理策略，就需要将 log.cleanup.policy 设置为 compact。需要注意的是，采用日志压缩的清理策略参数 log.cleaner.enable 需要设定为 true。通过将参数 log.cleanup.policy 设置为 delete，compact

还可以同时支持日志删除和日志压缩两种策略。

日志清理的粒度可以控制到Topic级别，与参数log.cleanup.policy对应的主题级别的参数为cleanup.policy。

当Kafka日志信息需要被清理的时候，Kafka日志管理器中会有一个专门的周期性日志删除任务来检测不符合保留条件的日志文件，并执行相应的删除或压缩操作从而达到日志清理的目的。通过设置服务器端参数log.retention.check.interval.ms来设置这个周期性任务的检查间隔，其默认值为300s，即5min。

既然Kafka要周期性地检查日志的保留条件，那么什么样的日志可以保留下来呢？目前Kafka支持三种保留策略。下面分别来介绍它们。

（1）基于时间的保留策略。

周期性执行的日志删除任务会检查当前日志文件中是否有保留的时间超过设定的阈值，寻找可删除的日志文件内容。可以通过服务器端参数log.retention.hours、log.retention.minutes及log.retention.ms来配置。其中log.retention.ms的优先级最高，log.retention.minutes次之，log.retention.hours的优先级最低。

下面列出了在Kafka的conf目录下的server.properties参数文件中，有关日志保留的相关部分的参数说明。

```
01  ######### Log Retention Policy #########
02
03  # The following configurations control the disposal of log segments.
04  # The policy can be set to delete segments after a period of time,
05  # after a given size has accumulated.A segment will be deleted
06  # whenever *either* of these criteria are met. Deletion always happens
07  # from the end of the log.
08
09  # The minimum age of a log file to be eligible for deletion due to age
10  log.retention.hours=168
11
12  # A size-based retention policy for logs. Segments are pruned from
13  # the log unless the remaining segments drop below log.retention.bytes.
14  # Functions independently of log.retention.hours.
15  #log.retention.bytes=1073741824
16
17  # The maximum size of a log segment file. When this size is reached a
18  # new log segment will be created.
19  log.segment.bytes=1073741824
20
21  # The interval at which log segments are checked to see if they can
22  # be deleted according to the retention policies
23  log.retention.check.interval.ms=300000
```

从配置文件中可以看到，在默认情况下只配置了 log.retention.hours 参数，其值为 168。它表示日志分段文件的保留时间为 7 天。

（2）基于日志大小的保留策略。

在这种日志保留策略下，周期性执行的日志检查任务会检查当前日志的大小是否超过设定的阈值（Retention Size），从而寻找 Kafka 中可被执行删除的日志分段的文件集合（Deletable Segments）。日志保留的大小阈值可以通过服务器端参数 log.retention.bytes 来配置，其默认值为-1，表示无穷大。通过查看参数 server.properties 配置文件，可以看到该参数被注释了，即采用默认值-1。

```
01   # A size-based retention policy for logs. Segments are pruned from
02   # the log unless the remaining segments drop below log.retention.bytes.
03   # Functions independently of log.retention.hours.
04   #log.retention.bytes=1073741824
```

这里需要注意的是，参数 log.retention.bytes 配置的是日志文件的总大小，而不是单个的日志分段的大小，一个日志文件包含多个日志分段。

（3）基于日志起始偏移量的保留策略。

在一般情况下，日志文件的起始偏移量 logStartOffset 等于第一个日志分段的 baseOffset，但这并不是绝对的，日志文件的起始偏移量 logStartOffset 的值可以通过 DeleteRecordsRequest 请求及日志的清理和截断等操作修改，如图 5.10 所示。

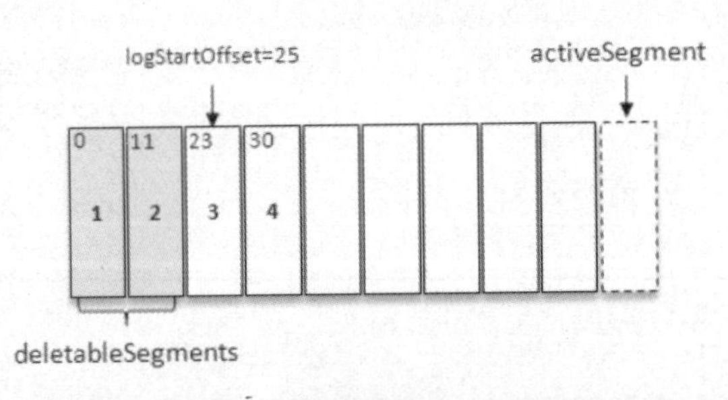


图 5.10 起始偏移量 logStartOffset

基于日志文件的起始偏移量的删除策略的判断依据，某日志分段的下一个日志分段的起始偏移量 baseOffset 是否小于等于 logStartOffset，若是则可以删除此日志分段，参考图 5.10。假设 logStartOffset 等于 25，日志分段 1 的起始偏移量为 0，日志分段 2 的起始偏移量为 11，日志分段 3 的起始偏移量为 23，我们通过如下的步骤可删除日志分段的文件集合 deletableSegments。

（1）从头开始遍历每个日志分段，日志分段 1 的下一个日志分段的起始偏移量为 11，小于 logStartOffset，将日志分段 1 加入 deletableSegments 中。

（2）日志分段 2 的下一个日志偏移量的起始偏移量为 23，也小于 logStartOffset 的大小，将日志分段 2 页加入 deletableSegments 中。

（3）日志分段 3 的下一个日志偏移量在 logStartOffset 的右侧，故从日志分段 3 开始的所有日志分段都不会被加入 deletableSegments 中。

这种日志保留策略用于判断 offset 是否比日志起始的偏移量小，如果小，就删除；否则就保留。

## 5.7　Kafka 与 ZooKeeper

在 Kafka 的体系架构中需要 ZooKeeper 的支持。Kafka 通过 ZooKeeper 管理集群配置选举 Leader，以及在 Consumer Group 发生变化时进行重平衡。

### 5.7.1　ZooKeeper 扮演的角色

ZooKeeper 用于分布式系统的协调，Kafka 使用 ZooKeeper 也是基于相同的原因。ZooKeeper 主要用来协调 Kafka 的各个 Broker，不但可以实现 Broker 的负载均衡，而且当增加 Broker 或某个 Broker 故障时，ZooKeeper 会通知生产者和消费者，这样可以保证整个系统正常运转。ZooKeeper 在 Kafka 集群中的作用主要体现在以下几个方面。

（1）存储 Kafka 的元数据。

注意，这里存储的是元数据而不是消息数据。

Kafka 可以在不更改生产者和消费者配置的条件下，通过使用 ZooKeeper 来实现动态的集群扩展。Kafka Broker 在启动过程中，会把所有元信息（例如，Topic 的信息、分区的信息等）在 ZooKeeper 注册并保持相关的元数据更新。同时，由于 ZooKeeper 提供了监听机制，生产者或消费者程序会在 ZooKeeper 上注册相关的监听器，一旦 ZooKeeper 中记录的元信息发生变化，生产者或消费者能及时感知并进行相应调整。这样就保证了在实现集群的动态扩容或缩容的过程中，各个 Broker 间仍能自动实现负载均衡，并能感知集群的变化。同时，ZooKeeper 利用监听的机制监听 Broker 和 Leader 副本的存活性。

（2）管理 Broker。

在 Kafka Broker 启动成功后，会向 ZooKeeper 注册 Broker 的信息，从而实现在服务器正常运行下的水平拓展。同时，当我们成功创建 Topic 后，ZooKeeper 也会维护 Topic 与 Broker 之间的对应关系，这是通过/brokers/topics/topic.name 节点来记录的。

（3）管理消费者。

消费者可以使用 Consumer Group 的形式消费 Kafka 集群中的消息数据。消费者在启动的过程中需要指定一个 Consumer Group 的 ID，这个 ID 会被 ZooKeeper 记录和维护，以保证同一份数据可以被同一个 Consumer Group 的不同消费者多次消费。

同时 ZooKeeper 管理消费者的偏移量用于跟踪当前消费者消费的位置。

（4）ZooKeeper 对生产者的意义。

生产者在启动过程中，会向 ZooKeeper 中注册监听器，从而帮助生产者了解 Topic 中的分区信息，包括分区的增加、减少、副本的选举等。同时生产者通过动态了解运行情况，实现负载均衡。

这里需要说明的是，ZooKeeper 并不直接对消息的生产者进行管理。生产者通过监听器感知 ZooKeeper 数据的变化。

## 5.7.2 Kafka 在 ZooKeeper 中存储的数据

在了解了 ZooKeeper 在 Kafka 集群中的作用后，Kafka 在 ZooKeeper 中存储了什么样的数据呢？存储的数据主要包含以下几个部分。

- Topic 的注册信息。
- Topic 配置。
- 分区的状态信息。
- Broker 的注册信息。
- Consumer 的注册信息。
- Consumer owner。
- Consumer offset。

图 5.11 展示了 Kafka 在 ZooKeeper 中存储的数据。

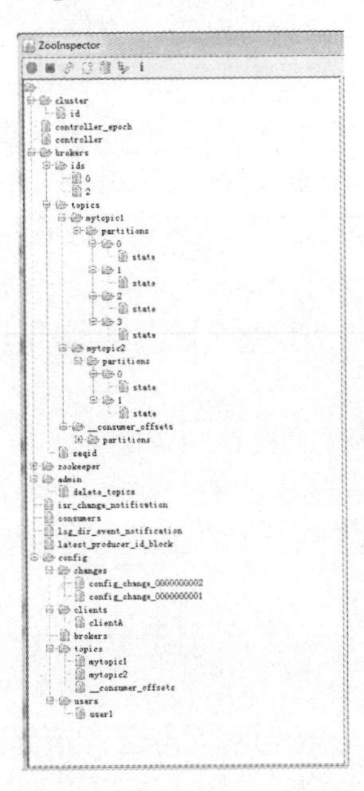


图 5.11　Kafka 在 ZooKeeper 中存储的数据

下面重点介绍其中几个比较主要的目录及其存储的数据。

- /brokers/topics/[topic]。

该目录用于存储 Topic 在 Broker 上的注册信息。

- /brokers/topics/[topic]/partitions/[0...N]，其中[0...N]表示 partition 索引号。

该目录用于存储 Topic 分区的状态信息。

- /brokers/ids/[0...N]。

该目录用于存储 Broker 的注册信息。在 server.properties 中，每个 Broker 都需要指定一个数字类型的 id 号，这个 id 号不能重复。这种目录节点其实是 ZooKeeper 的临时节点类型。如果 Broker 出现了宕机，这个目录节点将会自动被删除。

- /consumers/[groupId]/ids/[consumerId]。

每创建一个消费者实例，就会在该目录下创建一个 consumerId 节点，用于跟踪这个消费者实例。

- /consumers/[groupId]/offsets/[topic]/[partitionId] -> long (offset)。

该目录用来跟踪每个消费者所消费的分区数据中最大的 offset。

- /config/topics/[topic_name]。

该目录用于存储在 Topic 中的配置信息。

## 5.8　服务器端参数设置

Kafka 服务器端的配置参数都在 server.properties 中进行了定义。这里我们给出完整的 server.properties 文件内容。

```
01  # Licensed to the Apache Software Foundation (ASF) under one or more
02  # contributor license agreements.  See the NOTICE file distributed with
03  # this work for additional information regarding copyright ownership.
04  # The ASF licenses this file to You under the Apache License, Version 2.0
05  # (the "License"); you may not use this file except in compliance with
06  # the License.  You may obtain a copy of the License at
07  #
08  #    http://www.apache.org/licenses/LICENSE-2.0
09  #
10  # Unless required by applicable law or agreed to in writing, software
11  # distributed under the License is distributed on an "AS IS" BASIS,
12  # WITHOUT WARRANTIES OR CONDITIONS OF ANY KIND, either express or implied.
13  # See the License for the specific language governing permissions and
14  # limitations under the License.
15  # see kafka.server.KafkaConfig for additional details and defaults
16
17  ############################# Server Basics #############################
```

```
18  # The id of the broker. This must be set to a unique integer for each broker.
19  broker.id=0
20
21  ############################# Socket Server Settings ####################
22  # The address the socket server listens on. It will get the value returned from
23  # java.net.InetAddress.getCanonicalHostName() if not configured.
24  #   FORMAT:
25  #     listeners = listener_name://host_name:port
26  #   EXAMPLE:
27  #     listeners = PLAINTEXT://your.host.name:9092
28  #listeners=PLAINTEXT://:9092
29
30  # Hostname and port the broker will advertise to producers and consumers. If
not set,
31  # it uses the value for "listeners" if configured.  Otherwise, it will use
the value
32  # returned from java.net.InetAddress.getCanonicalHostName().
33  #advertised.listeners=PLAINTEXT://your.host.name:9092
34
35  # Maps listener names to security protocols, the default is for them to be
the same.
36  # See the config documentation for more details
37  #listener.security.protocol.map=PLAINTEXT:PLAINTEXT,
38  # SSL:SSL,SASL_PLAINTEXT:SASL_PLAINTEXT,SASL_SSL:SASL_SSL
39
40  # The number of threads that the server uses for receiving requests from
41  # the network and sending responses to the network
42  num.network.threads=3
43
44  # The number of threads that the server uses for processing requests,
45  # which may include disk I/O
46  num.io.threads=8
47
48  # The send buffer (SO_SNDBUF) used by the socket server
49  socket.send.buffer.bytes=102400
50
51  # The receive buffer (SO_RCVBUF) used by the socket server
52  socket.receive.buffer.bytes=102400
53
54  # The maximum size of a request that the socket server will accept
55  # (protection against OOM)
56  socket.request.max.bytes=104857600
57
58
```

```
59   ############################# Log Basics #############################
60   # A comma separated list of directories under which to store log files
61   log.dirs=/root/training/kafka_2.11-2.4.0/logs/broker0
62
63   # The default number of log partitions per topic. More partitions allow greater
64   # parallelism for consumption, but this will also result in more files across
65   # the brokers.
66   num.partitions=1
67
68   # The number of threads per data directory to be used for log recovery
69   # at startup and flushing at shutdown.
70   # This value is recommended to be increased for installations with
71   # data dirs located in RAID array.
72   num.recovery.threads.per.data.dir=1
73
74   ############################# Internal Topic Settings  #############################
75   # The replication factor for the group metadata internal topics
76   # "__consumer_offsets" and "__transaction_state"
77   # For anything other than development testing, a value greater than 1
78   # is recommended to ensure availability such as 3.
79   offsets.topic.replication.factor=1
80   transaction.state.log.replication.factor=1
81   transaction.state.log.min.isr=1
82
83              #############################      Log      Flush      Policy
#############################
84   # Messages are immediately written to the filesystem but by default we only
fsync() to sync
85   # the OS cache lazily. The following configurations control the flush of data
to disk.
86   # There are a few important trade-offs here:
87   #  1. Durability: Unflushed data may be lost if you are not using replication.
88   # 2. Latency: Very large flush intervals may lead to latency spikes when the
89   #              flush does occur as there will be a lot of data to flush.
90   #   3. Throughput: The flush is generally the most expensive operation, and a
91   #                 small flush interval may lead to excessive seeks.
92   # The settings below allow one to configure the flush policy to flush data after
93   #  a period of time or
94   # every N messages (or both). This can be done globally and overridden on
95   # a per-topic basis.
96
97   # The number of messages to accept before forcing a flush of data to disk
98   #log.flush.interval.messages=10000
99
```

```
100  # The maximum amount of time a message can sit in a log before we force a flush
101  #log.flush.interval.ms=1000
102
103  ############### Log Retention Policy ##############################
104  # The following configurations control the disposal of log segments. The
policy can
105  # be set to delete segments after a period of time, or after a given size
has accumulated.
106  # A segment will be deleted whenever *either* of these criteria are met.
Deletion
107  # always happens from the end of the log.
108
109  # The minimum age of a log file to be eligible for deletion due to age
110  log.retention.hours=168
111
112  # A size-based retention policy for logs. Segments are pruned from the log
113  # unless the remaining segments drop below log.retention.bytes. Functions
114  # independently of log.retention.hours.
115  #log.retention.bytes=1073741824
116
117  # The maximum size of a log segment file. When this size is reached a new log
118  # segment will be created.
119  log.segment.bytes=1073741824
120
121  # The interval at which log segments are checked to see if they can be
deleted according
122  # to the retention policies
123  log.retention.check.interval.ms=300000
124
125  ############################# Zookeeper #############################
126  # Zookeeper connection string (see zookeeper docs for details).
127  # This is a comma separated host:port pairs, each corresponding to a zk
128  # server. e.g. "127.0.0.1:3000,127.0.0.1:3001,127.0.0.1:3002".
129  # You can also append an optional chroot string to the urls to specify the
130  # root directory for all kafka znodes.
131  zookeeper.connect=kafka101:2181,kafka102:2181,kafka103:2181
132
133  # Timeout in ms for connecting to zookeeper
134  zookeeper.connection.timeout.ms=6000
135
136  ############################# Group Coordinator Settings ##############
137  # The following configuration specifies the time, in milliseconds,
138  # that the GroupCoordinator will delay the initial consumer rebalance.
```

```
139 # The rebalance will be further delayed by the value of group.initial.
rebalance.delay.ms
140 # as new members join the group, up to a maximum of max.poll.interval.ms.
141 # The default value for this is 3 seconds.
142 # We override this to 0 here as it makes for a better out-of-the-box
experience
143 # for development and testing.
144 # However, in production environments the default value of 3 seconds is more
145 # suitable as this will help to avoid unnecessary, and potentially expensive,
rebalances
146 # during application startup.
147 group.initial.rebalance.delay.ms=0
```

# 第 6 章 流处理引擎 Kafka Stream

从 Apache Kafka 0.10 版本开始引入了一个新的特性：Kafka Stream，它是一个轻量级的流式处理引擎。通过 Kafka Stream 可以对存储于 Kafka Topic 内的数据进行处理与分析。

Kafka Stream 的特点如下。

- 相对于其他流式计算引擎，Kafka Stream 是一个非常轻量级的开发库，并且已经集成在 Kafka 的环境中。除了 Kafka，Kafka Stream 无任何外部依赖。
- 由于 Kafka Stream 集成在 Kafka 的环境中，可以很方便地实现集群的水平扩展和消息的顺序处理。
- 支持高效的状态管理及其相关操作，例如，窗口计算、聚合操作等。
- 支持 Exectly-Once 的处理方式。这里需要说明的是，消息的处理方式有两种：Exectly-Once 和 At-Least-Once，即正好处理一次和至少处理一次。
- 由于 Kafka 本身提供高效的消息处理机制，因此 Kafka Stream 支持毫秒级的低延迟。
- 支持事件的 window 操作。
- 提供底层处理的 APIProcessor；同时支持 DSL 语句。DSL 语句类似 Spark 中的算子。

## 6.1 Kafka Stream 的体系架构

### 6.1.1 为什么需要 Kafka Stream

前面提到 Kafka Stream 是一个轻量级的流式计算框架，相对于批处理的离线计算而言，在流式计算中，数据源的输入是源源不断的，是一种时间上数据无边的数据集合。只要数据源产生了数据，分析计算的结果就是源源不断的。流式计算一般对实时性要求较高，要求具有较低的时间延迟。目前的流式计算引擎除了 Kafka Stream，还有 Apache Storm、Spark Streaming 和 Flink 中的 DataStream。我们会在后续的章节中详细介绍这些流式计算引擎。

那么，我们为什么需要 Kafka Stream 进行流式计算呢？

- 第一，Storm、Spark 与 Flink 都是一个处理框架，尤其是 Spark 和 Flink 还包含一个完整的生态圈系统，在 Spark 和 Flink 中，不仅支持流式计算，还支持批处理的离线计算。开发者很难了解完整框架的具体运行方式，从而使得调试成本高，并且使用受限。Kafka Stream 可以看成是一个开放库，直接提供具体的类给开发者调用，方便开发者使用和调试。
- 第二，尽管现在有很多集成环境方便 Storm、Spark 和 Flink 的部署，但是这些框架的部署相对复杂。而 Kafka Stream 作为类库，已经被集成在 Kafka 中，它对应用的打包和部署基本没有任何要求，可以非常方便地嵌入应用程序中。
- 第三，在大数据的流式计算体系中，一般都需要 Kafka 作为输入的数据源来缓存数据。这就决定了其他流式计算引擎中都是支持 Kafka 的，例如，Storm 中提供 Kafka Spout；Spark 中提供 Spark Streaming Kafka 模块；而在 Flink 中也提供了相应的 Source 组件，这就决定了我们使用 Kafka Stream 的成本非常低，因为大数据流式计算体系中都支持。
- 第四，在前面的章节中提到，由于 Kafka 本身提供了基于日志的数据持久化特性，因此 Kafka Stream 提供滚动部署和滚动升级及重新计算的能力。
- 最后，由于 Kafka 消费者的再平衡机制，Kafka Stream 可以方便地实现调整任务的并行度，而这种调整可以是在线动态调整。

## 6.1.2　Kafka Stream 的体系架构

Kafka Stream 的体系架构如图 6.1 所示。

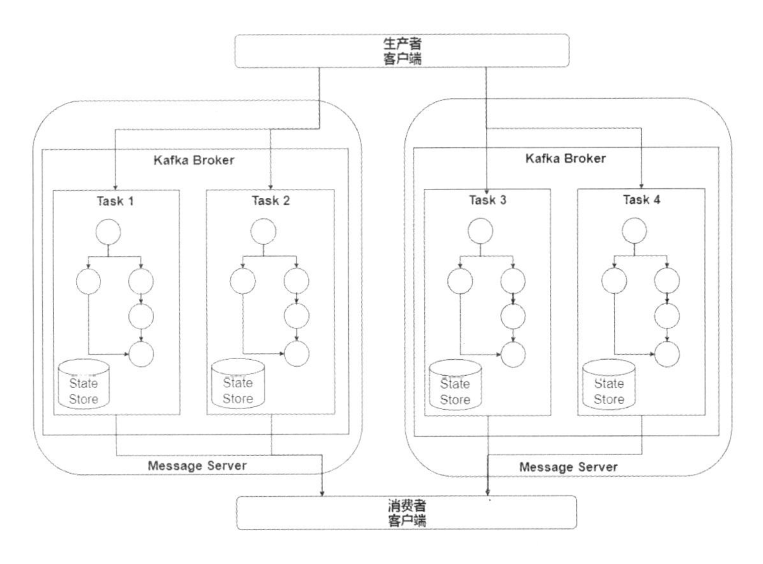


图 6.1　Kafka Stream 的体系架构

图 6.1 展示了 Kafka Streams 的基本体系架构，下面我们来介绍在 Kafka Stream 中需要了解的相关知识及其术语。

（1）拓扑。

拓扑（Topology）就是一个流式计算任务，类似 MapReduce 中的 Job。它定义了如何处理数据源的数据，最终输出结果。

Topology 中包含源算子和输出算子。源算子可以看成是上游 Topic 消息的消费者，它通过消费 Topic 中的数据，并将其转发给下游流处理算子进行处理；输出算子接收上游算子处理的数据结构，并且将数据写到下游的 Kafka Topic 中。所以输出算子可以看成是下游 Kafka Topic 的消息生产者。

（2）分区与任务。

分区与任务是 Kafka Stream 数据并发处理模型中的逻辑单元。Kafka Stream 中的分区类似于 Kafka Topic 的分区。当一个 Topology 中的算子处理数据时，它可以被并发执行，也就是 Kafka 会创建多个相应的实例并发处理数据，这就形成了一定数量的任务（Task），每个任务处理 Kafka Topic 中的一个分区。

（3）任务与线程。

Kafka Streams 允许自定义任务执行的线程数，从而达到数据的并行处理，每个线程可以执行一个或多个任务。同时，线程与线程之间不会共享状态，所以也不需要线程之间的协作处理。

### 6.1.3 执行 Kafka Stream 示例程序

Kafka 提供了一个示例程序（WordCountDemo）演示 Kafka Stream 的功能。可以在之前部署好的环境上直接执行这个示例程序。

（1）启动 kafka101、kafka102 和 kafka103 上的 ZooKeeper 集群。

```
zkServer.sh start
```

（2）启动 kafka101、kafka102 和 kafka103 上的 Kafka 集群。

```
bin/kafka-server-start.sh config/server.properties  &
```

（3）创建 Topic：streams-plaintext-input。

```
bin/kafka-topics.sh --create --zookeeper kafka101:2181 \
--replication-factor 1 --partitions 1 --topic streams-plaintext-input
```

后面我们会使用生产者的 Console 命令行工具，将数据发送到这个 Topic 中，这个 Topic 将作为 WordCountDemo 输入的数据源。

（4）创建 Topic：streams-wordcount-output。

```
bin/kafka-topics.sh --create --zookeeper kafka101:2181 \
--replication-factor 1 --partitions 1 --topic streams-wordcount-output
```

这个 Topic 将作为 WordCountDemo 的输出。在执行 WordCountDemo 的时候，使用消

费者的 Console 命令行工具，从这个 Topic 中接收处理的结果，并将其输出到屏幕上。

（5）启动应用程序 WordCountDemo。

```
bin/kafka-run-class.sh
org.apache.kafka.streams.examples.wordcount.WordCountDemo
```

WordCountDemo 示例程序将从 streams-plaintext-input topic 中接收数据，并统计每个单词出现的频率，并将结果输出到 streams-wordcount-output topic 中。启动 WordCountDemo 如图 6.2 所示。

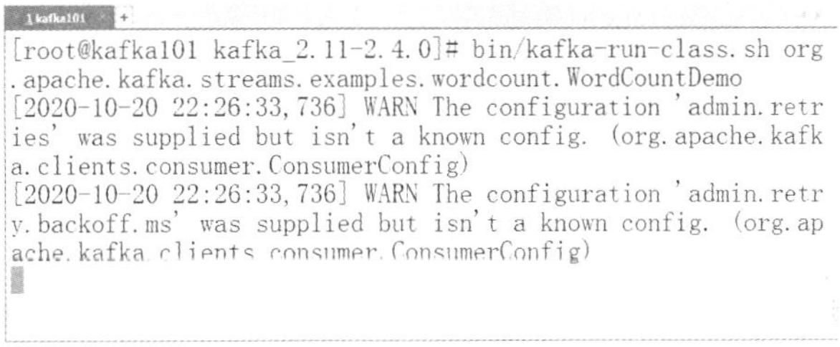

```
[root@kafka101 kafka_2.11-2.4.0]# bin/kafka-run-class.sh org
.apache.kafka.streams.examples.wordcount.WordCountDemo
[2020-10-20 22:26:33,736] WARN The configuration 'admin.retr
ies' was supplied but isn't a known config. (org.apache.kafk
a.clients.consumer.ConsumerConfig)
[2020-10-20 22:26:33,736] WARN The configuration 'admin.retr
y.backoff.ms' was supplied but isn't a known config. (org.ap
ache.kafka.clients.consumer.ConsumerConfig)
```

图 6.2　启动 WordCountDemo

（6）启动生产者客户端。

```
bin/kafka-console-producer.sh --broker-list kafka101:9092 --topic streams-
plaintext-input
```

这里将消息发送到之前创建的 streams-plaintext-input 主题中。

（7）启动消费者客户端。

```
bin/kafka-console-consumer.sh --bootstrap-server kafka101:9092 \
--topic streams-wordcount-output \
--from-beginning \
--property print.key=true \
--property print.value=true \
--property
key.deserializer=org.apache.kafka.common.serialization.StringDeserializer \
--property
value.deserializer=org.apache.kafka.common.serialization.LongDeserializer
```

我们从 streams-wordcount-output topic 中接收处理的结果，并直接打印在屏幕上。这里的--from-beginning 表示 Topic 的开始消费数据；同时，通过参数 key.deserializer 和 value.deserializer 指定了数据的序列化方式为 String 和 Long。

（8）在消息的生产者命令行终端中输入一些数据，如图 6.3 所示。

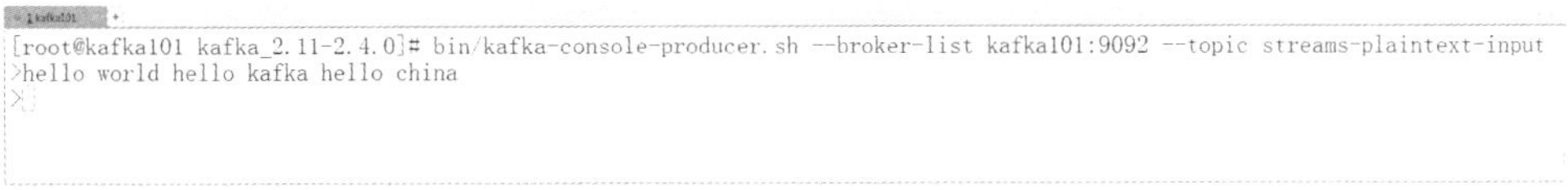

```
[root@kafka101 kafka_2.11-2.4.0]# bin/kafka-console-producer.sh --broker-list kafka101:9092 --topic streams-plaintext-input
>hello world hello kafka hello china
>
```

图 6.3　在 Producer Console 中输入数据

这里输入的数据是“hello world hello kafka hello china”。

（9）图 6.4 展示了 WordCountDemo 执行完成的处理结果。

```
[root@kafka101 kafka_2.11-2.4.0]# bin/kafka-console-producer.sh --broker-list kafka101:9092 --topic streams-plaintext-input
>hello world hello kafka hello china
>
```

```
[root@kafka101 kafka_2.11-2.4.0]# bin/kafka-console-consumer.sh --bootstrap-server kafka101:9092 \
> --topic streams-wordcount-output \
> --from-beginning \
> --property print.key=true \
> --property print.value=true \
> --property key.deserializer=org.apache.kafka.common.serialization.StringDeserializer \
> --property value.deserializer=org.apache.kafka.common.serialization.LongDeserializer
hello   1
world   1
hello   2
kafka   1
hello   3
china   1
```

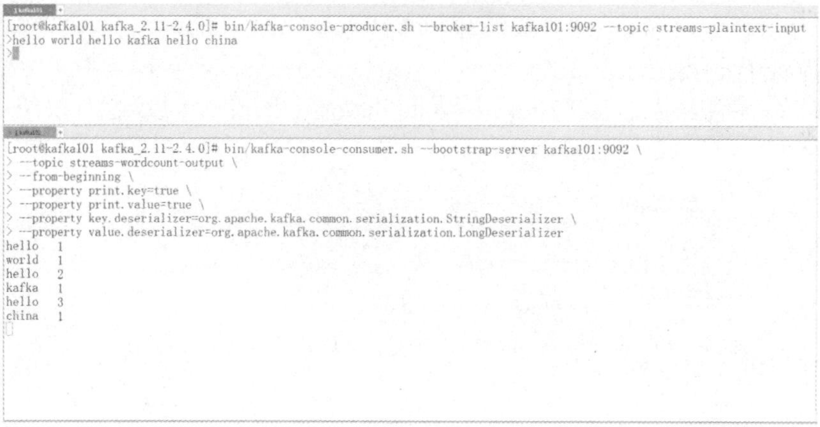

图 6.4　WordCountDemo 执行完成的处理结果

# 6.2　开发自己的 Kafka Stream 应用程序

在 6.1 节中，我们演示了官方提供的示例程序 WordCountDemo。本节来开发自己的 Kafka Stream 应用程序。

开发 Kafka Stream 应用程序，需要将下面的依赖加入我们的 Maven 工程中。

```
01  <dependencies>
02      <dependency>
03          <groupId>org.apache.kafka</groupId>
04          <artifactId>kafka-streams</artifactId>
05          <version>2.4.0</version>
06      </dependency>
07      <dependency>
08          <groupId>org.apache.kafka</groupId>
09          <artifactId>kafka-clients</artifactId>
10          <version>2.4.0</version>
11      </dependency>
12  </dependencies>
```

首先，我们分析一下 WordCountDemo 数据处理的流程。为了测试的方便，我们每次随机从下面的数据中采集一条数据作为数据源的数据。

```
I love Beijing
I love China
Beijing is the capital of China
```

图 6.5 展示了 WordCountDemo 的数据处理流程。

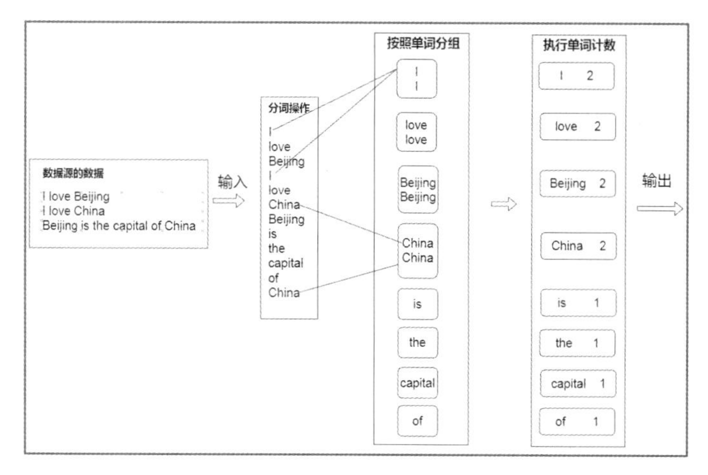


图 6.5　WordCountDemo 数据处理的流程

在了解了 WordCountDemo 数据处理的过程后，我们首先使用 Java 语言开发对应的应用程序。

（1）开发消息的生产者，代码如下。

```
01  import java.util.Properties;
02  import java.util.Random;
03  import org.apache.kafka.clients.producer.KafkaProducer;
04  import org.apache.kafka.clients.producer.Producer;
05  import org.apache.kafka.clients.producer.ProducerRecord;
06
07  public class WordCountProducer {
08
09      //构造测试数据
10      private static String[] data= {"I love Beijing","I love China",
11                                      "Beijing is the capital of China"};
12
13      private static String topic = "streams-plaintext-input";
14
15      public static void main(String[] args) throws InterruptedException {
16          Properties props = new Properties();
17          props.put("bootstrap.servers", "kafka101:9092");
18          props.put("acks", "all");
19
20          props.put("retries", 0);
21          props.put("batch.size", 16384);
22          props.put("linger.ms", 1);
23          props.put("buffer.memory", 33554432);
24
25          props.put("key.serializer",
```

```
26                 "org.apache.kafka.common.serialization.StringSerializer");
27         props.put("value.serializer",
28                 "org.apache.kafka.common.serialization.StringSerializer");
29
30         Producer<String, String> producer = new KafkaProducer<String,
String>(props);
31         for(int i=0;i<50;i++) {
32             //随机从 data 中采集一条数据
33             int random = new Random().nextInt(3);
34
35             System.out.println("发送的数据是: " + data[random]);
36
37             producer.send(new ProducerRecord<String, String>
38                     (topic, "key"+i, data[random]));
39             Thread.sleep(1000);
40         }
41         producer.close();
42
43     }
44 }
```

（2）开发消息的处理程序 WordCountDemo，代码如下。

```
01 import org.apache.kafka.common.serialization.Serdes;
02 import org.apache.kafka.streams.KafkaStreams;
03 import org.apache.kafka.streams.StreamsBuilder;
04 import org.apache.kafka.streams.StreamsConfig;
05 import org.apache.kafka.streams.Topology;
06 import org.apache.kafka.streams.kstream.*;
07
08 import java.util.Arrays;
09 import java.util.Properties;
10
11 public class WordCountDemo {
12
13     public static void main(String[] args) {
14         //指定 Kafka 的相关配置
15         Properties props = new Properties();
16         props.put(StreamsConfig.APPLICATION_ID_CONFIG, "WordCountDemo");
17         props.put(StreamsConfig.BOOTSTRAP_SERVERS_CONFIG, "kafka101:9092");
18         props.put(StreamsConfig.DEFAULT_KEY_SERDE_CLASS_CONFIG,
19                 Serdes.String().getClass());
20         props.put(StreamsConfig.DEFAULT_VALUE_SERDE_CLASS_CONFIG,
21                 Serdes.String().getClass());
22
23         final StreamsBuilder builder = new StreamsBuilder();
```

```
24
25          /**
26           * 指定数据源 topic
27           */
28          KStream<String, String> source = builder.stream("streams-plaintext-
input");
29
30          /*
31           * flatMapValues 将执行分词操作，其中的泛型，
32           * String：表示输入的数据类型，即接收到的每一句文本
33           * Iterable<String>：表示处理完成后的结果，即文本中的每个单词
34           */
35          source.flatMapValues(new ValueMapper<String, Iterable<String>>() {
36
37              @Override
38              public Iterable<String> apply(String data) {
39                  return Arrays.asList(data.split(" "));
40              }
41          }).groupBy(new KeyValueMapper<String,String,String>() {
42              /*
43               * groupBy 执行分组的操作
44               * 这里直接返回文本中的每个单词作为分组的条件
45               * 为了很好地看到结果，我们将拆分后的单词打印出来
46               */
47              @Override
48              public String apply(String key, String value) {
49                  //System.out.println("value 数据是："+value);
50                  return value;
51              }
52          })
53          .count()  //执行单词的计数
54          .toStream()
55          .to("streams-wordcount-output",  //将结果输出
56             Produced.with(Serdes.String(), Serdes.Long()));;
57
58          final Topology topology = builder.build();
59          final KafkaStreams streams = new KafkaStreams(topology, props);
60          streams.start();
61      }
62  }
```

其中第35～56行代码，也可以简化成下面的形式。

```
01  source.flatMapValues(data ->
02                               Arrays.asList(data.split(" ")))
03          .groupBy((key, value) -> value)
```

```
04            .count(Materialized.as("counts-store"))
05            .toStream()
06            .to("streams-wordcount-output",
07                Produced.with(Serdes.String(), Serdes.Long()));
```

（3）程序开发完成后，我们就可以进行测试了。这里可以直接使用 Consumer Console 的命令行工具来打印处理完成的结果。

（4）启动 Consumer Console 命令行工具，命令如下。

```
bin/kafka-console-consumer.sh --bootstrap-server kafka101:9092 \
--topic streams-wordcount-output \
--from-beginning \
--property print.key=true \
--property print.value=true \
--property
key.deserializer=org.apache.kafka.common.serialization.StringDeserializer \
--property
value.deserializer=org.apache.kafka.common.serialization.LongDeserializer
```

（5）启动 WordCountProducer 程序，程序执行的效果如图 6.6 所示。

图 6.6　启动 WordCountProducer 程序

（6）观察 Consumer Console 命令行工具的输出结果，如图 6.7 所示。

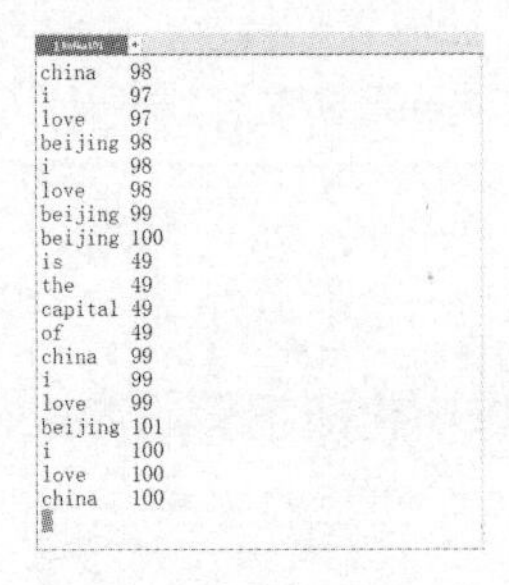


图 6.7　在 Console 上观察 WordCountDemo 的处理结果

（7）在前面的演示中，直接使用了 Consumer Console 命令行工具打印处理结果，也可以开发 Consumer 应用程序来接收数据。

# 6.3　Kafka Stream 中的数据模型

## 6.3.1　KStream 与 KTable

Kafka Stream 中两个非常重要的概念就是 KStream 和 KTable，可以把它们理解为数据集合的抽象。下面我们分别对其进行解释。

- KStream 是一个无边界的数据流集合，在前面开发的 WordCountDemo 中就创建了这样的数据流来代表源源不断的数据。它类似 Spark Streaming 中 DStream（离散流）的概念。对 KStream 进行的处理，就可以理解为 Kafka Stream 的流式计算。
- KTable 可以理解为数据库中的一张“表”，“表”中每条记录都是<key,value>的格式。Key 相当于“表”的主键；Value 则代表对应的记录。这种数据模型就类似 HBase 中的 row 和列族的关系，并且 KTable 只维护同一个 Key 的最新记录。

下面举一个简单的例子说明 KStream 和 KTable 数据结构的区别。图 6.8 说明了 KStream 的数据模型。图 6.9 说明了 KTable 的数据模型。

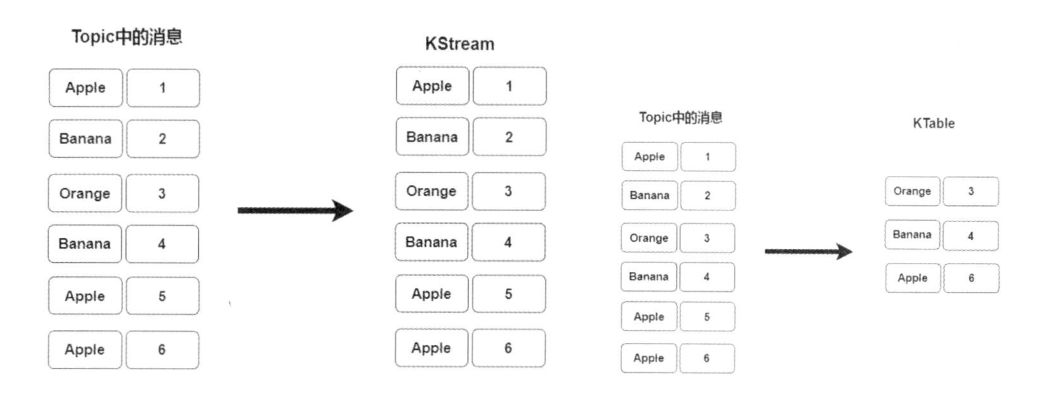


图 6.8　KStream 的数据模型　　　　图 6.9　KTable 的数据模型

如果对上面的 KStream 和 KTable 分别基于 Key 进行分组操作，再对 Value 进行求和，将会得到不同的结果，对 KStream 的计算结果是<Apple，12>，<Banana，6>，<Orange，3>，对 Ktable 的计算结果是<Orange，3>，<Banana，4>，<Apple，6>。

为了说明上面的区别，可以开发如下的程序代码来进行一个简单的测试。

（1）开发 KStream 程序。

前面开发的 WordCountDemo 示例程序使用的就是 KStream，其核心代码如下。

```
01   final StreamsBuilder builder = new StreamsBuilder();
```

```
02
03   /**
04    * 指定数据源 topic
05    */
06   KStream<String, String> source = builder.stream("streams-plaintext-input");
```

这里的 source 就是 KStream，其中接收的<Key,Value>类型是<String,String>，都为字符串类型。

（2）开发 KTable 程序。

可以直接通过 StreamBuilder 构建 KTable，其核心代码如下。

```
01   final StreamsBuilder builder = new StreamsBuilder();
02
03   /**
04    * 指定数据源 topic
05    */
06   KTable<String, String> table = builder.table("streams-plaintext-input");
07                                 table.toStream().print(Printed.<String,
String>toSysOut().withLabel("WordCount-KTable"));
```

这里的代码从 streams-plaintext-input 中接收数据，并创建 KTable 对象，将接收到的数据直接打印在屏幕上。

当然，也可以将一个 KStream 转换为 KTable。例如，可以将之前的 WordCountDemo 程序改写，下面展示了完整的代码。

```
01   import org.apache.kafka.common.serialization.Serdes;
02   import org.apache.kafka.common.utils.Bytes;
03   import org.apache.kafka.streams.KafkaStreams;
04   import org.apache.kafka.streams.StreamsBuilder;
05   import org.apache.kafka.streams.StreamsConfig;
06   import org.apache.kafka.streams.Topology;
07   import org.apache.kafka.streams.kstream.*;
08   import org.apache.kafka.streams.state.KeyValueStore;
09
10   import java.util.Arrays;
11   import java.util.Properties;
12
13   public class WordCountKTable {
14
15       public static void main(String[] args) {
16           //指定 Kafka 的相关配置
17           Properties props = new Properties();
18           props.put(StreamsConfig.APPLICATION_ID_CONFIG,
19                   "WordCountDemo");
20           props.put(StreamsConfig.BOOTSTRAP_SERVERS_CONFIG,
21                   "kafka101:9092");
```

```
22            props.put(StreamsConfig.DEFAULT_KEY_SERDE_CLASS_CONFIG,
23                    Serdes.String().getClass());
24            props.put(StreamsConfig.DEFAULT_VALUE_SERDE_CLASS_CONFIG,
25                   Serdes.String().getClass());
26
27            final StreamsBuilder builder = new StreamsBuilder();
28
29            /**
30             * 指定数据源 topic
31             */
32            KStream<String, String> source = builder.stream("streams-plaintext-
input");
33
34            //执行单词计数，并将结果转换为 KTable
35            KTable<String, Long> counts =
36               source.flatMapValues(new ValueMapper<String, Iterable<String>>()
{
37                     @Override
38                     public Iterable<String> apply(String value) {
39                         return Arrays.asList(value.split(" "));
40                     }
41                 }).groupBy(new KeyValueMapper<String, String, String>() {
42                     @Override
43                     public String apply(String key, String value) {
44                         return value;
45                     }
46                 })
47                .count(Materialized.<String, Long, KeyValueStore<Bytes, byte[]>>
48                      as("counts-store"));
49
50            counts.toStream().to("streams-wordcount-output",
51                          Produced.with(Serdes.String(), Serdes.Long()));
52
53            final Topology topology = builder.build();
54            final KafkaStreams streams = new KafkaStreams(topology, props);
55            streams.start();
56        }
57    }
```

## 6.3.2 状态管理

在数据的流式计算中，在某些情况下需要记录计算的中间状态，例如，窗口计算和 aggregate 聚合操作等。前面介绍过一些常见的流式计算引擎，表 6.1 对比了它们之间的差别。

表 6.1 常见的流式计算引擎

| 流式计算引擎 | 保证次数 | 容错机制 | 状态管理 | 延时 | 吞吐量 |
|---|---|---|---|---|---|
| Storm | At-least-once（至少一次） | ACK 机制 | 无 | 低 | 低 |
| Spark Streaming | Exactly-once（仅一次） | CheckPoint | 基于 DStream | 中等 | 高 |
| Flink DataStream | Exactly-once（仅一次） | CheckPoint | 基于操作 | 低 | 高 |
| Kafka Stream | Exactly-once（仅一次） | 基于 Kafka | State Store | 低 | 中等 |

在 Kafka Stream 中，State Store 用来存储中间状态。可以把 State Store 存储在内存中，也可以进行基于 Key-Value 的持久化存储，其底层依赖的是 Topic 的状态存储机制。因为数据在 Kafka Topic 中是以<Key,Value>的形式存储的，同时通过 Kafka 的日志将其进行持久化的存储，从而保证了数据的安全。

下面我们通过代码来演示如何使用 Kafka Stream 的 State Store。

```
01  import java.util.Arrays;
02  import java.util.Properties;
03
04  import org.apache.kafka.common.serialization.Serdes;
05  import org.apache.kafka.streams.KafkaStreams;
06  import org.apache.kafka.streams.StreamsBuilder;
07  import org.apache.kafka.streams.StreamsConfig;
08  import org.apache.kafka.streams.Topology;
09  import org.apache.kafka.streams.state.KeyValueStore;
10  import org.apache.kafka.streams.state.StoreBuilder;
11  import org.apache.kafka.streams.state.Stores;
12  import org.apache.kafka.streams.kstream.*;
13
14  public class StateStoreStreamDemo {
15      public static void main(String[] args) {
16          //指定 Kafka 的相关配置
17          Properties props = new Properties();
18          props.put(StreamsConfig.APPLICATION_ID_CONFIG,
19                  "StateStoreStreamDemo");
20          props.put(StreamsConfig.BOOTSTRAP_SERVERS_CONFIG,
21                  "kafka101:9092");
22          props.put(StreamsConfig.DEFAULT_KEY_SERDE_CLASS_CONFIG,
23                  Serdes.String().getClass());
24          props.put(StreamsConfig.DEFAULT_VALUE_SERDE_CLASS_CONFIG,
25                  Serdes.String().getClass());
```

```
26
27            //配置 State Store 的地址
28            props.put(StreamsConfig.STATE_DIR_CONFIG,
29                    "D:\\download\\kafka-state-store");
30
31            //创建一个自己的 State Store
32            StoreBuilder<KeyValueStore<String, String>> mystatestore =
33                    Stores.keyValueStoreBuilder(
34                        //指定 State Store 的名字
35                        //程序运行时，会在 STATE_DIR_CONFIG 目录下创建对应的子目录
36                        Stores.persistentKeyValueStore("mystatestore"),
37                        Serdes.String(),
38                        Serdes.String()
39                    ).withCachingEnabled();
40
41            final StreamsBuilder builder = new StreamsBuilder();
42
43            //向 builder 中添加 State Store
44            builder.addStateStore(mystatestore);
45
46            /**
47             * 指定数据源 topic
48             */
49            KStream<String, String> source = builder.stream("streams-plaintext-
input");
50            source.flatMapValues(new ValueMapper<String, Iterable<String>>() {
51
52                @Override
53                public Iterable<String> apply(String data) {
54                    return Arrays.asList(data.split(" "));
55                }
56            }).groupBy(new KeyValueMapper<String, String, String>() {
57                @Override
58                public String apply(String key, String value) {
59                    return value;
60                }
61            }).count()
62            .toStream()
63            .to("streams-wordcount-output", //将结果输出
64                Produced.with(Serdes.String(),
65                              Serdes.Long()));
66            ;
67
68            final Topology topology = builder.build();
```

```
69            final KafkaStreams streams = new KafkaStreams(topology, props);
70            streams.start();
71        }
72    }
```

运行程序，可以在目录“D:\download\kafka-state-store”中看到生成的持久化信息，如图 6.10 所示。

```
[D:\download\kafka-state-store]$ pwd
D:\download\kafka-state-store
[D:\download\kafka-state-store]$ tree
文件夹 PATH 列表
卷序列号为 00740073 6EDB:C922
D:.
└─StateStoreStreamDemo
    ├─0_0
    └─1_0
        └─rocksdb
            ├─counts-store
            └─KSTREAM-AGGREGATE-STATE-STORE-0000000003

[D:\download\kafka-state-store]$
```

图 6.10　State Store 的持久化信息

在默认的情况下，Kafka Streams 使用 RocksDB 来存储状态，除了 RocksDB，还可以使用下面这些静态工厂的方法来配置状态的持久化存储。

```
Stores.persistentKeyValueStore
Stores.lruMap
Stores.persistentWindowStore
Stores.persistentSessionStore
Stores.inMemoryKeyValueStore
```

# 6.4　Kafka Stream 中的窗口计算

## 6.4.1　时间

在流式计算框架中，时间是一个非常重要的特征。Kafka Topic 中的消息，除了基于<Key,Value>的形式，每条数据还带有一个时间戳的属性。目前，在 Kafka Stream 中有三种不同的时间。

（1）消息发生时间（Event Time）。

消息发生时间一般就是数据本身产生的时间。这个时间通常是在消息到达 Kafka Stream 之前就确定的，并且可以从每个消息中获取到消息时间戳。在消息发生时间中，时间取决于数据。

（2）消息接收时间（Ingestion Time）。

消息接收时间是消息进入 Kafka Stream 的时间。

（3）消息处理时间（Processing Time）。

消息处理时间是指消息被 Kafka Stream 处理时机器的系统时间。当应用程序在消息处理时间上运行时，所有基于时间的操作将使用当时机器的系统时间。

### 6.4.2　窗口

在了解了 Kafka Stream 中的时间以后，接下来我们看看 Kafka Stream 中的时间窗口。根据时间窗口做聚合，是流处理中非常重要的功能。比如，我们经常需要统计最近一段时间内的 count、sum、avg 等数据。一个典型的场景就是：网站定期统计网站的 PV 访问量。

在 Kafka Stream 中，共有四种类型的时间窗口，表 6.2 对它们进行了简要的描述。

表 6.2　四种类型的时间窗口

| 时间窗口名称 | 窗口的行为 | 描述 |
|---|---|---|
| Hopping time window（跳跃时间窗口） | 基于时间 | 窗口大小固定，窗口重叠 |
| Tumbling time window（滚动时间窗口） | 基于时间 | 窗口大小固定，窗口不重叠，窗口间无缝隙 |
| Sliding time window（滑动时间窗口） | 基于时间 | 窗口大小固定，窗口重叠，通过处理窗口时间戳的差异支持窗口重叠 |
| Session window（会话窗口） | 基于会话 | 窗口大小可动态调整，窗口不重叠，通过数据进行驱动 |

下面我们分别对这四种类型的窗口进行介绍。

（1）Hopping time window。

跳跃时间窗口由两个属性定义。

- 窗口的大小。
- 窗口前进的间隔。

前进的间隔指定一个窗口相对于前一个窗口向前移动多少的时间。由于跳跃窗口可以重叠，因此，同一条数据可能属于多个窗口。

图 6.11 是 Kafka 官方网站上提供的示例，即窗口大小为 5min、前进的间隔为 1min 的跳跃时间窗口。

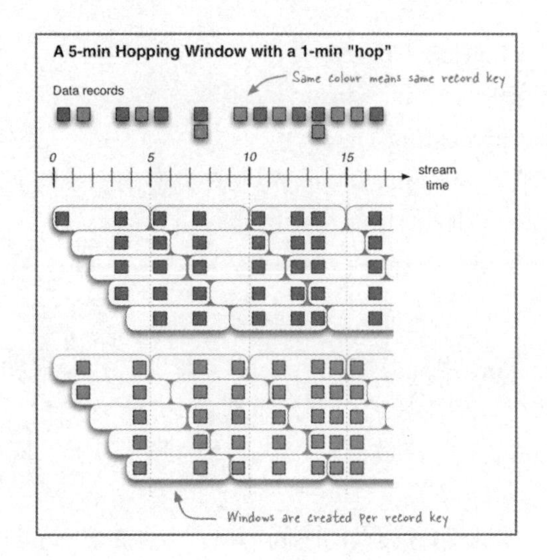


图 6.11　跳跃时间窗口

下面的代码展示了一个跳跃时间窗口的使用过程，这里我们定义窗口的大小为 5min，窗口的前进间隔为 1min。

```
01  import org.apache.kafka.common.serialization.Serdes;
02  import org.apache.kafka.streams.KafkaStreams;
03  import org.apache.kafka.streams.StreamsBuilder;
04  import org.apache.kafka.streams.StreamsConfig;
05  import org.apache.kafka.streams.Topology;
06  import org.apache.kafka.streams.kstream.*;
07
08  import java.time.Duration;
09  import java.time.Instant;
10  import java.util.Properties;
11
12  public class HoppingTimeWindowsDemo {
13
14      //窗口大小为 5min
15      private static final long TIME_WINDOW_MINUTES = 5L;
16      //前进间隔为 1min
17      private static final long ADVANCED_BY_MINUTES = 1L;
18
19      public static void main(String[] args) {
20          //指定 Kafka 的相关配置
21          Properties props = new Properties();
22          props.put(StreamsConfig.APPLICATION_ID_CONFIG,
23                  "HoppingTimeWindows");
24          props.put(StreamsConfig.BOOTSTRAP_SERVERS_CONFIG,
```

```
25                     "kafka101:9092");
26          props.put(StreamsConfig.DEFAULT_KEY_SERDE_CLASS_CONFIG,
27                     Serdes.String().getClass());
28          props.put(StreamsConfig.DEFAULT_VALUE_SERDE_CLASS_CONFIG,
29                     Serdes.String().getClass());
30
31          final StreamsBuilder builder = new StreamsBuilder();
32
33          KStream<String, String> source = builder.stream("streams-plaintext-
input");
34         Instant initTime = Instant.now();
35
36         source
37         .groupByKey()
38         //设置跳跃时间窗口的大小和向前滑动的距离
39         .windowedBy(TimeWindows.of(Duration.ofSeconds(TIME_WINDOW_MINUTES))
40                   .advanceBy(Duration.ofSeconds(ADVANCED_BY_MINUTES))
41                   .grace(Duration.ZERO))
42         .count(Materialized.with(Serdes.String(), Serdes.Long()))
43         .suppress(Suppressed.untilWindowCloses(Suppressed.BufferConfig.unbou
nded()))
44         .toStream()
45         .filterNot(((windowedKey, value) -> {
46          //删除太旧的时间窗口
47          Instant windowEnd = windowedKey.window().endTime();
48            return windowEnd.isBefore(initTime);
49         }))
50         .foreach((windowedKey, value) -> {
51          System.out.println("结果是: key=" + windowedKey + ",value=" + value);
52         });
53
54          final Topology topology = builder.build();
55          final KafkaStreams streams = new KafkaStreams(topology, props);
56          streams.start();
57      }
58  }
```

（2）Tumbling time window。

滚动时间窗口是跳跃时间窗口的一种特殊情况，滚动时间窗口也是基于时间间隔的。但它是固定大小、不重叠、无间隙的窗口。滚动时间窗口只由一个属性定义，即窗口的大小。

图 6.12 是 Kafka 的示例，即窗口大小为 5min 的滚动时间窗口。

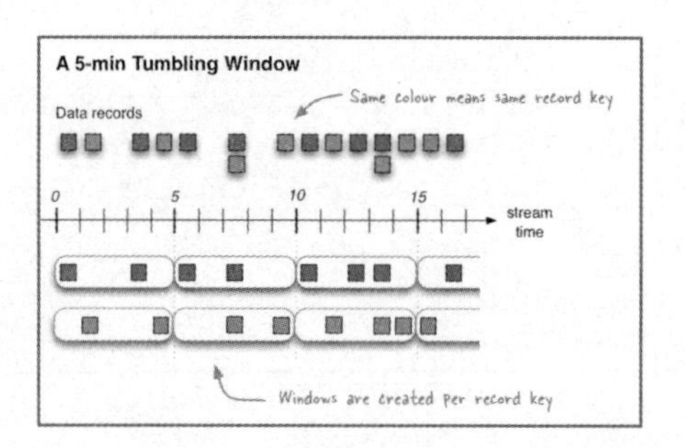


图 6.12　滚动时间窗口

滚动时间窗口实际上是一种跳跃时间窗口，窗口的大小与其前进间隔相等。由于滚动时间窗口从不重叠，所以数据记录只属于一个窗口。

下面的代码展示了一个滚动时间窗口的使用过程，这里我们定义的窗口滚动的大小为 5min。

```
01  import org.apache.kafka.common.serialization.Serdes;
02  import org.apache.kafka.streams.KafkaStreams;
03  import org.apache.kafka.streams.StreamsBuilder;
04  import org.apache.kafka.streams.StreamsConfig;
05  import org.apache.kafka.streams.Topology;
06  import org.apache.kafka.streams.kstream.*;
07
08  import java.time.Duration;
09  import java.time.Instant;
10  import java.util.Properties;
11
12  public class TumblingTimeWindowsDemo {
13
14      public static void main(String[] args) {
15          //指定 Kafka 的相关配置
16          Properties props = new Properties();
17          props.put(StreamsConfig.APPLICATION_ID_CONFIG,
"TumblingTimeWindows");
18          props.put(StreamsConfig.BOOTSTRAP_SERVERS_CONFIG, "kafka101:9092");
19          props.put(StreamsConfig.DEFAULT_KEY_SERDE_CLASS_CONFIG,
20                  Serdes.String().getClass());
21          props.put(StreamsConfig.DEFAULT_VALUE_SERDE_CLASS_CONFIG,
22                  Serdes.String().getClass());
23
24          final StreamsBuilder builder = new StreamsBuilder();
25
26          /**
27           * 指定数据源 topic
```

```
28           */
29          KStream<String, String> source = builder.stream("streams-plaintext-
input");
30
31          Instant initTime = Instant.now();
32
33          source
34          .groupByKey()
35          .windowedBy(TimeWindows.of(Duration.ofMinutes(5L)))
36          .count(Materialized.with(Serdes.String(), Serdes.Long()))
37          .toStream()
38          .filterNot(((windowedKey, value) -> {
39           //删除太旧的时间窗口，程序二次启动时，会重新读取历史数据进行整套流处理
40           //为了不影响观察，这里过滤掉历史数据
41           Instant windowEnd = windowedKey.window().endTime();
42             return windowEnd.isBefore(initTime);
43          }))
44          .foreach((windowedKey, value) -> {
45           System.out.println("结果是: key=" + windowedKey + ",value=" + value);
46          });
47
48           final Topology topology = builder.build();
49           final KafkaStreams streams = new KafkaStreams(topology, props);
50           streams.start();
51      }
52  }
```

（3）Sliding time window。

滑动时间窗口只用于KStream进行Join计算时候。该窗口的大小定义了KStream的数据记录执行Join时，在同一个窗口的最大时间差。假设滑动时间窗口的大小为5min，则参与Join的2个KStream数据，记录时间差小于5min的数据，被认为在同一个窗口中可以直接进行连接操作。

下面的代码是Kafka官方网站上提供的一个示例。该示例程序是在Inner Join中使用了Sliding time window的设置方式。

```
01  import java.time.Duration;
02  KStream<String, Long> left = ...;
03  KStream<String, Double> right = ...;
04
05  // Java 8+ example, using lambda expressions
06  KStream<String, String> joined =
07      left.join(right,(leftValue, rightValue)
08                -> "left=" + leftValue + ", right=" + rightValue,
09      JoinWindows.of(Duration.ofMinutes(5)),
```

```
10      Joined.with(
11        Serdes.String(), /* key */
12        Serdes.Long(),   /* left value */
13        Serdes.Double())  /* right value */
14    );
15
16  // Java 7 example
17  KStream<String, String> joined = left.join(right,
18      new ValueJoiner<Long, Double, String>() {
19        @Override
20        public String apply(Long leftValue, Double rightValue) {
21          return "left=" + leftValue + ", right=" + rightValue;
22        }
23      },
24      JoinWindows.of(Duration.ofMinutes(5)),
25      Joined.with(
26        Serdes.String(), /* key */
27        Serdes.Long(),   /* left value */
28        Serdes.Double())  /* right value */
29    );
```

除了可以在 Inner Join 中使用 Sliding time window，也可以在 Left Join 和 Outer Join 中使用 Sliding time window。

（4）Session window。

Session window 用于对 Key 进行分组操作后的聚合。该窗口首先需要对 Key 进行分组，然后对组内的数据记录定义一个窗口的开始和结束。假设网站要统计某个用户访问网站的时间，这里就可以将用户作为 Key。当用户登录网站时，该用户的窗口开始，当用户退出后，该用户的窗口结束。窗口结束时，可计算该用户的访问时间或单击次数等。

下面的代码是 Kafka 官方网站上提供的 Session window 的设置方式，这里设置的窗口大小为 5min。

```
01  import java.time.Duration;
02  import org.apache.kafka.streams.kstream.SessionWindows;
03
04  //将会话的时间窗口设置为 5min
05  SessionWindows.with(Duration.ofMinutes(5));
```

# 第 7 章　监控 Kafka

一个功能健全的 Kafka 集群可以处理相当大的数据量，由于消息系统是很多大型应用的基石，因此 Broker 集群在性能上的缺陷，都会引起整个应用栈的各种问题。

Kafka 的度量指标主要有以下三类。

- Kafka 服务器指标。
- 生产者指标。
- 消费者指标。

另外，由于 Kafka 的状态靠 ZooKeeper 来维护，对 ZooKeeper 性能的监控也成为整个 Kafka 监控计划中一个必不可少的组成部分。

## 7.1　Kafka 的监控指标

Kafka 的服务器端度量指标是为了监控 Broker，也是整个消息系统的核心，如表 7.1 所示。因为所有消息都通过 Kafka Broker 传递，然后被消费，所以对 Broker 集群上出现问题的监控和告警就尤为重要。Broker 性能指标有以下三类。

- Kafka 本身的指标。
- 主机层面的指标。
- JVM 垃圾回收指标。

表 7.1　Kafka 的服务器端度量指标

| 指　　标 | 作　　用 |
| --- | --- |
| UnderReplicatedPartitions | 在一个运行健康的集群中，处于同步状态的副本数（ISR）应该与总副本数（简称 AR，Assigned Repllicas）完全相等，如果分区的副本远远落后于 Leader，则这个 Follower 将被 ISR 池删除，随之而来的是 IsrShrinksPerSec（可理解为 isr 的缩水情况，后面章节会讲）的增加。由于 Kafka 的高可用性必须通过副本来满足，所以有必要重点关注这个指标，让它长期处于大于 0 的状态 |

续表

| 指　标 | 作　用 |
| --- | --- |
| IsrShrinksPerSec<br>IsrExpandsPerSec | 任意一个分区处于同步状态的副本数（ISR）应该保持稳定，只有一种例外，就是当你扩展 Broker 节点或删除某个 Partition 的时候。为了保证高可用性，健康的 Kafka 集群必须要保证最小 ISR 数，以防在某个 Partiton 的 Leader 挂掉时它的 Follower 可以接管。如果 IsrShrinksPerSec（ISR 缩水）增加了，但并没有随之而来的 IsrExpandsPerSec（ISR 扩展）增加，将引起重视并让人工介入 |
| ActiveControllerCount | Controller 的职责是维护 Partition Leader 的列表，当遇到这个值等于 0 且持续了一小段时间（<1 秒）的时候，必须发出明确的告警 |
| OfflinePartitionsCount | 这个指标报告了没有活跃 Leader 的 Partition 数 |
| LeaderElectionRateAndTimeMs | Leader 选举的频率（每秒多少次）和集群中无 Leader 状态的时长（以毫秒为单位） |
| UncleanLeaderElectionsPerSec | 这个指标如果存在就很糟糕，这说明 Kafka 集群在寻找 Partition Leader 节点上出现了故障 |
| TotalTimeMs | 这个指标是由 4 个其他指标的总和构成的。<br>● queue：处于请求队列中的等待时间。<br>● local：Leader 节点处理的时间。<br>● remote：等待 Follower 节点响应的时间。<br>● response：发送响应的时间。 |
| BytesInPerSec<br>BytesOutPerSec | Kafka 的吞吐量 |

生产者、消费者度量指标如表 7.2 和表 7.3 所示。

**表 7.2　生产者度量指标**

| 指　标 | 作　用 |
| --- | --- |
| Response rate | 响应的速率是指数据从生产者发送到 Broker 的速率 |
| Request rate | 请求的速率是指数据从生产者发送到 Broker 的速率 |
| Request latency avg | 平均请求延迟 |
| Outgoing byte rate | 生产者的网络吞吐量 |
| IO wait time ns avg | 生产者的 I/O 等待的时间 |

**表 7.3　消费者度量指标**

| 指　标 | 作　用 |
| --- | --- |
| ConsumerLag MaxLag | 指消费者当前的日志偏移量相对生产者的日志偏移量 |
| BytesPerSec | 消费者的网络吞吐量 |
| MessagesPerSec | 消息的消费速度 |

续表

| 指　　标 | 作　　用 |
| --- | --- |
| ZooKeeperCommitsPerSec | 当 ZooKeeper 处于高写负载的时候，会成为性能瓶颈，从而导致从 Kafka 管道抓取数据变得缓慢。随着时间推移跟踪这个指标，可以帮助定位到 ZooKeeper 的性能问题，如果发现有大量发往 ZooKeeper 的 Commit 请求，你需要考虑要不要对 ZooKeeper 集群进行扩展 |
| MinFetchRate | 消费者的最小拉取速率 |

通过官方网站的说明可以查看 Kafka 提供的所有监控指标参数，本节只是列出了部分主要的参数指标。

# 7.2　使用 Kafka 客户端监控工具

Kafka 常用的客户端管理、监控工具主要有以下几种。

- Kafka Manager
- Kafka Tool
- KafkaOffsetMonitor
- JConsole

其中，前三个工具都是专门用于 Kafka 集群管理与监控的；而 JConsole（Java Monitoring and Management Console）是一种基于 JMX 的可视化监视与管理工具，安装好 JDK 以后，Java 就提供了 JConsole 的客户端工具。利用它也可以监控 Kafka 的各项指标。

这里我们简单介绍一下 JMX。JMX 的全称为 Java Management Extensions。JMX 可以管理、监控正在运行中的 Java 程序，常用于管理线程、内存、日志 Level、服务重启、系统环境等。Kafka 底层也是基于 Java 的，所以我们也可以使用 JMX 的标准来管理和监控运行中的 Kafka。

下面我们分别介绍它们的使用方法。

## 7.2.1　Kafka Manager

Kafka Manager 的监控框架的好处在于监控内容相对丰富，既能够实现 Broker 级常见的 JMX 监控（比如出入站流量监控），又能对消费者的消费进度进行监控（比如 lag 等）。另外，用户还能在页面上直接对集群进行管理，比如分区重分配或创建 Topic——当然这是一把双刃剑，好在 Kafka Manager 自己提供了只读机制，允许用户禁掉这些管理功能。

这里我们使用的版本是：kafka-manager-2.0.0.2.zip，其安装和配置非常简单，按照下面的步骤配置 Kafka Manager。

（1）在启动 Kafka 集群的命令脚本中，增加 JMX 的相关参数，否则无法使用客户端工具管理和监控 Kafka 集群。这里我们以 kafka101 主机上运行的 Broker 0 和 Broker 1 为例，进入 Kafka 安装目录下的 bin 目录。

```
cd /root/training/kafka_2.11-2.4.0/bin/
```

（2）修改 kafka-run-class.sh 文件，找到"JMX settings"的位置（第 176 行），增加 JMX Server 的配置信息，如图 7.1 所示。

```
-Djava.rmi.server.hostname=kafka101
```

```
176 # JMX settings
177 if [ -z "$KAFKA_JMX_OPTS" ]; then
178   KAFKA_JMX_OPTS="-Dcom.sun.management.jmxremote -Dcom.sun.management.jmxremote.authenticate=false  -Dcom.sun.management.jmxremote.ssl=false -Djava.rmi.server.hostname=kafka101"
179 fi
```

图 7.1 修改 Kafka Manager 的 JMX settings

由于在 kafka101 主机上将会启动两个 Broker，为了方便可以在命令终端使用 export 命令设置 JMX 的端口地址；也可以把 JMX 的端口写到 kafka-server-start.sh 脚本中，如图 7.2 所示，修改第 30 行。

```
export JMX_PORT="9999"
```

```
28 if [ "x$KAFKA_HEAP_OPTS" = "x" ]; then
29     export KAFKA_HEAP_OPTS="-Xmx1G -Xms1G"
30     export JMX_PORT="9999"
31 fi
```

图 7.2 设置 JMX 的端口

（3）启动 Kafka Broker 0。

```
export JMX_PORT="9990"
bin/kafka-server-start.sh config/server.properties &
```

（4）重新开启一个命令行终端，启动 Kafka Broker 1。

```
export JMX_PORT="9991"
bin/kafka-server-start.sh config/server1.properties &
```

（5）将 Kafka Manager 的压缩包解压至/root/training 目录。

```
unzip kafka-manager-2.0.0.2.zip -d ~/training/
```

（6）进入 Kafka Manager 的 conf 目录，并修改 application.conf 文件。

```
#这里我们指定 ZooKeeper 集群的地址
kafka-manager.zkhosts="kafka101:2181,kafka102:2181,kafka103:2181"

#将下面的这一行注释掉
#kafka-manager.zkhosts=${?ZK_HOSTS}
```

（7）采用 nohup 的方式启动 Kafka Manager。

```
nohup bin/kafka-manager &
```

也可以在启动 Kafka Manager 的时候，指定相关参数。

```
nohup bin/kafka-manager -Dconfig.file=conf/application.conf -Dhttp.port=8080 &
```

（8）启动成功后，输出如下日志信息，如图 7.3 所示。

```
[root@kafka101 kafka-manager-2.0.0.2]# tail -f nohup.out
2020-10-08 06:06:13,903 - [INFO] o.a.z.ClientCnxn - Session establishment complete on server kafka101/192.168.157.101
:2181, sessionid = 0x17507aa07960000, negotiated timeout = 40000
2020-10-08 06:06:14,113 - [INFO] k.m.a.KafkaManagerActor - Started actor akka://kafka-manager-system/user/kafka-manag
er
2020-10-08 06:06:14,127 - [INFO] k.m.a.KafkaManagerActor - Starting delete clusters path cache...
2020-10-08 06:06:14,153 - [INFO] k.m.a.DeleteClusterActor - Started actor akka://kafka-manager-system/user/kafka-mana
ger/delete-cluster
2020-10-08 06:06:14,154 - [INFO] k.m.a.DeleteClusterActor - Starting delete clusters path cache...
2020-10-08 06:06:14,172 - [INFO] k.m.a.KafkaManagerActor - Starting kafka manager path cache...
2020-10-08 06:06:14,209 - [INFO] k.m.a.DeleteClusterActor - Adding kafka manager path cache listener...
2020-10-08 06:06:14,209 - [INFO] k.m.a.DeleteClusterActor - Scheduling updater for 10 seconds
2020-10-08 06:06:14,209 - [INFO] k.m.a.KafkaManagerActor - Adding kafka manager path cache listener...
2020-10-08 06:06:14,539 - [INFO] play.api.Play - Application started (Prod)
2020-10-08 06:06:15,255 - [INFO] k.m.a.KafkaManagerActor - Updating internal state...
2020-10-08 06:06:15,708 - [INFO] p.c.s.AkkaHttpServer - Listening for HTTP on /0:0:0:0:0:0:0:0:9000
2020-10-08 06:06:25,241 - [INFO] k.m.a.KafkaManagerActor - Updating internal state...
2020-10-08 06:06:35,241 - [INFO] k.m.a.KafkaManagerActor - Updating internal state...
2020-10-08 06:06:45,240 - [INFO] k.m.a.KafkaManagerActor - Updating internal state...
```

图 7.3　启动 Kafka Manager

可以看到，Kafka Manager 将运行在 9000 端口上。

（9）通过浏览器访问 9000 端口，可以打开 Kafka Manager 的 Web 控制台，如图 7.4 所示。

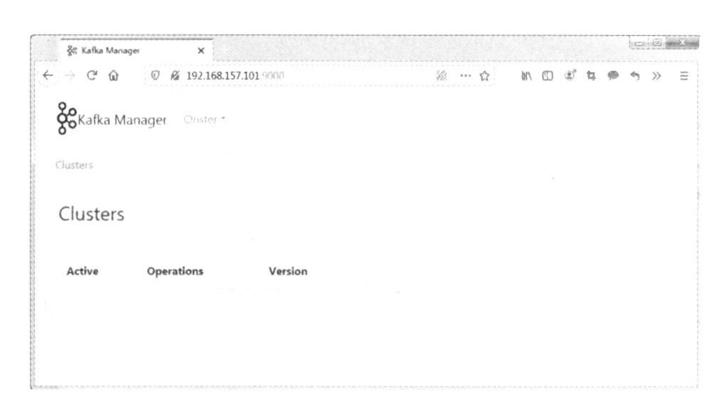


图 7.4　Kafka Manager 的 Web 控制台

（10）选择“Cluster”→“Add Cluster”选项，添加一个新的 Kafka 集群。勾选“Enable JMX Polling”复选框，如图 7.5 所示。

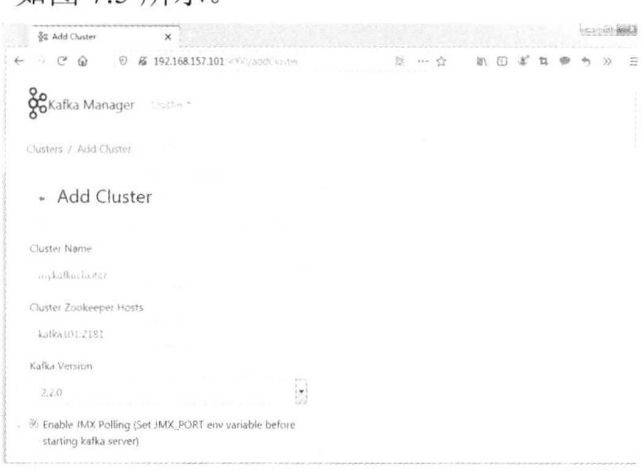


图 7.5　添加 Kafka 集群

（11）添加成功后，单击“Go to Cluster View”命令，跳转到 Kafka 集群的首页，如图 7.6 所示。

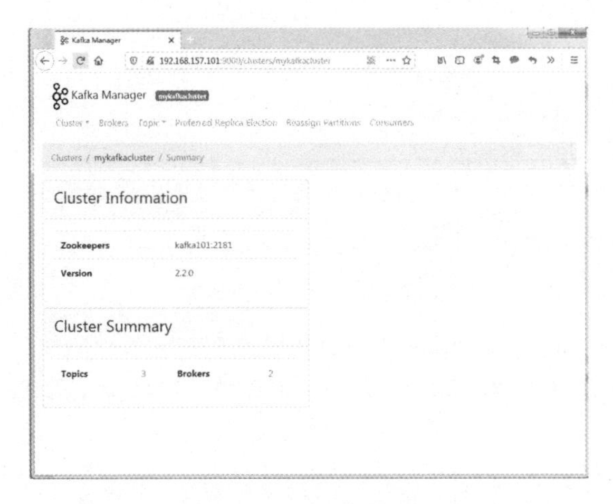


图 7.6　Kafka 集群的首页

在这里可以看到当前的 Kafka 集群中共存在 2 个 Broker，即 Broker 0 和 Broker 1；还有 3 个 Topics。

（12）单击 Brokers 的数字“2”，跳转到 Broker 的监控页面上。在这里就可以实时监控 Kafka 集群 Broker 的相关信息，如 Kafka 集群的吞吐量（Bytes in /sec、Bytes out /sec）等，如图 7.7 所示。

（13）图 7.8 展示了 Kafka 集群 Topic 的监控信息。

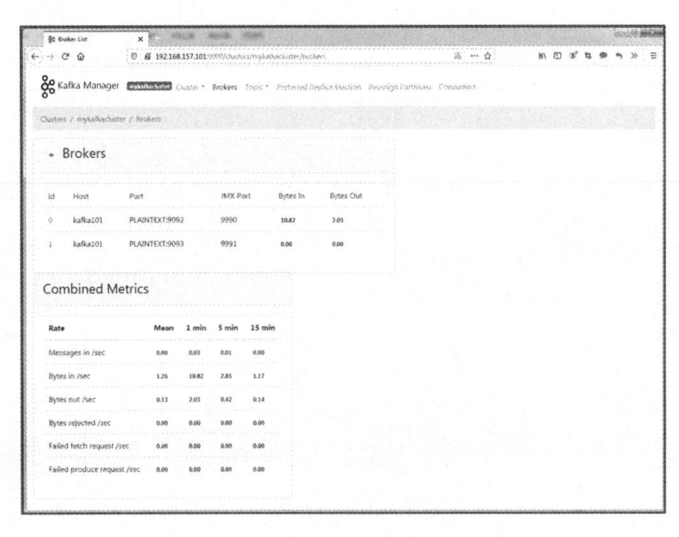


图 7.7　监控 Kafka Broker

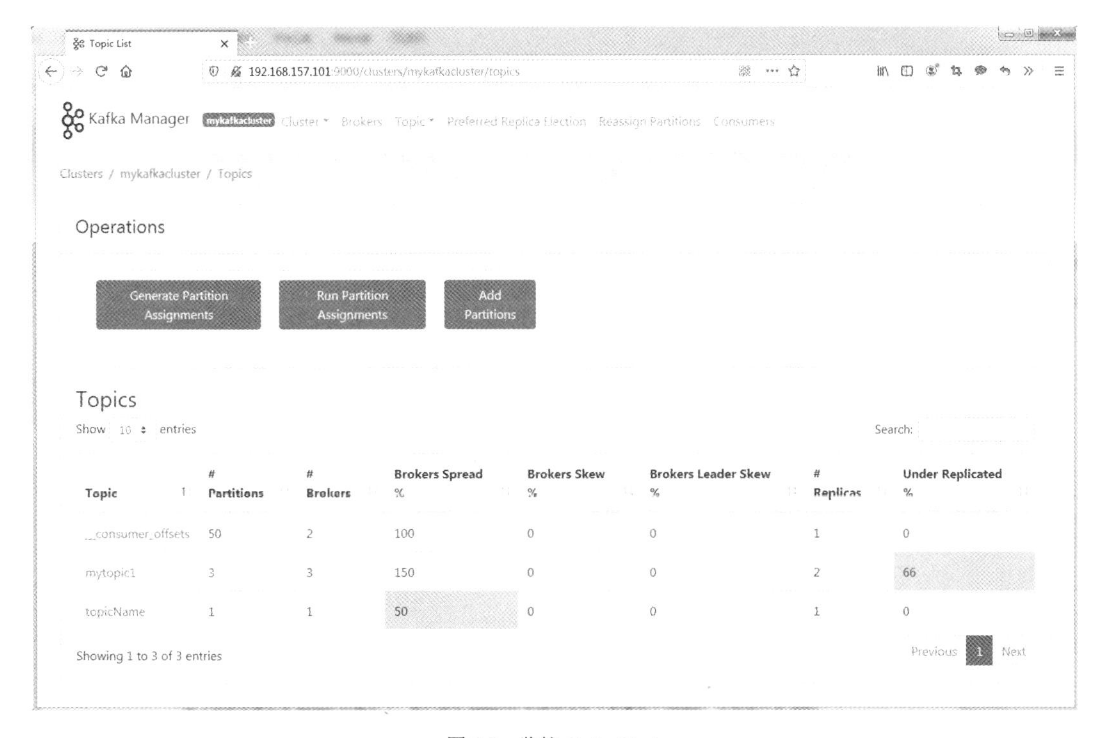


图 7.8　监控 Kafka Topic

## 7.2.2　Kafka Tool

Kafka Tool 是用于管理和使用 Apache Kafka 集群的图形应用程序。它提供了一种直观的界面，可让用户快速查看 Kafka 集群中的对象及集群主题中存储的消息。它包含面向开发人员和管理员的功能，一些关键功能如下。

- 快速查看所有 Kafka 集群，包括其 Broker、主题和消费者。
- 查看分区中消息的内容并添加新消息。
- 查看消费者的偏移量，包括 Apache Storm 中的 Spout 消费者。
- 以良好的格式显示 JSON 和 XML 消息。
- 添加和删除主题及其他管理功能。
- 将单个消息从自己的分区保存到本地硬盘驱动器。
- 编写自己的插件，可以查看自定义数据格式。
- Kafka 工具可在 Windows、Linux 和 Mac OS 上运行。

从 Kafka Tool 的官方网站直接下载 Kafka Tool 2.0.8 的版本，如图 7.9 所示。下载完成后，直接安装启动即可。图 7.10 展示了 Kafka Tool 的启动界面。

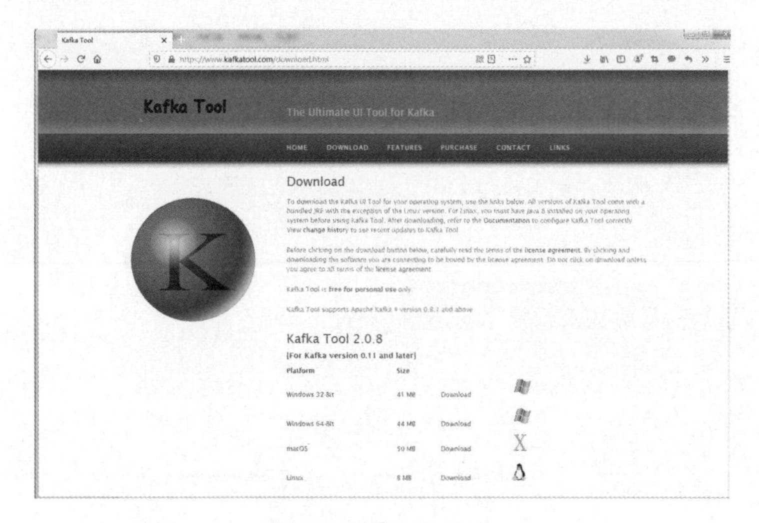


图 7.9　下载 Kafka Tool

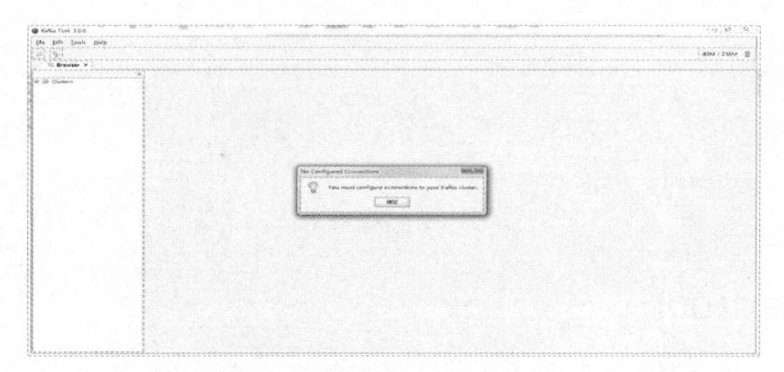

图 7.10　Kafka Tool 的启动界面

添加一个 Kafka Cluster 集群并测试，如图 7.11 所示。

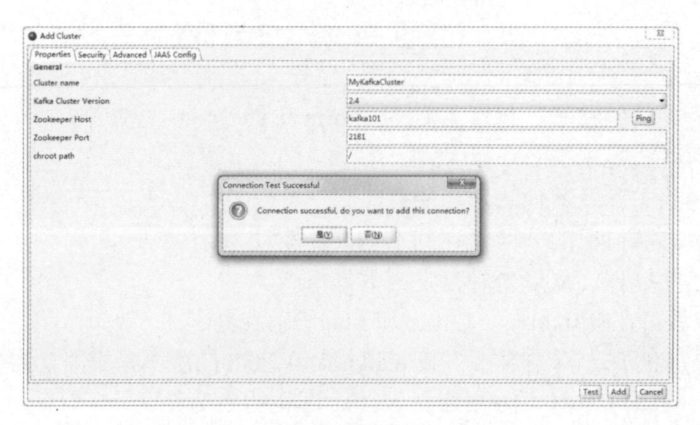


图 7.11　添加 Kafka Cluster 集群

单击“是”按钮，进入 Kafka 集群的首页，如图 7.12 所示。

图 7.12　Kafka 集群的首页

可以看到 Kafka 集群中的 Broker 信息、Topics 信息及 Consumers 信息。

现在我们使用 Kafka Tool 来创建一个新的 Topic。

（1）选择“Browsers”集群中的“Topics”节点，并在右边的界面上单击 [+] 按钮，添加一个新的 Topic。

（2）输入 Topic 的名称、分区数及每个分区的副本数。我们新创建的 Topic 名称是 mytopic2，它由两个分区组成，并且每个分区的副本数为 1，如图 7.13 所示。

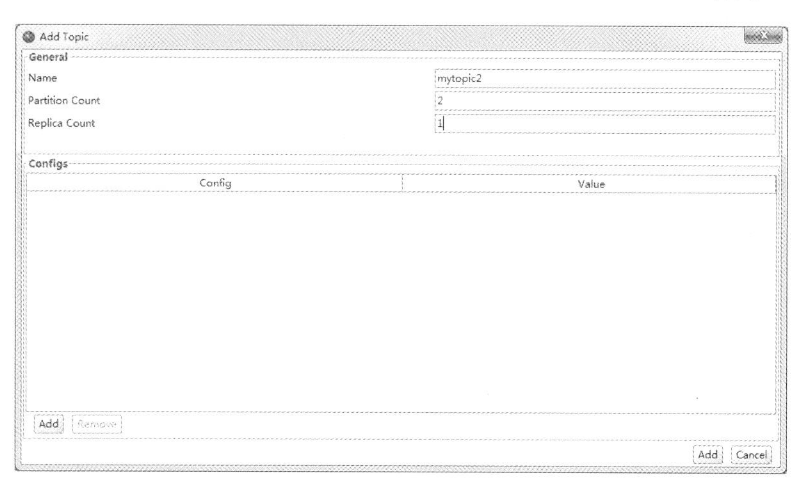


图 7.13　Add Topic

（3）单击“Add”按钮，成功创建 Topic，如图 7.14 所示。

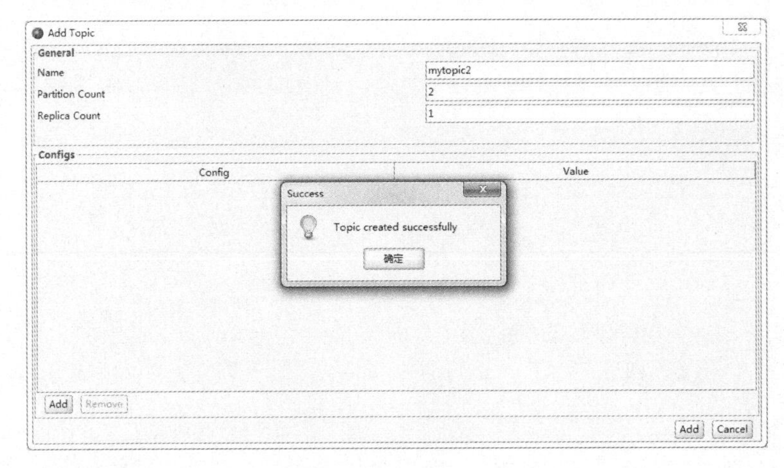


图 7.14　成功创建 Topic

（4）现在我们使用 Kafka Tool 来接收 mytopic2 上的消息数据。选择刚刚创建好的主题 mytopic2，并在右边的窗口中选择“Data”选项卡，如图 7.15 所示。

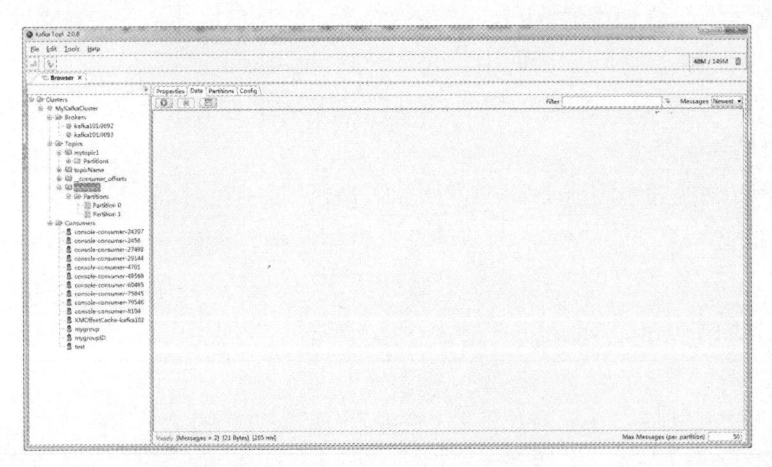

图 7.15　通过 Kafka Tool 接收数据

（5）启动一个 Kafka Producer 的命令行终端，并发送一些消息，如图 7.16 所示。

```
bin/kafka-console-producer.sh --broker-list kafka101:9092 --topic mytopic2
```

```
root@kafka101:~/training/kafka_2.11-2.4.0
[root@kafka101 kafka_2.11-2.4.0]# bin/kafka-console-producer.sh --broker-list ka
fka101:9092 --topic mytopic2
>hello kafka
>kafka tool
>123
>
```

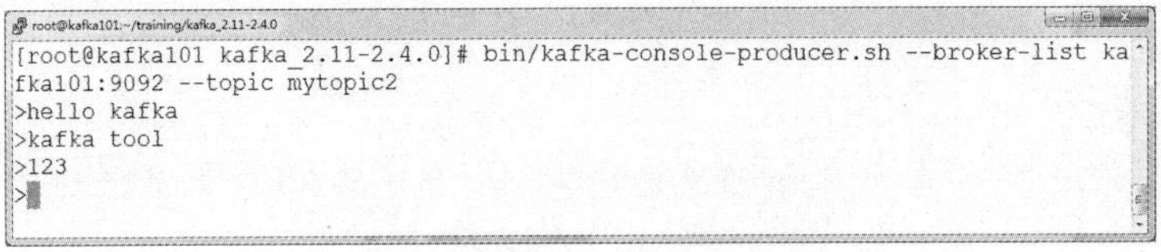

图 7.16　通过命令行发送数据

（6）在 Kafka Tool 上单击 按钮接收消息。这里就可以看到刚才在 Kafka Producer 命令行上发送的消息，如图 7.17 所示。

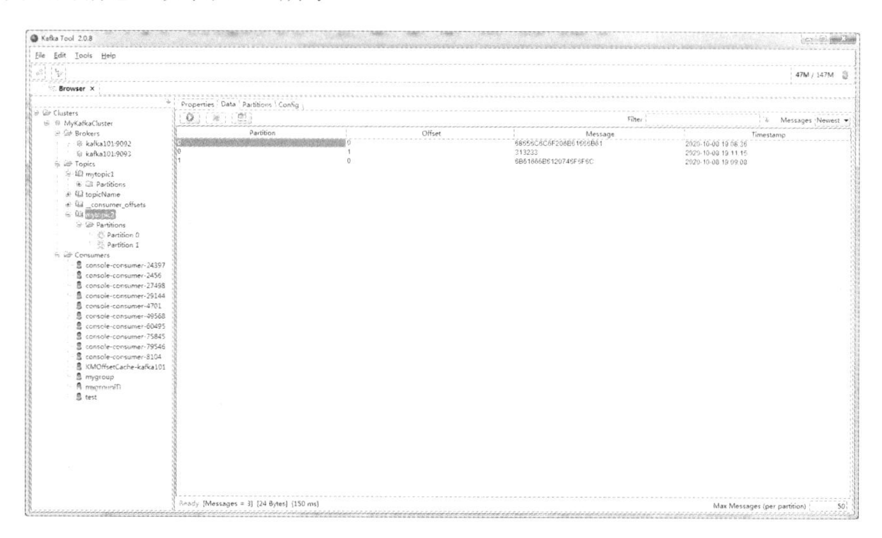


图 7.17　在 Kafka Tool 上接收消息

（7）这里的数据格式默认是“Byte Array”，可以在 Properties 的设置中将其修改为 String，并单击“Update”按钮，如图 7.18 所示。

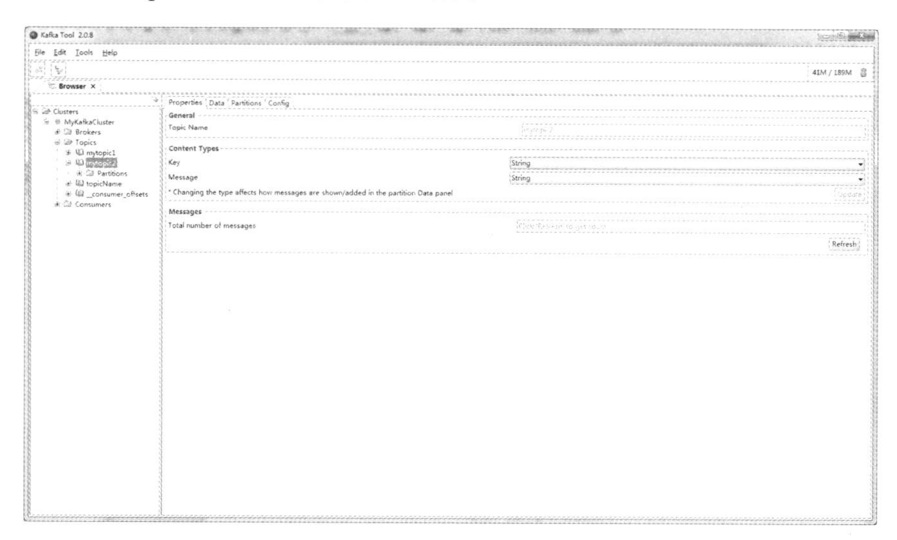


图 7.18　修改 Topic 的数据格式

（8）回到 Data 页面，这时数据将按照正确的格式显示，如图 7.19 所示。

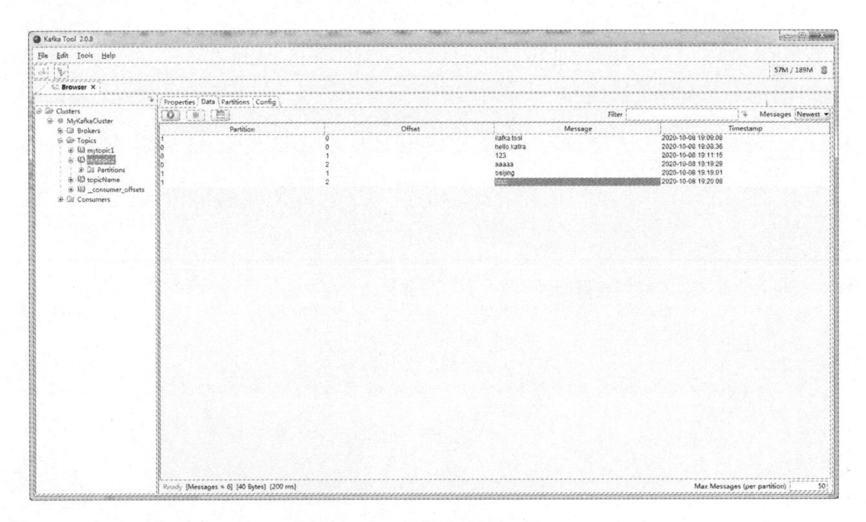

图 7.19　显示正确的数据格式

### 7.2.3 KafkaOffsetMonitor

KafkaOffsetMonitor 是一个基于 Web 界面的管理平台，可以用来实时监控 Kafka 服务的消费者及它们所在的 Partition 中的 Offset，可以浏览当前的消费者组，并且每个 Topic 的所有 Partition 的消费情况都可以进行实时监控。KafkaOffsetMonitor 可以从 github 上下载。这里我们使用的是 KafkaOffsetMonitor-assembly-0.2.0.jar。

KafkaOffsetMonitor 的安装启动比较简单，可以直接在 kafka101 的主机上执行下面的指令。

```
java -cp KafkaOffsetMonitor-assembly-0.2.0.jar \
com.quantifind.kafka.offsetapp.OffsetGetterWeb \
--zk kafka101:2181 \
--port 8089 \
--refresh 10.seconds \
--retain 1.days
```

其中，

- com.quantifind.kafka.offsetapp.OffsetGetterWeb 是运行 Web 监控的类。
- --zk 用于指定 ZooKeeper 的地址。
- --port 是 Web 运行端口。
- --refresh 和--retain 用于指定页面数据刷新的时间及保留数据的时间值。

打开浏览器访问 8089 端口，就可以打开 KafkaMonitor 的首页，如图 7.20 所示。

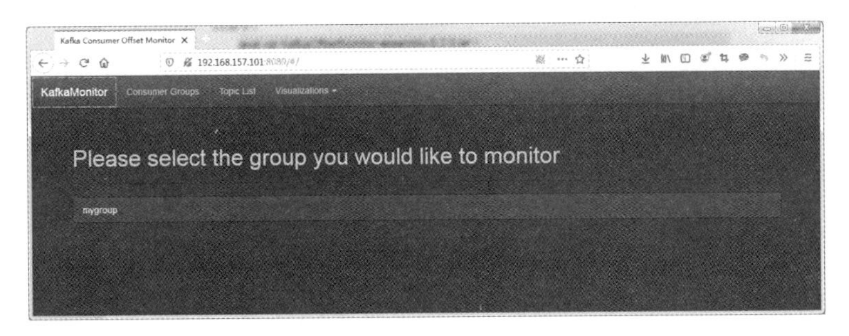


图 7.20　KafkaMonitor 首页

选择“Consumer Groups”选项卡，就可以监控某个 Topic 中具体的消费者信息，如图 7.21 所示。

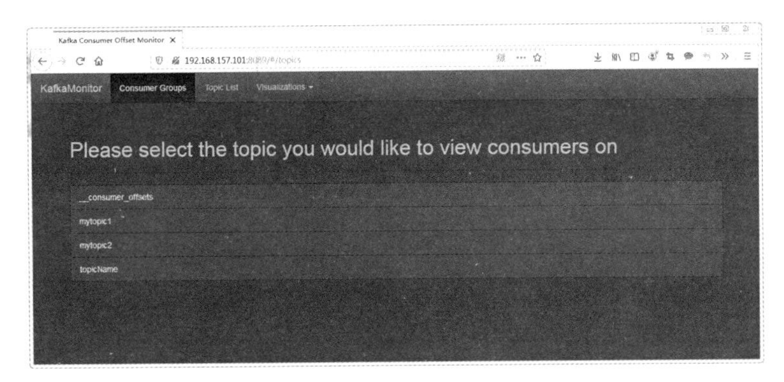


图 7.21　通过 KafkaOffsetMonitor 监控 Topic

## 7.2.4　JConsole

JConsole 是一种基于 JMX 的可视化监视与管理工具，从 Java 5 开始引入。JConsole 是用 Java 写的 GUI 程序，用来监控 VM，并可监控远程的 VM，易用且功能非常强。在命令行里输入 jconsole，就可以直接启动了。

这里我们直接在 Windows 上启动 JConsole。在 CMD 创建中直接输入 jconsole，如图 7.22 所示。

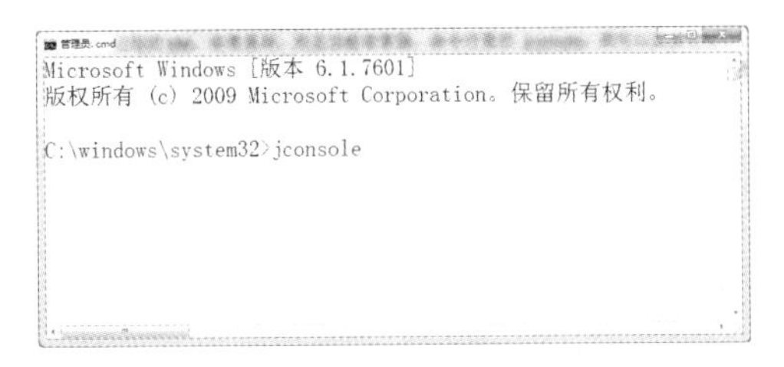


图 7.22　启动 JConsole

JConsole 的启动界面如图 7.23 所示。

由于在前面配置 Kafka Manager 时，我们已经启用了 Broker 0 和 Broker 1 的 JMX 配置，所以这里可以直接通过 JConsole 连接到 Broker 0 或 Broker 1 上。我们以 Broker 0 为例，选择“远程进程”单选按钮，并输入 Broker 0 的 JMX 地址，单击“连接”按钮，如图 7.24 所示。

```
kafka101:9990
```

图 7.23　JConsole 的启动界面

图 7.24　通过 JConsole 连接 Broker 0

弹出“安全连接失败”提示框，选择“不安全的连接”按钮，进入 JConsole 监控的主界面，如图 7.25 和图 7.26 所示。

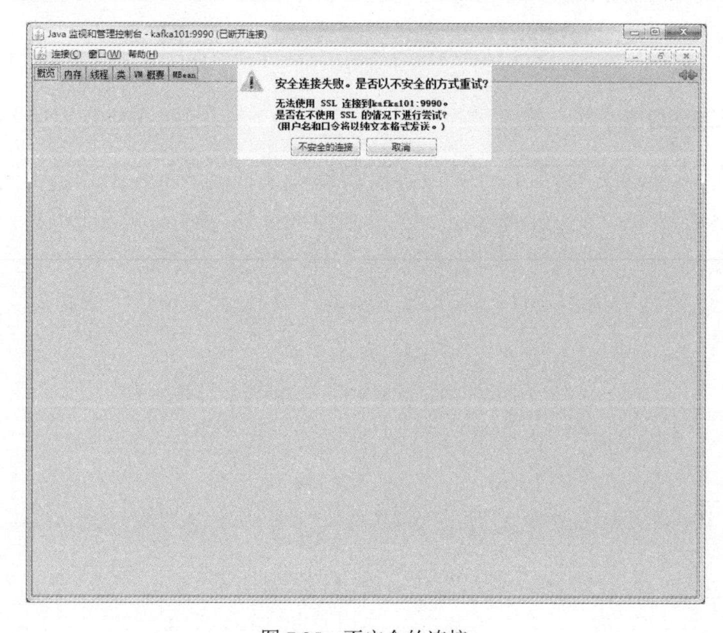


图 7.25　不安全的连接

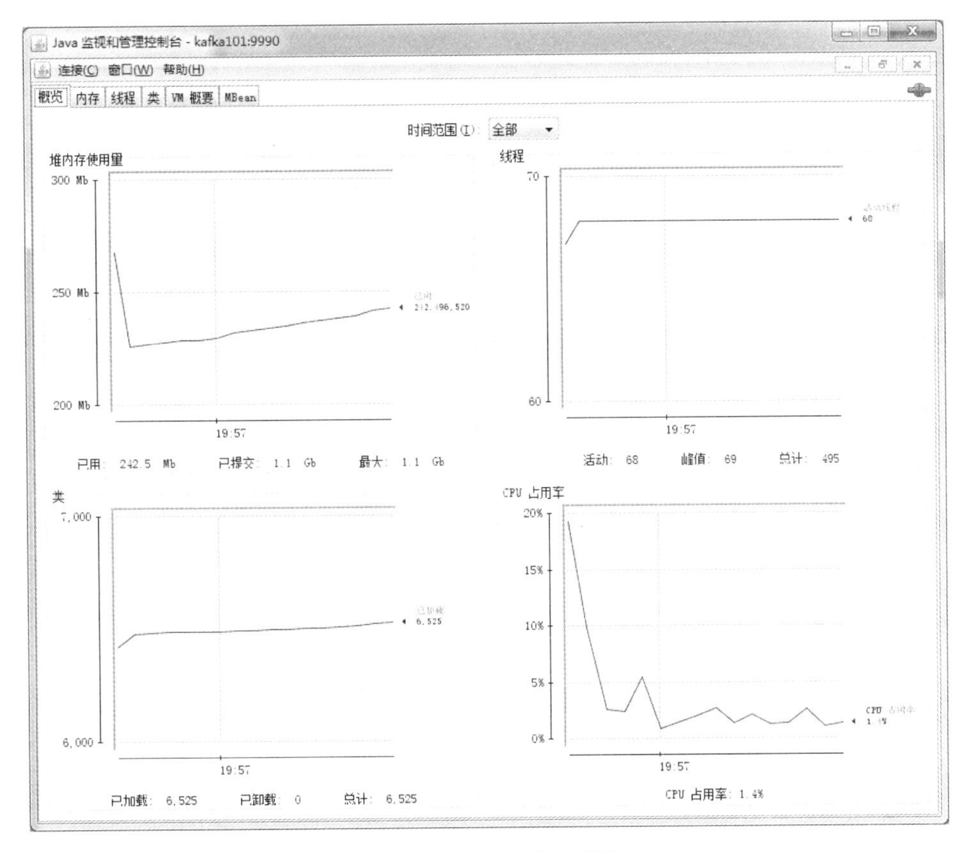


图 7.26　JConsole 的主界面

JConsole 提供 6 个选项卡显示应用信息。

（1）“概览”选项卡：提供内存使用的概述、运行的线程数量、创建的对象数量及 CPU 使用情况。

（2）“内存”选项卡：显示使用的内存数量。可以选择要监视的内存类型（堆、非堆或池）组合。

（3）“线程”选项卡：显示线程数量和每个线程的详细信息。

（4）“类”选项卡：显示加载的对象数量的信息。

（5）“VM 概要”选项卡：提供运行应用的 JVM 概要。

（6）“MBean”选项卡：显示有关应用的托管 bean 的信息。

这里我们选择“MBean”选项卡，就可以看到 Kafka 相关的 MBean 信息，如图 7.27 所示。

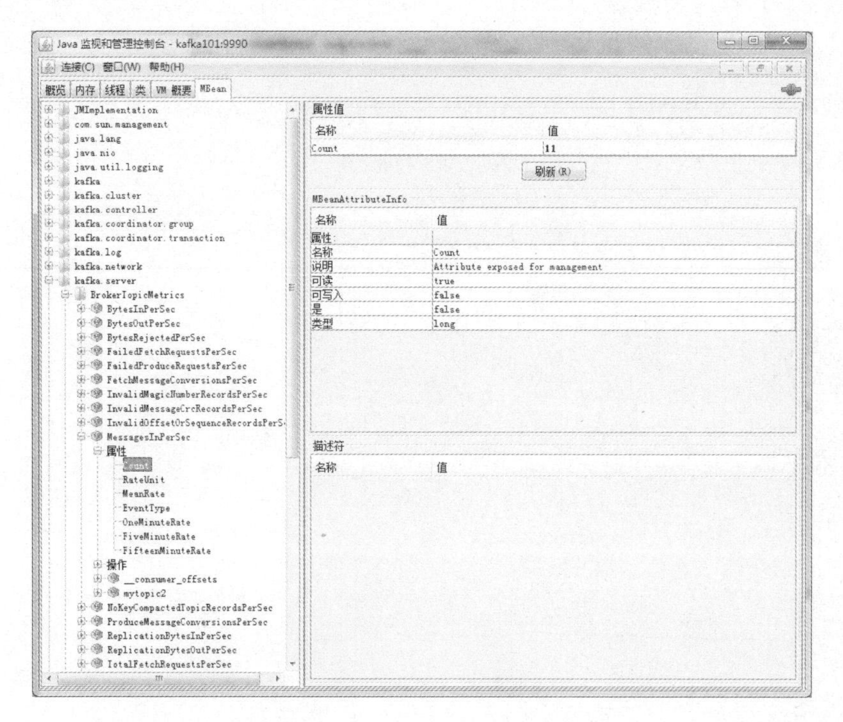


图 7.27 “MBean”选项卡

以图 7.27 监控的参数“MessagesInPerSec”为例，它表示的是 Kafka 集群消息的速率。关于所有 Kafka 监控的 MBean 信息，可以参考官方网站上的说明。

## 7.3 监控 ZooKeeper

前面提到，整个 Kafka 的状态靠 ZooKeeper 维护，因此 ZooKeeper 性能的监控也成为整个 Kafka 监控计划中一个必不可少的组成部分。在典型的 Kafka 集群中，Kafka 通过 ZooKeeper 管理集群配置，例如，选举 Leader，以及在 Consumer Group 发生变化时进行 Rebalance；生产者将消息发布到 Broker，消费者从 Broker 订阅并消费消息，这些操作都离不开 ZooKeeper。在 Kafka 集群的管理监控中，ZooKeeper 的监控也就成为非常重要的一部分。

由于 ZooKeeper 本身也是由 Java 开发的应用程序，我们也可以用前面提到的 JMX 的方式进行监控，例如，使用 JConsole 进行监控。图 7.28 展示了通过 JConsole 监控 ZooKeeper MBean 的监控信息。

这里也可以使用另一个客户端工具 ZooInspector 监控 ZooKeeper。图 7.29 展示了它的主界面。

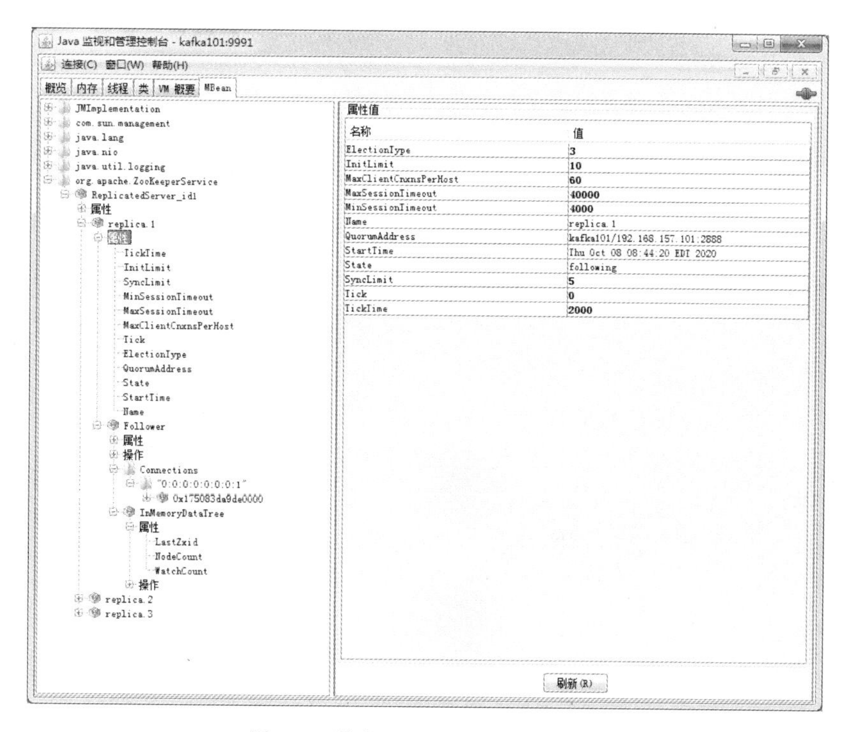


图 7.28　通过 JConsole 监控 ZooKeeper

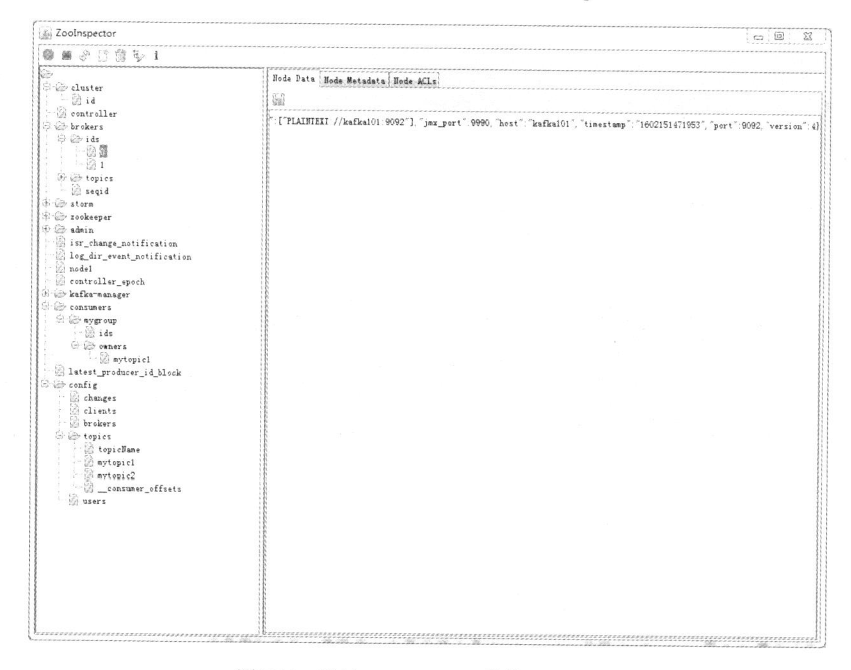


图 7.29　通过 ZooInspector 监控 ZooKeeper

# 第 8 章 Kafka 与 Flink 集成

Apache Flink 是一个用于对无边界和有边界数据流进行有状态计算的框架和分布式处理引擎。Flink 设计运行在所有常见的集群环境中，并且以内存速度和任意规模执行计算。

图 8.1 展示了 Flink 数据处理的流程及支持部署的平台。在图中可以看到 Flink 部署在 K8s、Yarn 或 Mesos 上。当然，也可以将 Flink 作为 Standalone 模式进行单独部署。关于 Flink 的部署，我们会在后续的章节中介绍。另外，通过图 8.1 也可以看到 Flink 支持使用 HDFS、S3 或 NFS 等文件系统作为存储。在早期的 Flink 版本中，Flink 直接集成 Hadoop，因此可以直接在 Flink 中访问 HDFS；但在 Flink 的最新版本中，取消了与 Hadoop 的集成，因此需要手动集成。

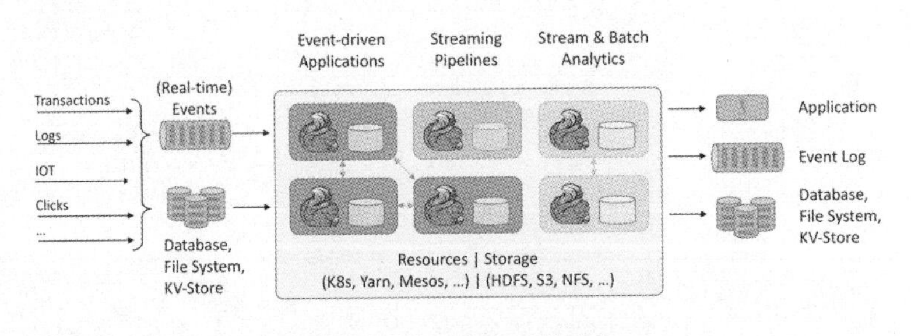


图 8.1 Flink 数据处理的流程

## 8.1 Flink 的体系架构

Flink 作为新一代的大数据处理引擎，其体系架构也是一种主从式的架构，可以利用 ZooKeeper 实现 Flink 的 HA 功能，这一点与其他主从式的大数据组件是完全一样的。

### 8.1.1 Flink 中的数据集

任何类型的数据都可以作为数据源事件流产生。例如，用户访问网站的信息、工业传感器的监控信息、用户在应用 App 上的交互信息等，可以被看成一种数据流，那么数据流就可以作为无边界的数据流或有边界的数据流进行处理。

无边界的数据流定义了开始但没有定义结束。它们不会在生成时终止提供数据，必须持续地处理无边界数据流，即必须在拉取到事件后立即处理它，无法等待所有输入数据到达后处理，因为输入是无边界的，并且在任何时间点都不会完成。处理无边界数据流通常要求以特定顺序（如事件发生的顺序）拉取事件，以便能够推断结果完整性。在 Flink 中，可以通过 Flink 的 DataStream API 来完成对无边界数据流的处理。

有边界数据流定义了开始和结束。可以在执行任何计算之前拉取到所有数据后处理有边界数据流。处理有边界数据流不需要有序拉取数据，因为可以随时对有界数据集进行排序。有边界数据流的处理也称批处理。在 Flink 中，可以通过 Flink 的 DataSet API 来完成对有边界数据流的处理。

## 8.1.2　Flink 的生态圈体系

Flink 以层级式系统形式组建其软件栈，不同层的软件栈建立在其下层基础上，并且各层接受程序不同层的抽象形式。图 8.2 展示了 Flink 的生态体系。

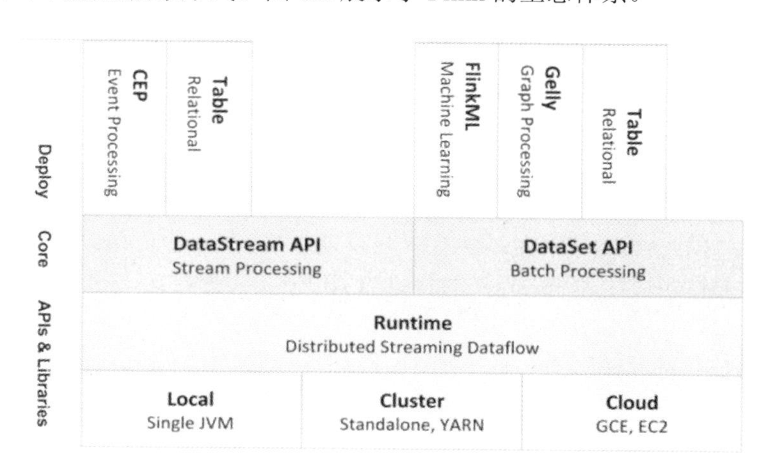


图 8.2　Flink 的生态圈系统

可以把 Flink 的生态圈系统划分成三层：平台部署层（Deploy）、执行引擎层（Core）、应用开发层（APIs & Library），下面我们分别对其进行介绍。

### 1. 平台部署层

Flink 支持的平台部署模式主要有以下三种。

- Local 模式：即单机部署环境。可以在集成化的开发工具中，比如，Eclipse、IDEA 直接基于 Flink 的 API 开发应用程序，并运行程序，这种模式跟运行一个普通的

Java 程序没有差别。Local 模式多用于开发和测试，并方便开发人员在 IDE 环境中调试自己的应用程序代码。

- Cluster 模式：即集群模式。在实际的生产上，我们不可能在 IDE 环境中运行应用程序，需要将其部署到集群模式上。而 Cluster 模式又可以具体划分成 Standalone 模式和 Yarn 模式。Standalone 模式是指 Flink 集群独立运行，不依赖其他任何组件，独立运行 Flink 应用程序，管理调度 Flink 的任务和管理 Flink 集群的资源。Yarn 是从 Hadoop 2.x 以后引入的资源管理器。Yarn 是一个通用资源管理系统，可为上层应用提供统一的资源管理和调度。例如，Hadoop MapReduce 运行在 Yarn 上，也可以将 Spark 和 Flink 运行在 Yarn 上，由 Yarn 统一管理和调度。Flink on Yarn 的模式也是很多大数据平台上实际采用的方式。
- Cloud 模式：将 Flink 运行在云环境下，发挥云的优势。云计算的核心是虚拟化技术，例如，可以使用 Docker、K8s 这样的技术创建虚拟化容器。云环境具备快速的硬件升级及扩容和缩容等优势。因此 Flink 上云后，将最大程度发挥其优势。除了 Docker 和 K8s，还有 Amazon AWS 中的 EC2 等。

### 2. 执行引擎层

Flink 的执行引擎负责计算逻辑（流计算和批处理）的执行，即 Flink 的 Runtime。在 Flink 中执行的所有技术，不管是 DataSet 的离线批处理操作，还是 DataStream 的流式实时计算，最终都将由 Flink 的执行引擎 Runtime 负责完成。

### 3. 应用开发层

Flink 的目标是批流一体，但是 API 层并没有做到统一。DataSet API 为批处理离线计算的 API，DataStrem API 为流处理实时计算的 API。可以使用 Java 语言或 Scala 语言完成应用程序的开发。

基于 API 的各种库（Library），用户在使用时需要感知流计算和批处理。下面主要提供了几个 Library，方便应用程序的开发。

（1）CEP（Complex Event Processing），复杂事件处理就是对行为数据进行一系列规则匹配。当数据匹配到了相应的处理规则后，再对数据进行处理。例如，电商购物的一个过程就可以用这样的规则去定义：用户购买商品并支付—商家收到订单并处理订单—商家发货—用户收到商品并确认评价，每个单元即一个模式，一系列模式联合即复杂事件处理。

（2）Table，在 Flink 中提供了两套处理结构化数据的接口：Table API 和 SQL。这两套接口既可以处理流式数据，又可以进行批处理的离线计算。Flink Table API 允许用户以一种直观的方式进行代码程序的开发，其既可以使用 Java 语言，又可以使用 Scala 语句。从而很方便地进行 select 、distinct、filter 和多表连接等操作；Flink SQL 实现了 SQL 标准。但在目前的 Flink 版本中，Table API 和 SQL 两套接口并不成熟，还处在开发的阶段，并不建议在实际生产环境中使用。下面是在 Flink 1.11 版本的官方文档中的描述。

```
Please note that the Table API and SQL are not yet feature complete and are being
actively developed. Not all operations are supported by every combination of
[Table API, SQL] and [stream, batch] input.
```

（3）Flink ML。Flink ML 是 Flink 的机器学习（Machine Learning）库。这是 Flink 社区的一项新工作，其中包含越来越多的算法和贡献者，而机器学习的本质就是一系列的算法。类似的框架还有 Hadoop 中的 Mahout 和 Spark 中的 MLlib。

（4）Gelly。Gelly 是 Flink 中用于支持图计算的 API，其中包含了一些用于图应用分析中的方法和接口。

## 8.1.3　Flink 的体系架构

Flink 集群启动后会有两种进程，一种是 JobManager（Master），另一种是 TaskManager（Worker），可以通过 jps 或 ps -ef | grep java 命令来查看 Flink 进程。图 8.3 展示了一个单节点 Flink 环境启动后的进程信息。

```
[root@kafka101 flink-1.11.0]# jps
49648 StandaloneSessionClusterEntrypoint
49942 TaskManagerRunner
49974 Jps
[root@kafka101 flink-1.11.0]#
```

图 8.3　Flink 的后台进程

Flink 的体系架构类似 Spark，是一个基于 Master-Slave 风格的架构，即主从式架构。图 8.4 展示了 Flink 的主从式架构。

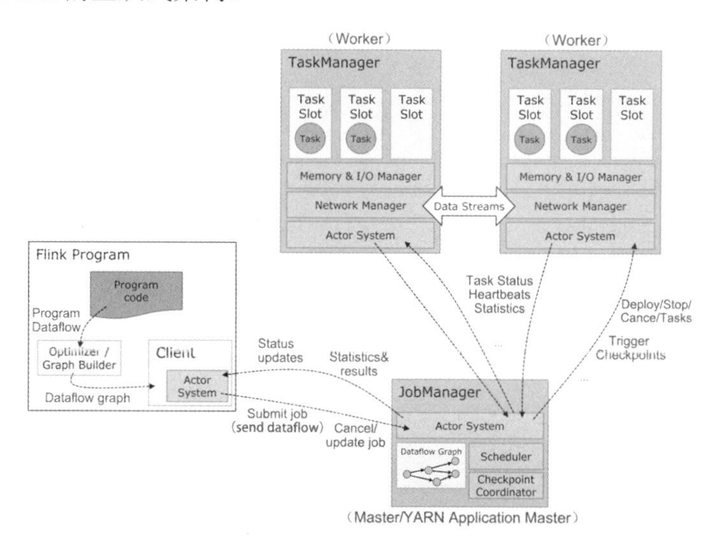


图 8.4　Flink 的体系架构

Flink 集群的 Master 进程（JobManager）和 Worker 进程（TaskManager）既可以通过容器技术部署启动，又可以在物理机上部署启动。当然也可以由 Yarn 资源管理框架启动。Worker 作为 Flink 的节点运行，连接到 Master 节点上等待任务的分配，同时向 Master 汇报自己的状态，如心跳信息等。

下面对 Master、Worker 和客户端 Client 分别进行说明。

- Master 进程。Flink 的主节点，用于分布式执行和调度任务、协调检查点（checkpoint）、协调失败恢复等，Flink 集群中至少有一个 Master 进程。为了保证高可用性，通常会有多个 Master 节点，选取其中一个作为 Leader，其余作为 Standby。Flink 高可用性需要借助 ZooKeeper 才能实现。
- Worker 进程。Flink 的从节点，用于执行 Data Flow 上的任务，同时缓存和交换数据流，至少有一个 TaskManager。而在实际的生产环境中，至少应该保证两个以上的 TaskManager，它们也与 JobManager 共同组成一个主从式的架构。
- Client，即 Flink 的客户端。它也叫作 Flink Program，不是 Flink 集群运行时的一部分，它作为客户端，用来准备和将数据流发送到 Master。任务提交成功后，客户端就可以断开，或者保持连接接收结果数据。客户端程序可以是 Java 或 Scala 程序，也可以用 bin/flink run 的方式来运行客户端程序。

## 8.2 安装部署 Flink Standalone 模式

本节将重点介绍 Flink Standalone 模式和 Flink on Yarn 模式的部署。关于这两种模式在前面的章节中已经进行了简单的介绍。

Flink 的核心配置文件是 conf 目录下的 flink-conf.yaml 文件，下面列出了这个文件 common 部分的配置内容。因为在 Standalone 和 Yarn 模式下，我们只会用到这部分的配置参数。

```
01
#==============================================================================
02  # Common
03
#==============================================================================
04  # The external address of the host on which the JobManager runs and can be
05  # reached by the TaskManagers and any clients which want to connect. This
setting
06  # is only used in Standalone mode and may be overwritten on the JobManager
side
07  # by specifying the --host <hostname> parameter of the bin/jobmanager.sh
executable.
08  # In high availability mode, if you use the bin/start-cluster.sh script and
setup
09  # the conf/masters file, this will be taken care of automatically. Yarn/Mesos
```

```
10  # automatically configure the host name based on the hostname of the node
where the
11  # JobManager runs.
12  jobmanager.rpc.address: localhost
13
14  # The RPC port where the JobManager is reachable.
15  jobmanager.rpc.port: 6123
16
17  # The total process memory size for the JobManager.
18  # Note this accounts for all memory usage within the JobManager process,
including JVM metaspace and other overhead.
19  jobmanager.memory.process.size: 1600m
20
21  # The total process memory size for the TaskManager.
22  # Note this accounts for all memory usage within the TaskManager process,
including JVM metaspace and other overhead.
23  taskmanager.memory.process.size: 1728m
24
25  # To exclude JVM metaspace and overhead, please, use total Flink memory size
instead of 'taskmanager.memory.process.size'.
26  # It is not recommended to set both 'taskmanager.memory.process.size' and
Flink memory.
27  # taskmanager.memory.flink.size: 1280m
28  # The number of task slots that each TaskManager offers. Each slot runs one
parallel pipeline.
29  taskmanager.numberOfTaskSlots: 1
30
31  # The parallelism used for programs that did not specify and other parallelism.
32  parallelism.default: 1
33
34  # The default file system scheme and authority.
35  # By default file paths without scheme are interpreted relative to the local
36  # root file system 'file:///'. Use this to override the default and interpret
37  # relative paths relative to a different file system,
38  # for example 'hdfs://mynamenode:12345'
39  # fs.default-scheme
```

其中的参数说明如下。

- jobmanager.rpc.address 表示 JobManager 的 IP 地址。
- jobmanager.rpc.port 表示 Flink 客户端通过 RPC 协议可以连接 JobManager 的端口。
- jobmanager.memory.process.size 表示 JobManager JVM Heap 的内存大小。
- taskmanager.memory.process.size 表示 TaskManager JVM Heap 的内存大小。
- taskmanager.numberOfTaskSlots 表示每个 TaskManager 提供的任务 Slots 数量大小。

- parallelism.default 表示程序默认并行计算的个数。

下面我们通过具体的步骤学习如何配置 Flink Standalone 和 Flink on Yarn 的模式。

## 8.2.1　Flink Standalone 模式的部署

### 1. Flink Standalone 伪分布模式的部署

可以在 kafka101 的主机上部署一个伪分布模式的 Flink Standalone 模式，即只有一个 JobManager 和一个 TaskManager。

（1）首先，将压缩包解压至/root/training 目录。

```
tar -zxvf flink-1.11.0-bin-scala_2.12.tgz -C ~/training/
```

这里我们使用的是 flink-1.11.0-bin-scala_2.12.tgz 的版本。

（2）进入 Flink 的 conf 目录，并修改 flink-conf.yaml 文件。

```
cd training/flink-1.11.0/conf
vi flink-conf.yaml
```

（3）下面列出了需要修改的 flink-conf.yaml 参数。

```
jobmanager.rpc.address: kafka101
```

在前面的内容中提到 jobmanager.rpc.address 参数是指 JobManager 运行的 RPC 端口。Flink Client 可以通过这个端口连接到 JobManager 上。

（4）修改 conf/workers 文件，指定 Flink 从节点运行的地址。这里我们把 TaskManager 运行在当前主机上。

```
kafka101
```

（5）启动 FlinkKafka Broker，支持下面的命令。

```
cd ..
bin/start-cluster.sh
```

启动成功后，将输出如下的日志信息，如图 8.5 所示。

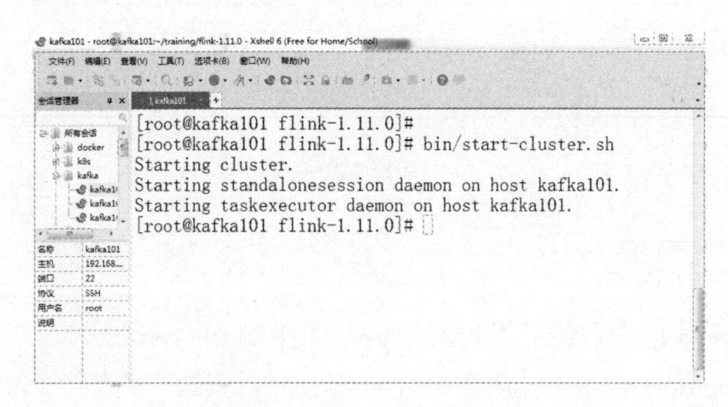


图 8.5　Flink 启动成功日志

也可以通过 Java 的 jps 命令查看后台的 Java 进程，如图 8.6 所示。

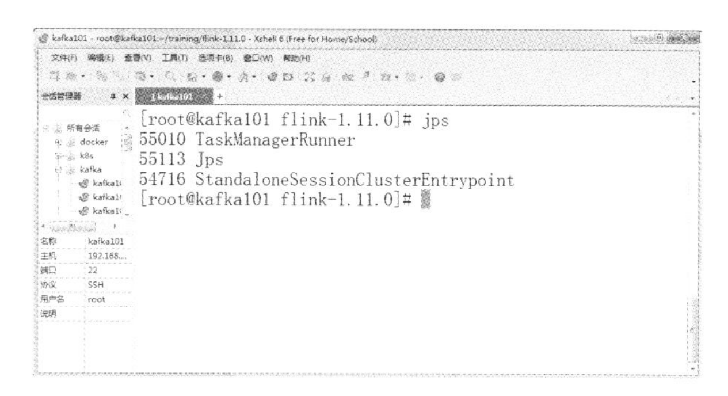


图 8.6　Flink 的后台进程信息

（6）Flink 启动成功后，提供了一个 Web Console 的界面。通过 8081 的端口可以访问这个 Web Console，从而监控 Flink 任务运行的状态，如图 8.7 所示。

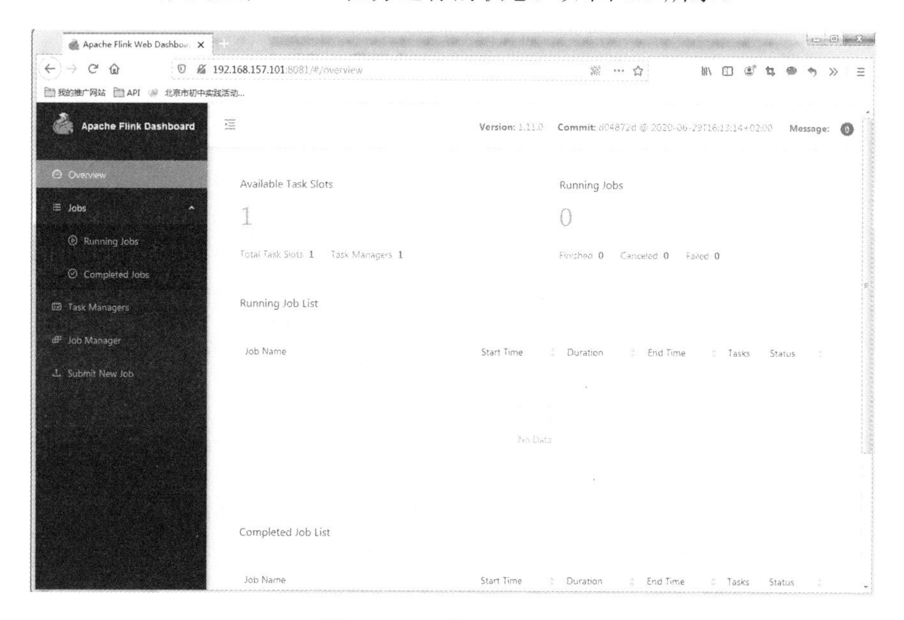


图 8.7　Flink 的 Web Console

由于目前没有任务在运行，所以在这个界面上看不到任务执行的状态。接下来我们就可以使用 Flink 官方提供的 Example 示例程序来执行批处理的离线计算和流处理的实时计算。

另外，由于当前的 Flink 环境是一个单节点的伪分布模式，所以在 Web Console 的界面上我们只能看到一个 Task Manager。

### 2．Flink Standalone 全分布模式的部署

在实际的生产环境中，Flink 集群至少应该是一个全分布的模式，包含一主两从的架构，如图 8.8 所示。

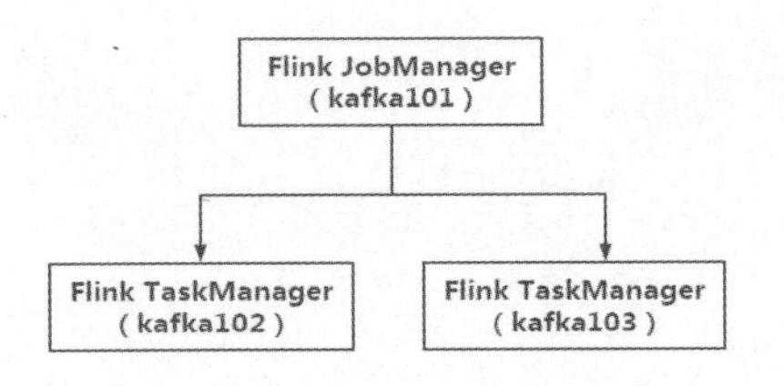


图 8.8　Flink 的全分布模式

在前面的章节中，我们已经完成了 Flink 伪分布模式的搭建，可以在此基础上继续进行 Flink 全分布模式的搭建。

（1）修改 conf/workers 文件，指定 Flink 从节点运行的地址。这里我们把 TaskManager 运行在 kafka102 和 kafka103 的虚拟机上。

```
kafka102
kafka103
```

（2）使用 scp 命令将 kafka101 的虚拟机上的 flink 目录复制到 kafka102 和 kafka103 的虚拟机上。执行下面的命令。

```
scp -r flink-1.11.0/ root@kafka102:/root/training
scp -r flink-1.11.0/ root@kafka103:/root/training
```

需要注意的是，由于没有配置免密码登录，这一步执行 scp 复制的时候需要输入 kafka102 和 kafka103 虚拟机的密码。

（3）在主节点 kafka101 的虚拟机上启动 Flink 集群，启动的信息如图 8.9 所示。

```
bin/start-cluster.sh
```

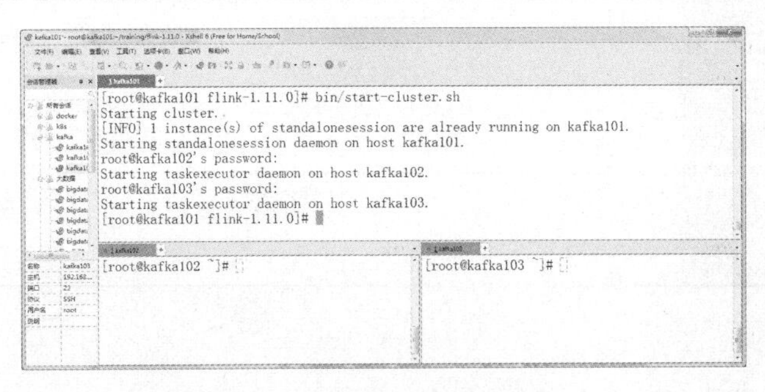


图 8.9　Flink 全分布模式的启动

可以看到在 kafka101 的虚拟机上启动了主节点的信息，而在 kafka102 和 kafka103 的虚拟机上各启动了一个 Task Executor 的从节点信息。这里需要注意的问题是，由于没有配置免密码登录，在集群启动和停止的过程中，都需要输入 kafka102 和 kafka103 的虚拟机的密码。可以按照下面的步骤进行免密码登录的配置。

```
首先，在 kafka101 上产生公钥和私钥信息。执行下面的命令，每步操作直接按 Enter 键即可。
ssh-keygen -t rsa
```

将 kafka101 的虚拟机公钥文件/root/.ssh/id_rsa.pub 复制到 kafka102 和 kafka103 的虚拟机上，并输入 kafka102 和 kafka103 的虚拟机密码。

```
ssh-copy-id -i .ssh/id_rsa.pub root@kafka102
ssh-copy-id -i .ssh/id_rsa.pub root@kafka103
```

至此，启动 Flink 集群或停止 Flink 集群都不需要再次输入密码。读者可以自行测试，执行如下的命令：

```
bin/stop-cluster.sh
```

说明：免密码登录采用的是不对称加密的方式。这里的第一步 ssh-keygen 命令就是用于产生本台主机的公钥信息和私钥信息，即 id_rsa 文件和 id_rsa.pub 文件。

（4）在 kafka101、kafka102、kafka103 的虚拟机上分别执行 jps 命令，查看每台主机的 Java 进程信息，如图 8.10 所示。

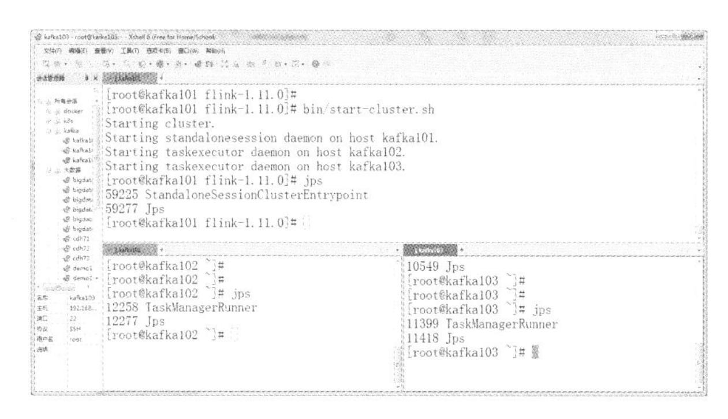


图 8.10　Flink 全分布模式的进程信息

（5）使用浏览器查看 kafka101 的虚拟机 8081 端口，并查看 Flink 集群的信息，可以看到有两个 Task Manager 在运行，如图 8.11 所示。

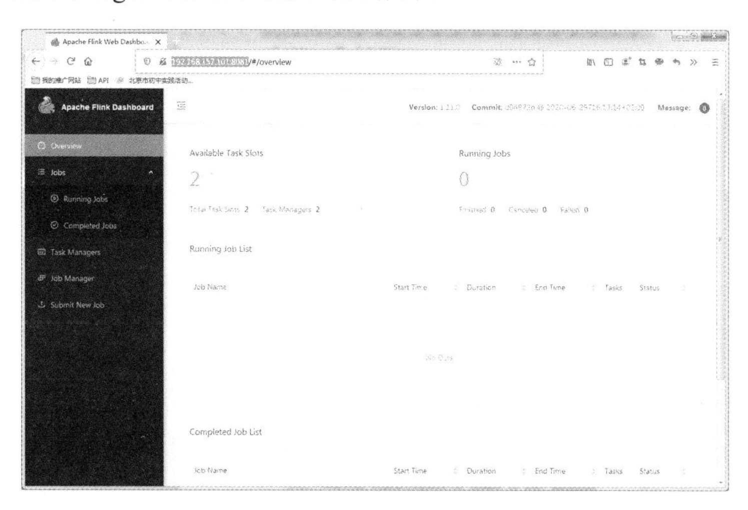


图 8.11　Flink 全分布模式的 Web Console 界面

## 8.2.2 在 Standalone 模式上执行 Flink 任务

在前面的内容中提到，Flink 支持 DataSet API 和 DataStream API 用于执行离线计算和流式计算。在部署完成的 Flink 环境中，Flink 官方提供了相应的 Example 示例程序。可以通过下面的命令分别执行。

### 1. 执行 WordCount 批处理的离线计算

（1）准备测试的数据。将使用 Flink 直接计算出本地文件系统上的数据。处理完成后的结果也直接存放在本地文件系统上。执行下面的操作。

```
mkdir -p ~/temp/input
vi ~/temp/input/data.txt
```

（2）在 data.txt 文件中输入以下文本，保存并退出。

```
I love Beijing
I love China
Beijing is the capital of China
```

（3）进入 Flink 的安装目录执行下面的命令。

```
cd ~/training/flink-1.11.0/
bin/flink run examples/batch/WordCount.jar -input /root/temp/input/data.txt -output /root/temp/wc
```

图 8.12 展示了运行过程中的输出信息。

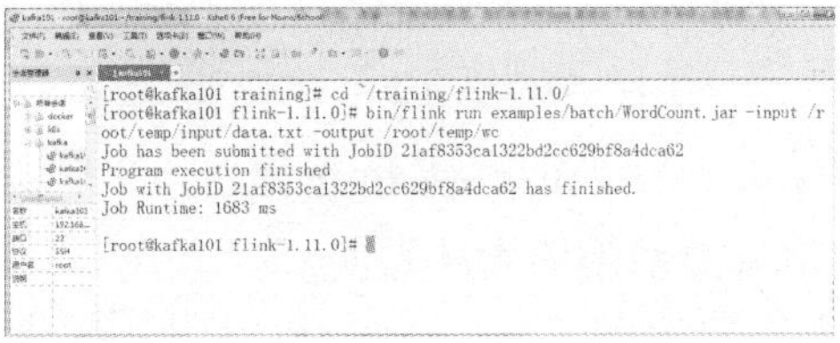

```
[root@kafka101 training]# cd ~/training/flink-1.11.0/
[root@kafka101 flink-1.11.0]# bin/flink run examples/batch/WordCount.jar -input /root/temp/input/data.txt -output /root/temp/wc
Job has been submitted with JobID 21af8353ca1322bd2cc629bf8a4dca62
Program execution finished
Job with JobID 21af8353ca1322bd2cc629bf8a4dca62 has finished.
Job Runtime: 1683 ms

[root@kafka101 flink-1.11.0]#
```

图 8.12 WordCount 离线计算的输出信息

（4）查看输出文件/root/temp/wc 的内容。

```
cat /root/temp/wc

beijing 2
capital 1
china 2
i 2
is 1
love 2
of 1
the 1
```

可以看到每个单词在数据文件中出现的总频率就统计了出来。这里我们处理的数据是本地文件系统的数据，也可以是 Hadoop HDFS 文件系统上的数据。需要注意的是，由于在新版本的 Flink 中并没有集成 Hadoop，因此要访问 HDFS 或运行 Flink On Yarn，需要将图 8.13 中的 jar 包复制至 Flink 的 lib 目录下。该 jar 包可以从 Flink 的官方网站上下载。

图 8.13　集成 Flink 与 Hadoop 的 jar 包

### 2. 执行 WordCount 流式计算

在流式计算中，数据源将源源不断地产生数据。数据也将源源不断地送到 Flink 中进行实时处理。为了模拟这样一个过程，我们使用 Linux 下的 netcat 工具作为数据源来产生数据。netcat 被誉为网络安全界的"瑞士军刀"，简称 nc.，它是一个简单而有用的工具，使用 TCP 或 UDP 协议的网络连接去读写数据。

（1）单独开启一个命令行终端，并执行以下命令，如图 8.14 所示。

```
nc -l -p 1234
```

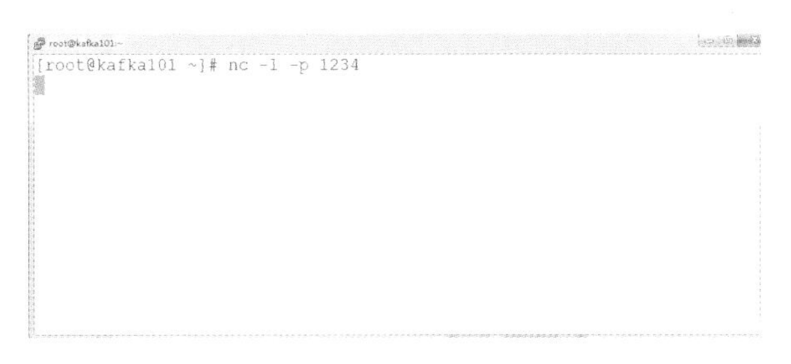


图 8.14　启动 netcat 工具

其中，

- -l 表示 netcat 的监听模式，用于客户端接入连接。这里的客户端其实就是 Flink 的 DataStream，它将从 netcat 中接收数据。
- -p 表示 netcat 运行的端口。

（2）进入 Flink 的安装目录，执行 SocketWindowWordCount 示例程序，如图 8.15 所示。

```
cd ~/training/flink-1.11.0/
bin/flink run examples/streaming/SocketWindowWordCount.jar --port 1234
```

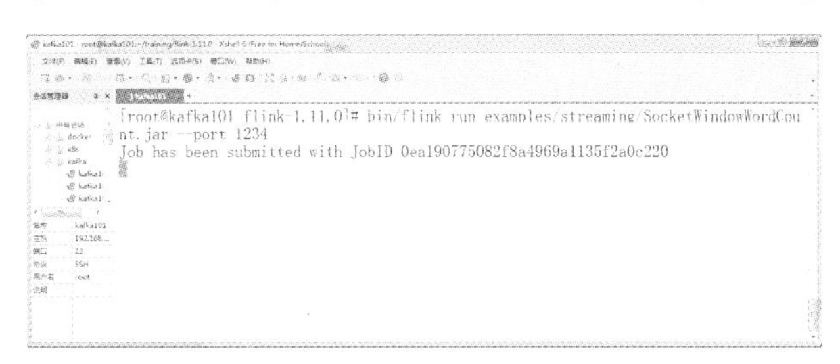


图 8.15　执行 Flink 的 SocketWindowWordCount 程序

需要注意的是，在流式计算中，只要任务不被人为终止，任务程序将永远在这里运行。这一点也可以通过 Flink 的 Web Console 确定，如图 8.16 所示。

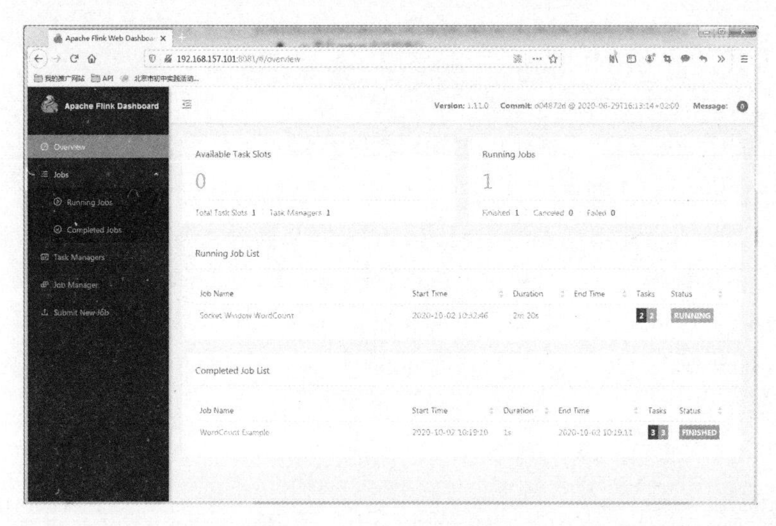


图 8.16　通过 Flink Web Console 监控 Flink 任务

从 Web Console 的界面上可以看到有两个任务的信息：一个是刚才执行的离线计算任务，它已经处于 FINISHED 状态；另一个是正在运行的实时流式计算任务，它一直处于 RUNNING 状态。

（3）在第一个命令行终端 netcat 中输入一个测试的字符串，并按 Enter 键，如图 8.17 所示。

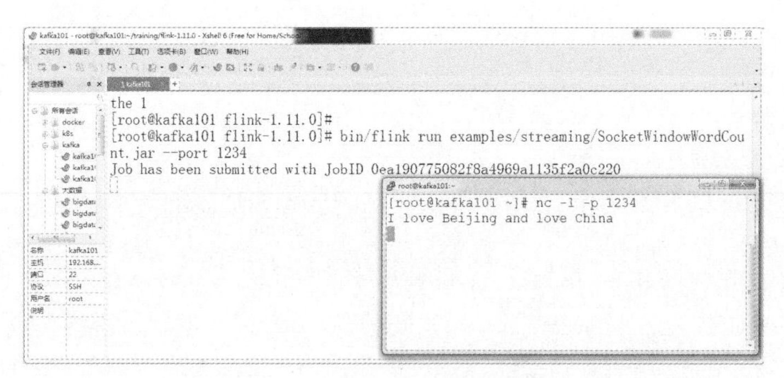


图 8.17　在 netcat 中输入测试数据

观察后面的 Flink 窗口似乎没有发生变化。这里的 Flink 其实已经完成了计算，并把计算结果写入日志文件中。

（4）进入 Flink 的日志目录，并使用 tail -f 命令查看 flink-root-taskexecutor-0-kafka101.out 文件，如图 8.18 所示。

```
cd training/flink-1.11.0/log/
ls
tail -f flink-root-taskexecutor-0-kafka101.out
```

```
[root@kafka101 log]# cd ~
[root@kafka101 ~]#
[root@kafka101 ~]# cd training/flink-1.11.0/log/
[root@kafka101 log]# ls
flink-root-client-kafka101.log                   flink-root-taskexecutor-0-kafka101.log
flink-root-standalonesession-0-kafka101.log      flink-root-taskexecutor-0-kafka101.log.1
flink-root-standalonesession-0-kafka101.log.1    flink-root-taskexecutor-0-kafka101.log.2
flink-root-standalonesession-0-kafka101.log.2    flink-root-taskexecutor-0-kafka101.out
flink-root-standalonesession-0-kafka101.out      flink-root-taskexecutor-0-kafka101.out.1
flink-root-standalonesession-0-kafka101.out.1    flink-root-taskexecutor-0-kafka101.out.2
flink-root-standalonesession-0-kafka101.out.2
[root@kafka101 log]# tail -f flink-root-taskexecutor-0-kafka101.out
I : 1
China : 1
and : 1
Beijing : 1
love : 2
```

图 8.18　查看 Flink 日志文件中的结果

从图 8.18 中可以看到统计出来的结果，每个单词的出现频率被统计了出来。例如，I 出现了一次；love 出现了两次，等等。

（5）试着再在 netcat 窗口中输入一个数据，并按 Enter 键，再观察 Flink 日志的信息，如图 8.19 所示。

```
[root@kafka101 log]# ls
flink-root-client-kafka101.log                   flink-root-taskexecutor-0-kafka101.log
flink-root-standalonesession-0-kafka101.log      flink-root-taskexecutor-0-kafka101.log.1
flink-root-standalonesession-0-kafka101.log.1    flink-root-taskexecutor-0-kafka101.log.2
flink-root-standalonesession-0-kafka101.log.2    flink-root-taskexecutor-0-kafka101.out
flink-root-standalonesession-0-kafka101.out      flink-root-taskexecutor-0-kafka101.out.1
flink-root-standalonesession-0-kafka101.out.1    flink-root-taskexecutor-0-kafka101.out.2
flink-root-standalonesession-0-kafka101.out.2
[root@kafka101 log]# tail -f flink-root-taskexecutor-0-kafka101.out
I : 1
China : 1
and : 1
Beijing : 1
love : 2
Hello : 2
Flink : 1
World : 1
```

```
[root@kafka101 ~]# nc -l -p 1234
I love Beijing and love China
Hello World Hello Flink
```

图 8.19　Flink 流式计算的输出结果

# 8.3　Flink DataSet API 算子

Flink 提供 DataSet API 用户处理批量数据。Flink 先将接入数据转换成 DataSet 数据集，并行分布在集群的每个节点上；然后将 DataSet 数据集进行各种转换操作（map、filter 等），最后通过 DataSink 操作将结果数据集输出到外部系统。表 8.1 列举了 Flink DataSet API 中常见的算子，以及它们的功能特性。

表 8.1 Flink DataSet 中的算子

| 算子 | 功能说明 |
| --- | --- |
| Map | Map 操作为 DataSet 提供了一个用户自定义的 Map 函数，它实现了一对一的映射，并且该函数只能返回一个元素。这里可以实现 MapFunction 接口或 RichMapFunction 定义用户自己的逻辑 |
| FlatMap | FlatMap 操作可以为 DataSet 的每一个元素定义一个用户自定义的 FlatMap 函数，它和 Map 类似，不同的是它的返回结果可以是任意个（包括 0 个）。可以实现 FlatMapFunction 接口或 RichFlatMapFucntion 接口 |
| MapPartition | MapPartition 操作可以使用单个函数调用处理一个并行的分区，它可以自定义 MapPartitionFunction 获得每一个分区的 Iterable 集合，然后通过处理输出任意个数的结果值，每一个分区操作的元素依赖于它的并行度和之前的操作。 |
| Filter | Filter 操作可以让用户使用自定义的 FilterFunction 函数实现自定义逻辑，最后保留该函数中返回 true 的元素 |
| Distinct | 依据元素的所有字段或字段的子集，删除 DataSet 中所有重复的数据 |
| Join | Join 操作和 DataStream 中的操作类似，它可以分为 cross join、innner join、outer join 等几大类 |
| First-N | 无论是批处理还是流处理，该操作都非常有用，可以根据自己实现的一些逻辑获取前 *N* 个需要的元素 |
| Outer-Join | 外连接操作，类似 SQL 语句汇总的外连接操作 |

在了解上面的算子及其功能特性后，我们通过具体的代码示例展示它们的功能。这里主要使用 Java 语言开发的示例代码，当然也可以使用 Scala 开发 Flink 程序，需要把下面的依赖加入工程中。

```
01  <dependency>
02      <groupId>org.apache.flink</groupId>
03      <artifactId>flink-streaming-java_2.11</artifactId>
04      <version>1.11.0</version>
05      <scope>provided</scope>
06  </dependency>
07
08  <dependency>
09      <groupId>org.apache.flink</groupId>
10      <artifactId>flink-clients_2.11</artifactId>
11      <version>1.11.0</version>
12  </dependency>
```

### 1. Map 算子

```
01  //创建 Flink 的运行环境
```

```
02    ExecutionEnvironment env = ExecutionEnvironment.getExecutionEnvironment();
03
04    //创建测试数据
05    ArrayList<String> data = new ArrayList<String>();
06    data.add("I love Beijing");
07    data.add("I love China");
08    data.add("Beijing is the capital of China");
09    DataSource<String> text = env.fromCollection(data);
10
11          //执行 Map 算子操作
12          DataSet<List<String>>  mapData  =  text.map(new  MapFunction<String,
List<String>>() {
13
14        public List<String> map(String data) throws Exception {
15            String[] words = data.split(" ");
16
17            //创建一个 List
18            List<String> result = new ArrayList<String>();
19            for(String w:words){
20                result.add(w);
21            }
22            return result;
23        }
24    });
25
26    //输出结果
27    mapData.print();
```

程序运行的结果输出如图 8.20 所示。

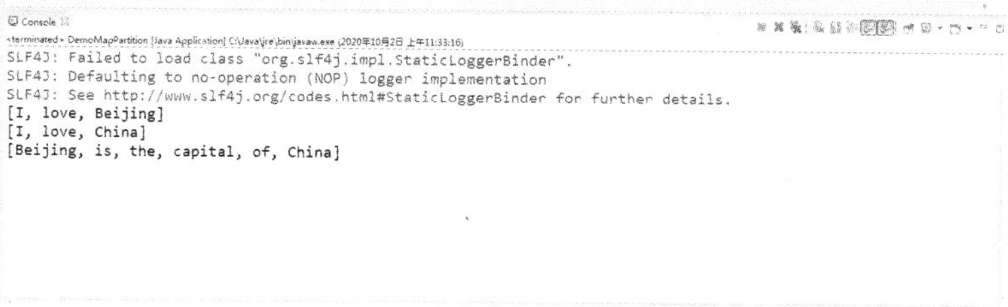

```
Console
<terminated> DemoMapPartition [Java Application] C:\Java\jre\bin\javaw.exe (2020年10月2日 上午11:33:16)
SLF4J: Failed to load class "org.slf4j.impl.StaticLoggerBinder".
SLF4J: Defaulting to no-operation (NOP) logger implementation
SLF4J: See http://www.slf4j.org/codes.html#StaticLoggerBinder for further details.
[I, love, Beijing]
[I, love, China]
[Beijing, is, the, capital, of, China]
```

图 8.20　Map 算子的运行结果

## 2. FlatMap 算子

```
01      DataSet<String>  flatMapData  =  text.flatMap(new  FlatMapFunction<String,
String>() {
02
03       public  void  flatMap(String  data,  Collector<String>  collection)  throws
Exception {
```

```
04          String[] words = data.split(" ");
05          for(String w:words){
06              collection.collect(w);
07          }
08      }
09  });
10  flatMapData.print();
```

程序运行的结果如图 8.21 所示。

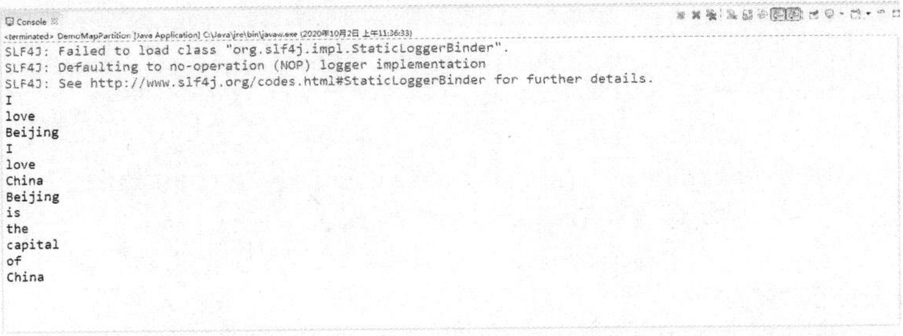

```
SLF4J: Failed to load class "org.slf4j.impl.StaticLoggerBinder".
SLF4J: Defaulting to no-operation (NOP) logger implementation
SLF4J: See http://www.slf4j.org/codes.html#StaticLoggerBinder for further details.
I
love
Beijing
I
love
China
Beijing
is
the
capital
of
China
```

图 8.21　FlatMap 算子的运行结果

### 3. MapPartition 算子

```
01      DataSet<String> flatMapData = text.flatMap(new FlatMapFunction<String,
String>() {
02
03      public void flatMap(String data, Collector<String> collection) throws
Exception {
04          String[] words = data.split(" ");
05          for(String w:words){
06              collection.collect(w);
07          }
08      }
09  });
10  flatMapData.print();
```

程序运行的结果如图 8.22 所示。

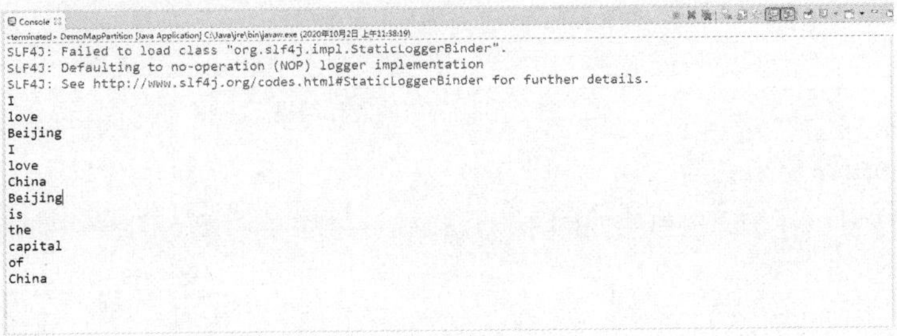

```
SLF4J: Failed to load class "org.slf4j.impl.StaticLoggerBinder".
SLF4J: Defaulting to no-operation (NOP) logger implementation
SLF4J: See http://www.slf4j.org/codes.html#StaticLoggerBinder for further details.
I
love
Beijing
I
love
China
Beijing
is
the
capital
of
China
```

图 8.22　MapPartition 算子的运行结果

### 4. Filter 算子

```
01      DataSet<String> flatMapData = text.flatMap(new FlatMapFunction<String,
String>() {
02
03       public void flatMap(String data, Collector<String> collection) throws
Exception {
04           String[] words = data.split(" ");
05           for(String w:words){
06               collection.collect(w);
07           }
08       }
09   });
10
11   //选出长度大于 3 的单词
12   flatMapData.filter(new FilterFunction<String>() {
13
14       @Override
15       public boolean filter(String value) throws Exception {
16           int length = value.length();
17           return length>3?true:false;
18       }
19   }).print();
```

程序运行的结果如图 8.23 所示。

```
Console
<terminated> DemoMapPartition [Java Application] C:\Java\jre\bin\javaw.exe (2020年10月2日 上午11:44:00)
SLF4J: Failed to load class "org.slf4j.impl.StaticLoggerBinder".
SLF4J: Defaulting to no-operation (NOP) logger implementation
SLF4J: See http://www.slf4j.org/codes.html#StaticLoggerBinder for further details.
love
Beijing
love
China
Beijing
capital
China
```

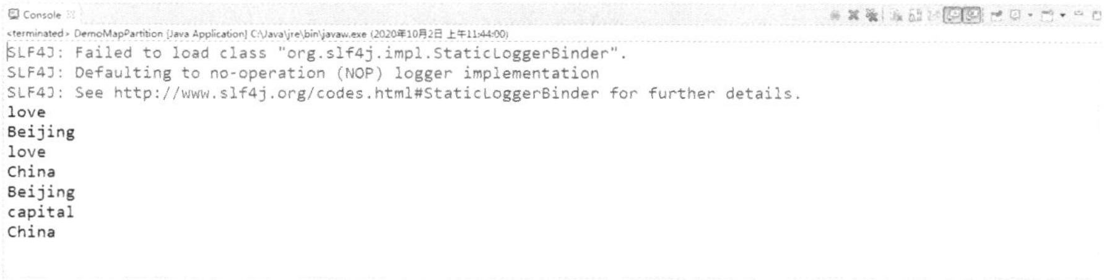

图 8.23　Filter 算子的运行结果

### 5. Distinct 算子

```
01      DataSet<String> flatMapData = text.flatMap(new FlatMapFunction<String,
String>() {
02
03       public void flatMap(String data, Collector<String> collection) throws
Exception {
04           String[] words = data.split(" ");
05           for(String w:words){
06               collection.collect(w);
07           }
```

```
08        }
09    });
10
11    //去掉重复的单词
12    flatMapData.distinct().print();
```

程序运行的结果如图 8.24 所示。

```
Console
<terminated> DemoMapPartition [Java Application] C:\Java\jre\bin\javaw.exe (2020年10月2日 上午11:44:47)
SLF4J: Defaulting to no-operation (NOP) logger implementation
SLF4J: See http://www.slf4j.org/codes.html#StaticLoggerBinder for further details.
China
is
of
Beijing
I
love
the
capital
```

图 8.24　Distinct 算子的运行结果

### 6. Join 算子

```
01    import java.util.ArrayList;
02    import org.apache.flink.api.common.functions.JoinFunction;
03    import org.apache.flink.api.java.DataSet;
04    import org.apache.flink.api.java.ExecutionEnvironment;
05    import org.apache.flink.api.java.tuple.Tuple2;
06    import org.apache.flink.api.java.tuple.Tuple3;
07
08    public class DemoJoin {
09        public static void main(String[] args) throws Exception {
10            //获取运行的环境
11            ExecutionEnvironment env = ExecutionEnvironment.
getExecutionEnvironment();
12
13            //创建第一张表：用户 ID  姓名
14            ArrayList<Tuple2<Integer, String>> data1 =
15                    new ArrayList<Tuple2<Integer,String>>();
16            data1.add(new Tuple2(1,"Tom"));
17            data1.add(new Tuple2(2,"Mike"));
18            data1.add(new Tuple2(3,"Mary"));
19            data1.add(new Tuple2(4,"Jone"));
20
21            //创建第二张表：用户 ID  所在的城市
22            ArrayList<Tuple2<Integer, String>> data2 =
23                    new ArrayList<Tuple2<Integer,String>>();
24            data2.add(new Tuple2(1,"北京"));
25            data2.add(new Tuple2(2,"上海"));
26            data2.add(new Tuple2(3,"广州"));
```

```
27              data2.add(new Tuple2(4,"重庆"));
28
29              //实现join的多表查询：用户ID  姓名  所在的程序
30              DataSet<Tuple2<Integer, String>> table1 = env.fromCollection(data1);
31              DataSet<Tuple2<Integer, String>> table2 = env.fromCollection(data2);
32
33              table1.join(table2).where(0).equalTo(0)
34             /*第一个Tuple2<Integer,String>：表示第一张表
35              *第二个Tuple2<Integer,String>：表示第二张表
36              * Tuple3<Integer,String, String>：多表join连接查询后的返回结果  */

37            .with(new JoinFunction<Tuple2<Integer,String>,
38                                Tuple2<Integer,String>,
39                                    Tuple3<Integer,String, String>>() {
40                     public  Tuple3<Integer,  String,  String>  join(Tuple2<Integer,
String> table1,
41                             Tuple2<Integer, String> table2) throws Exception {
42                         return          new          Tuple3<Integer,          String,
String>(table1.f0,table1.f1,table2.f1);
43                     }
44                 })
45            .print();
46        }
47    }
```

程序运行的结果如图 8.25 所示。

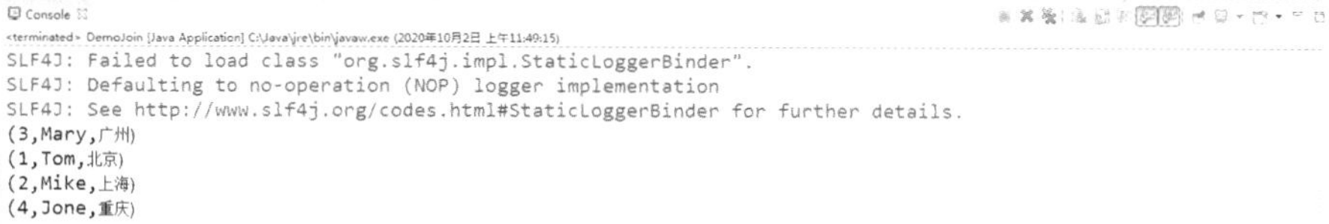

```
Console
<terminated> DemoJoin [Java Application] C:\Java\jre\bin\javaw.exe (2020年10月2日 上午11:49:15)
SLF4J: Failed to load class "org.slf4j.impl.StaticLoggerBinder".
SLF4J: Defaulting to no-operation (NOP) logger implementation
SLF4J: See http://www.slf4j.org/codes.html#StaticLoggerBinder for further details.
(3,Mary,广州)
(1,Tom,北京)
(2,Mike,上海)
(4,Jone,重庆)
```

图 8.25　Join 算子的运行结果

### 7．First-N 算子

```
01  import org.apache.flink.api.common.operators.Order;
02  import org.apache.flink.api.java.DataSet;
03  import org.apache.flink.api.java.ExecutionEnvironment;
04  import org.apache.flink.api.java.tuple.Tuple3;
05
06  public class DemoFirstN {
07
08      public static void main(String[] args) throws Exception {
```

```
09          ExecutionEnvironment env = ExecutionEnvironment.
getExecutionEnvironment();
10
11          //这里的数据是：员工姓名、薪水、部门号
12          DataSet<Tuple3<String, Integer,Integer>> grade =
13                 env.fromElements(
14                 new Tuple3<String, Integer,Integer>("Tom",1000,10),
15                 new Tuple3<String, Integer,Integer>("Mary",1500,20),
16                 new Tuple3<String, Integer,Integer>("Mike",1200,30),
17                 new Tuple3<String, Integer,Integer>("Jerry",2000,10));
18
19          //按照插入顺序取前三条记录
20          grade.first(3).print();
21          System.out.println("**********************");
22
23          //先按照部门号排序，再按照薪水排序
24          grade.sortPartition(2, Order.ASCENDING).sortPartition(1, Order.
ASCENDING)
25              .print();
26          System.out.println("**********************");
27
28          //按照部门号分组，求每组的第一条记录
29          grade.groupBy(2).first(1).print();
30      }
31  }
```

程序运行的结果输出如图 8.26 所示。

```
Console
<terminated> DemoFirstN [Java Application] C:\Java\jre\bin\javaw.exe (2020年10月2日 上午11:53:00)
SLF4J: Failed to load class "org.slf4j.impl.StaticLoggerBinder".
SLF4J: Defaulting to no-operation (NOP) logger implementation
SLF4J: See http://www.slf4j.org/codes.html#StaticLoggerBinder for further details.
(Tom,1000,10)
(Mary,1500,20)
(Mike,1200,30)
**********************
(Tom,1000,10)
(Jerry,2000,10)
(Mary,1500,20)
(Mike,1200,30)
**********************
(Mike,1200,30)
(Tom,1000,10)
(Mary,1500,20)
```

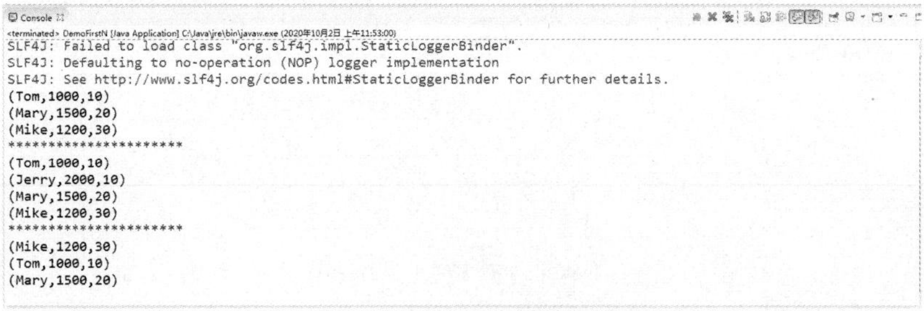

图 8.26　First-N 算子的运行输出

## 8. Outer-Join 算子

```
01  import java.util.ArrayList;
02  import org.apache.flink.api.common.functions.JoinFunction;
03  import org.apache.flink.api.java.DataSet;
04  import org.apache.flink.api.java.ExecutionEnvironment;
05  import org.apache.flink.api.java.tuple.Tuple2;
```

```
06  import org.apache.flink.api.java.tuple.Tuple3;
07
08  public class DemoOutJoin {
09      public static void main(String[] args) throws Exception {
10          ExecutionEnvironment env = ExecutionEnvironment.
getExecutionEnvironment();
11
12          //创建第一张表：用户 ID  姓名
13          ArrayList<Tuple2<Integer, String>> data1 = new ArrayList<Tuple2<
Integer,String>>();
14          data1.add(new Tuple2(1,"Tom"));
15          data1.add(new Tuple2(3,"Mary"));
16          data1.add(new Tuple2(4,"Jone"));
17
18          //创建第二张表：用户 ID  所在的城市
19          ArrayList<Tuple2<Integer, String>> data2 = new ArrayList<Tuple2<
Integer,String>>();
20          data2.add(new Tuple2(1,"北京"));
21          data2.add(new Tuple2(2,"上海"));
22          data2.add(new Tuple2(4,"重庆"));
23
24          //实现join的多表查询：用户 ID  姓名  所在的程序
25          DataSet<Tuple2<Integer, String>> table1 = env.fromCollection(data1);
26          DataSet<Tuple2<Integer, String>> table2 = env.fromCollection(data2);
27
28          //左外连接
29          table1.leftOuterJoin(table2).where(0).equalTo(0)
30               .with(new JoinFunction<Tuple2<Integer,String>,
31                                   Tuple2<Integer,String>,
32                                    Tuple3<Integer,String,String>>() {
33
34          public Tuple3<Integer, String, String> join(Tuple2<Integer, String>
table1,
35              Tuple2<Integer, String> table2) throws Exception {
36              //左外连接表示等号左边的信息会被包含
37              if(table2 == null){
38                  return new Tuple3<Integer, String, String>(table1.f0,table1.
f1,null);
39              }else{
40                  return new Tuple3<Integer, String, String>(table1.f0,table1.
f1,table2.f1);
41              }
42          }
43              }).print();
```

```
44          System.out.println("**********************************");
45
46          //右外连接
47          table1.rightOuterJoin(table2).where(0).equalTo(0)
48              .with(new JoinFunction<Tuple2<Integer,String>,
49                              Tuple2<Integer,String>,
50                               Tuple3<Integer,String,String>>() {
51
52          public Tuple3<Integer, String, String> join(Tuple2<Integer, String>
table1,
53                 Tuple2<Integer, String> table2) throws Exception {
54              //右外连接表示等号右边的信息会被包含
55              if(table1 == null){
56                 return new Tuple3<Integer, String, String>(table2.f0,null,
table2.f1);
57              }else{
58                return new Tuple3<Integer, String, String>(table2.f0,table1.
f1,table2.f1);
59              }
60          }
61              }).print();
62          System.out.println("**********************************");
63
64          //全外连接
65          table1.fullOuterJoin(table2).where(0).equalTo(0)
66          .with(new JoinFunction<Tuple2<Integer,String>,
67                          Tuple2<Integer,String>,
68                              Tuple3<Integer,String,String>>() {
69
70          public Tuple3<Integer, String, String> join(Tuple2<Integer, String>
table1,
71                                        Tuple2<Integer, String> table2)
72                 throws Exception {
73              if(table1 == null){
74                 return new Tuple3<Integer, String, String>(table2.f0,null,
table2.f1);
75              }else if(table2 == null){
76                return new Tuple3<Integer, String, String>(table1.f0,table1.
f1,null);
77              }else{
78                return new Tuple3<Integer, String, String>(table1.f0,table1.
f1,table2.f1);
79              }
80          }
```

```
81              }).print();
82          }
83    }
```

程序运行的结果如图 8.27 所示。

```
Console
<terminated> DemoOutJoin (1) [Java Application] C:\Java\jre\bin\javaw.exe (2020年10月2日 上午11:57:37)
SLF4J: Defaulting to no-operation (NOP) logger implementation
SLF4J: See http://www.slf4j.org/codes.html#StaticLoggerBinder for further details.
(3,Mary,null)
(1,Tom,北京)
(4,Jone,重庆)
**********************************
(1,Tom,北京)
(2,null,上海)
(4,Jone,重庆)
**********************************
(3,Mary,null)
(1,Tom,北京)
(2,null,上海)
(4,Jone,重庆)
```

图 8.27　外连接算了的运行结果

## 8.4　Flink DataStream API 算子

DataStream 是 Flink 中可以在数据流的基础上实现各种 Transformation 操作的程序，例如，filtering、updating state、defining windows、aggregating 等。这些数据流最初的来源可以有很多种，比如消息队列、socket 流、文件等，计算的结果通过 sinks 途径返回，也可以将这些数据写到一个文件或标准的输出系统中，例如，命令行控制台。

Flink 的 DataStream 由以下三部分组成。

（1）DataSources 数据源。source 是程序的数据源输入，可以通过 StreamExecutionEnvironment.addSource()为程序添加一个 source。Flink 提供了大量已经实现好的 source 方法，可以自定义 source 通过实现 sourceFunction 接口来自定义无并行度的 source，也可以通过实现 ParallelSourceFunction 接口，或者继承 RichParallelSource-Function 自定义有并行度的 source。

在后面的章节中，我们使用 Kafka 作为 Flink DataStream 的 source，将 Kafka 转发来的消息由 Flink DataStream 进行处理。

（2）DataStream Transformation 转换操作。表 8.2 列举了一些 Flink DataStream 中常见的 Transformation 的转换操作。稍后我们会通过具体的代码展示其中部分算子的使用方式。

表 8.2　常见的 Transformation 的转换操作

| 操作 | 说明 |
| --- | --- |
| map | 输入一个元素，然后返回一个元素，中间可以进行一些清洗转换等操作 |
| flatmap | 输入一个元素，可以返回零个、一个或多个元素 |

续表

| 操作 | 说明 |
| --- | --- |
| filter | 过滤函数，对传入的数据进行判断，符合条件的数据会被留下 |
| keyBy | 根据指定的 key 进行分组，相同 key 的数据会进入同一个分区 |
| reduce | 对数据进行聚合操作，结合当前元素和上一次 reduce 返回的值进行聚合操作，然后返回一个新的值 |
| aggregations | sum()、min()、max()等 |
| union | 合并多个流，新的流会包含所有流中的数据，但是 union 是一个限制，就是所有合并的流类型必须是一致的 |
| connect | 和 union 类似，但是只能连接两个流，两个流的数据类型可以不同，对两个流中的数据应用不同的处理方法 |
| coMap, coFlatMap | 在 ConnectedStreams 中需要使用这种函数，类似于 map 和 flatmap |
| split | 根据规则把一个数据流切分为多个流 |
| select | 和 split 配合使用，选择切分后的流 |

（3）Data Sinks 将结果输出。当 Flink DataStream 完成了数据流的处理后，可以使用 Sink 组件将结果进行保存。这些 Sink（包括前面的 Source）在 Flink 中都是以 Connector 方式提供出来的。下面列举了一些常见的 Connector。

- Apache Kafka (source/sink)。
- Apache Cassandra (sink)。
- Elasticsearch (sink)。
- Hadoop FileSystem (sink)。
- RabbitMQ (source/sink)。
- Apache ActiveMQ (source/sink)。
- Redis (sink)。

在了解了 Flink DataStream API 中的算子后，我们通过以下具体的示例代码来展示它们的功能特性。

### 1. 创建一个单并行度的数据源 source

```
01  import org.apache.flink.streaming.api.functions.source.SourceFunction;
02
03  //自定义实现并行度为 1 的 source
04  //注意，SourceFunction 和 SourceContext 都需要指定数据类型
05  //如果不指定，代码运行的时候会报错
06  public class MyNoParalleSource implements SourceFunction<Long>{
07      private long count = 1;
08      private boolean isRunning = true;
09
10      //主要的方法，启动一个 source
```

```
11      public void run(SourceContext<Long> ctx) throws Exception {
12         while(isRunning){
13            ctx.collect(count);
14            count++;
15            //每秒产生一条数据
16            Thread.sleep(1000);
17         }
18      }
19
20      //取消一个 cancel 的时候会调用的方法
21      public void cancel() {
22         isRunning = false;
23      }
24   }
```

为了测试上面的数据源，通过下面的测试代码进行测试。

```
01   import org.apache.flink.api.common.functions.MapFunction;
02   import org.apache.flink.streaming.api.datastream.DataStream;
03   import org.apache.flink.streaming.api.datastream.DataStreamSource;
04                                                                    import
org.apache.flink.streaming.api.environment.StreamExecutionEnvironment;
05   import org.apache.flink.streaming.api.windowing.time.Time;
06
07   public class MyNoParalleSourceMain {
08
09       public static void main(String[] args) throws Exception {
10           StreamExecutionEnvironment env =
11               StreamExecutionEnvironment.getExecutionEnvironment();
12
13           //注意，针对此 source，并行度为 1
14          DataStreamSource<Long> source = env.addSource(new MyNoParalleSource());
15
16           DataStream<Long> data = source.map(new MapFunction<Long, Long>() {
17
18               public Long map(Long value) throws Exception {
19                   System.out.println("接收到的数据是: " + value);
20                   return value;
21               }
22           });
23
24           //每 2 秒处理一次数据：求和
25           DataStream<Long> result = data.timeWindowAll(Time.seconds(2)).sum(0);
26           result.print().setParallelism(1);
27
28           env.execute("MyNoParalleSourceMain");
```

```
29      }
30  }
```

注意，单并行度的数据源其并行度只能为 1。

## 2. union

这个算子可以连接多个流，但是流中的数据类型必须一致。

```
01  import org.apache.flink.api.common.functions.MapFunction;
02  import org.apache.flink.streaming.api.datastream.DataStream;
03  import org.apache.flink.streaming.api.datastream.DataStreamSource;
04                                                                  import
org.apache.flink.streaming.api.environment.StreamExecutionEnvironment;
05
06  public class MyUnionDemo {
07
08      public static void main(String[] args) throws Exception {
09          StreamExecutionEnvironment env = StreamExecutionEnvironment
10                                  .getExecutionEnvironment();
11
12          //注意，这里的 source1 是 Long 类型，source2 是 Long 类型
13      DataStreamSource<Long> source1 = env.addSource(new MyNoParalleSource());
14          DataStream<Long> source2 = env.addSource(
15             new MyNoParalleSource()).map(new MapFunction<Long, Long>() {
16               public Long map(Long value) throws Exception {
17                   //将第二个流中的数据*10
18                   return value*10;
19               }
20          });
21
22          //执行 union 运算
23          source1.union(source2).print();
24
25          env.execute();
26      }
27  }
```

## 3. connect

这个算子可以连接两个流，并且流的数据类型可以不一样，可以针对两个流中的数据分别进行处理。程序代码中使用了之前创建的单并行度的数据源。

```
01  import org.apache.flink.api.common.functions.MapFunction;
02  import org.apache.flink.streaming.api.datastream.ConnectedStreams;
03  import org.apache.flink.streaming.api.datastream.DataStream;
04  import org.apache.flink.streaming.api.datastream.DataStreamSource;
05   import org.apache.flink.streaming.api.environment.StreamExecutionEnvironment;
06  import org.apache.flink.streaming.api.functions.co.CoMapFunction;
```

```
07
08  public class MyConnectDemo {
09      public static void main(String[] args) throws Exception {
10          StreamExecutionEnvironment env = StreamExecutionEnvironment
11                                      .getExecutionEnvironment();
12
13      //注意，这里的 source1 是 Long 类型，source2 是 String 类型
14      DataStreamSource<Long> source1 = env.addSource(new MyNoParalleSource());
15          DataStream<String> source2 = env.addSource(
16              new MyNoParalleSource()).map(new MapFunction<Long, String>() {
17               public String map(Long value) throws Exception {
18                   //将其转换为 String 类型
19                   return "String " + value;
20               }
21          });
22
23          ConnectedStreams<Long, String> stream = source1.connect(source2);
24          //这里可以使用 CoMapFunction 对两个流中不同的数据类型分别进行处理
25          DataStream<Object> result = stream.map(
26              new CoMapFunction<Long, String, Object>() {
27               public Object map1(Long value) throws Exception {
28                   return "对 Long 类型的数据进行处理:" +value;
29               }
30
31               public Object map2(String value) throws Exception {
32                   return "对 String 类型的数据进行处理:" + value;
33               }
34          });
35
36          result.print().setParallelism(1);
37          env.execute("MyConnectDemo");
38      }
39  }
```

4．split

这个算子可以连接多个流，但是流中的数据类型必须一致，它可以根据规则把一个数据流切分成多个流。在实际的工作中，源数据流中可能混合了多种类型的数据，多种类型的数据处理规则不一样，所以就可以根据一定的规则，把一个数据流切分成多个数据流，这样每个数据流就可以使用不同的处理逻辑了。

```
01  import java.util.ArrayList;
02  import org.apache.flink.streaming.api.collector.selector.OutputSelector;
03  import org.apache.flink.streaming.api.datastream.DataStream;
04  import org.apache.flink.streaming.api.datastream.DataStreamSource;
```

```
05  import org.apache.flink.streaming.api.datastream.SplitStream;
06                                                                         import
org.apache.flink.streaming.api.environment.StreamExecutionEnvironment;
07
08  public class MySplitDemo {
09
10      public static void main(String[] args) throws Exception {
11          StreamExecutionEnvironment env = StreamExecutionEnvironment
12                                          .getExecutionEnvironment();
13          DataStreamSource<Long>      source      =      env.addSource(new
MyNoParalleSource());
14
15          //现在我们想把奇数和偶数分成两个流，分别对其进行处理
16          SplitStream<Long> stream = source.split(new OutputSelector<Long>()
{
17              public Iterable<String> select(Long value) {
18                  ArrayList<String> selector = new ArrayList<String>();
19
20                  //返回值就是将其分成几个流
21                  if(value%2 == 0){
22                      selector.add("even");//偶数
23                  }else{
24                      selector.add("odd"); //奇数
25                  }
26                  return selector;
27              }
28          });
29
30          DataStream<Long> stream1 =  stream.select("even");
31          stream1.print().setParallelism(1);
32
33          env.execute("MySplitDemo");
34      }
35  }
```

上面的代码使用了之前创建的单并行度的数据源，然后根据奇数和偶数对数据流进行切分，并在程序的最后选择了所有偶数流。

## 8.5 集成 Flink 与 Kafka

Apache Flink 是新一代的分布式流式数据处理框架，它的统一处理引擎既可以处理批数据（batch data），又可以处理流式数据（streaming data）。在实际场景中，Flink 利用 Apache Kafka 作为上下游的输入输出十分常见，本节内容将给出一个可运行的实际案例

来集成 Flink 与 Kafka。

在前面的内容中曾经提到，Kafka 可以作为 Connector 连接到 Flink 中，并且 Kafka 既可以作为 Flink 的 Source Connector，又可以作为 Flink 的 Sink Connector。为了在 Flink 中集成 Kafka，需要添加如下依赖信息。

```
01  <dependency>
02      <groupId>org.apache.flink</groupId>
03      <artifactId>flink-connector-kafka_2.11</artifactId>
04      <version>1.11.2</version>
05  </dependency>
```

## 8.5.1 将 Kafka 作为 Flink 的 Source Connector

在 Java 工程中添加了相应的依赖信息后，开发如下的 Java 代码来完成 Flink 和 Kafka 的集成，这时 Kafka 将作为 Flink 的 Source Connector 使用。当 Kafka 接收到消息生产者产生的消息后，将会转发给 Flink；从另一个角度看，这时候的 Flink 将作为 Kafka 的消息消费者使用。

```
01 import java.util.Properties;
02 
03 import org.apache.flink.streaming.api.environment.StreamExecutionEnvironment;
04 import org.apache.flink.streaming.connectors.kafka.FlinkKafkaConsumer;
05 import org.apache.flink.streaming.util.serialization.SimpleStringSchema;
06 
07 public class FlinkWithKafkaSource {
08 
09     public static void main(String[] args) throws Exception {
10         StreamExecutionEnvironment env = StreamExecutionEnvironment
11                                          .getExecutionEnvironment();
12 
13         Properties props = new Properties();
14 
15         //指定 Kafka 的 Broker 地址。如果有多个 Broker，可以使用逗号分隔
16         props.setProperty("bootstrap.servers", "kafka101:9092");
17         //指定 ZooKeeper 的地址。如果是 ZooKeeper 集群，可以使用逗号分隔
18         props.setProperty("zookeeper.connect", "kafka101:2181");
19         //指定 Kafka 的消费者组的信息
20         props.setProperty("group.id", "test");
21 
22         //创建 FlinkKafkaConsumer
23         FlinkKafkaConsumer<String> source =
24                 new FlinkKafkaConsumer<String>
25                ("mytopic1", new SimpleStringSchema(), props);
26 
```

```
27          //添加 Kafka Source 接收消息，并直接打印
28          env.addSource(source).print();
29
30          //执行任务
31          env.execute("FlinkWithKafkaSource");
32      }
33  }
```

对应的 Scala 代码如下所示。

```
01  import java.util.Properties
02  import org.apache.flink.streaming.connectors.kafka.FlinkKafkaConsumer
03  import org.apache.flink.api.common.serialization.SimpleStringSchema
04  import org.apache.flink.streaming.api.scala.StreamExecutionEnvironment
05  import org.apache.flink.streaming.api.scala._
06
07  object FlinkWithKafkaSource {
08    def main(args: Array[String]): Unit = {
09      val env: StreamExecutionEnvironment = StreamExecutionEnvironment
10                                      .getExecutionEnvironment
11
12      val properties = new Properties()
13      //指定 Kafka 的 Broker 地址。如果有多个 Broker，可以使用逗号分隔
14      properties.setProperty("bootstrap.servers", "kafka101:9092")
15      // only required for Kafka 0.8
16      //指定 ZooKeeper 的地址。如果是 ZooKeeper 集群，可以使用逗号分隔
17      properties.setProperty("zookeeper.connect", "kafka101:2181")
18      //指定 Kafka 的消费者组的信息
19      properties.setProperty("group.id", "test")
20
21      //创建 FlinkKafkaConsumer
22      val source = new FlinkKafkaConsumer[String]
23              ("mytopic1", new SimpleStringSchema(), properties)
24
25      //添加 Kafka Source 接收消息，并直接打印
26      env.addSource(source).print()
27
28      //执行任务
29      env.execute("FlinkWithKafka");
30    }
31  }
```

下面我们来进行一个简单的测试。

（1）启动 Kafka 的 kafka-console-producer.sh，作为消息的生产者来发送消息。

```
bin/kafka-console-producer.sh --broker-list kafka101:9092 --topic mytopic1
```

（2）在 IDE 环境中启动应用程序，如图 8.28 所示。

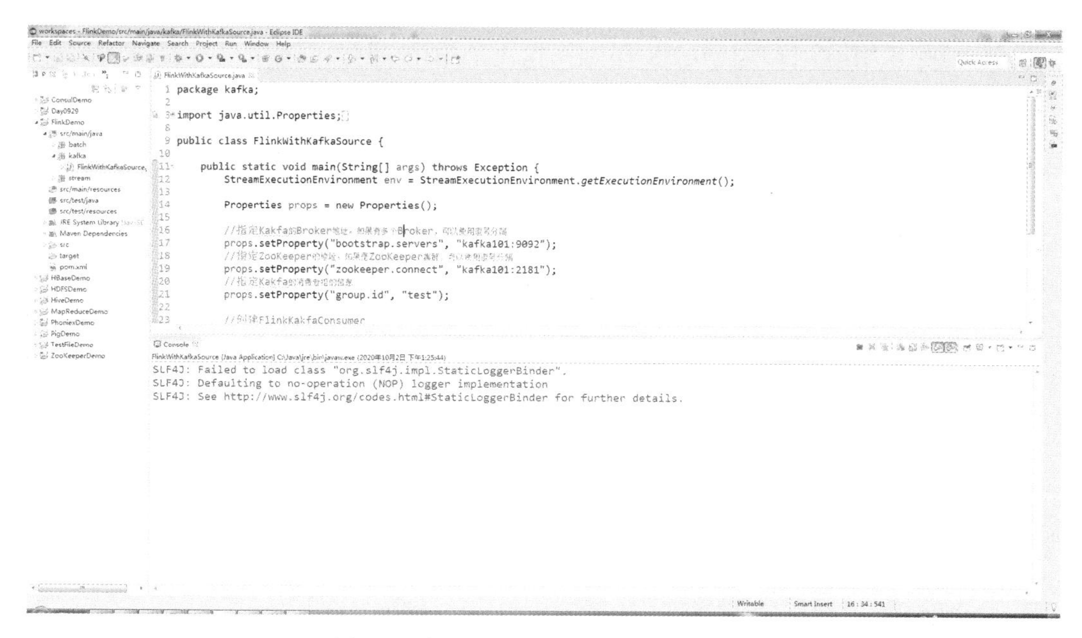


图 8.28　启动 Flink Source Connector 应用程序

（3）在 Kafka Producer 的命令行创建中输入“Hello World”，可以看到在 Flink 程序中成功接收生产者发送的消息，如图 8.29 所示。

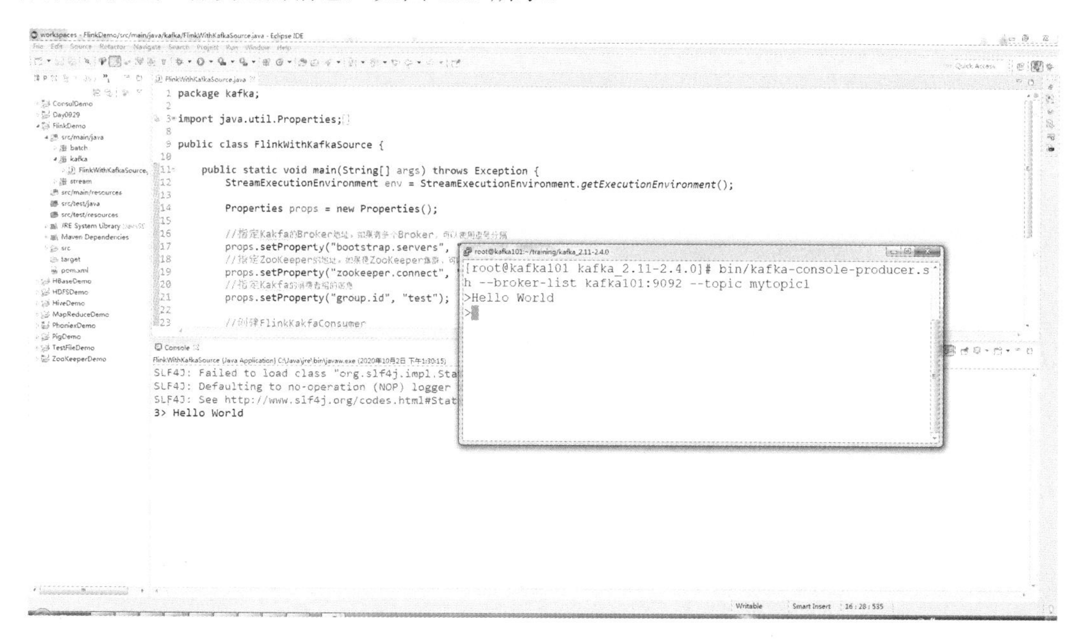


图 8.29　Flink Source Connector 接收 Kafka 的消息

### 8.5.2 将 Kafka 作为 Flink 的 Sink Connector

这时 Kafka 将作为 Flink 的 Sink Connector 使用。Flink 处理数据后，会通过 Kafka Sink Connector 输出到 Kafka。这时 Flink 将作为 Kafka 的消息生产者使用。

下面展示了相应的 Java 代码。在这段测试代码中，我们生成了一个 DataStream<String> 的数据流，并把里面的每个单词转换成大写，再输出到 Kafka 的 Sink 中。

```
01  import java.util.Properties;
02
03  import org.apache.flink.api.common.functions.MapFunction;
04  import org.apache.flink.streaming.api.datastream.DataStreamSource;
05  import org.apache.flink.streaming.api.environment.StreamExecutionEnvironment;
06  import org.apache.flink.streaming.connectors.kafka.FlinkKafkaProducer;
07  import org.apache.flink.streaming.util.serialization.SimpleStringSchema;
08
09  public class FlinkWithKafkaSink {
10
11      public static void main(String[] args) throws Exception {
12          Properties props = new Properties();
13
14          //指定 Kafka 的 Broker 地址。如果有多个 Broker，可以使用逗号分隔
15          props.setProperty("bootstrap.servers", "kafka101:9092");
16          //指定 ZooKeeper 的地址。如果是 ZooKeeper 集群，可以使用逗号分隔
17          props.setProperty("zookeeper.connect", "kafka101:2181");
18
19          //创建 Kafka Sink Connector
20          FlinkKafkaProducer<String> producer =
21                  new FlinkKafkaProducer<String>
22                  ("mytopic1", new SimpleStringSchema(), props);
23
24          StreamExecutionEnvironment env = StreamExecutionEnvironment
25                                          .getExecutionEnvironment();
26          //创建一个 DataStreamSource
27          DataStreamSource<String> dataStream =
28                  env.fromElements("Hello","World","Kafka");
29
30          //处理数据，并添加 Kafka Sink Connector
31          dataStream.map(new MapFunction<String, String>() {
32
33              @Override
34              public String map(String value) throws Exception {
35                  //将字符串转换成大写
36                  return value.toUpperCase();
37              }
```

```
38              }).addSink(producer);
39
40              env.execute("FlinkWithKafkaSink");
41          }
42      }
```

对应的 Scala 代码如下所示。

```
01
02  import org.apache.flink.api.common.serialization.SimpleStringSchema
03  import org.apache.flink.streaming.api.scala.StreamExecutionEnvironment
04  import org.apache.flink.streaming.api.scala._
05  import org.apache.flink.streaming.connectors.kafka.FlinkKafkaProducer
06
07  object FlinkWithKafkaSink {
08    def main(args: Array[String]): Unit = {
09      //创建 FlinkKafkaProducer
        //Kafka Broker 的地址
10      val sink = new FlinkKafkaProducer("kafka101:9092",
11                                  "mytopic1",        //topic 的信息
12                                  new SimpleStringSchema()) //数据序列化的方式
13
14      val env: StreamExecutionEnvironment = StreamExecutionEnvironment
15                                        .getExecutionEnvironment
16
17       //创建一个 DataStreamSource
18       val dataStream = env.fromElements("Hello","World","Kafka")
19
20       //将单词转换成大写，并输出到 Kafka Sink Connector 中
21       dataStream.map(_.toUpperCase()).addSink(sink)
22
23      //执行任务
24      env.execute("FlinkWithKafka");
25    }
26  }
```

下面我们来进行一个测试。

（1）首先启动 Kafka Consumer Console，执行下面的语句。

```
bin/kafka-console-consumer.sh --bootstrap-server kafka101:9092 --topic mytopic1
```

（2）在 IDE 环境中启动应用程序，并观察 Kafka Console 命令的输出，如图 8.30 所示，可以看到 Kafka 中成功接收 Flink Sink Connector 发送来的消息。

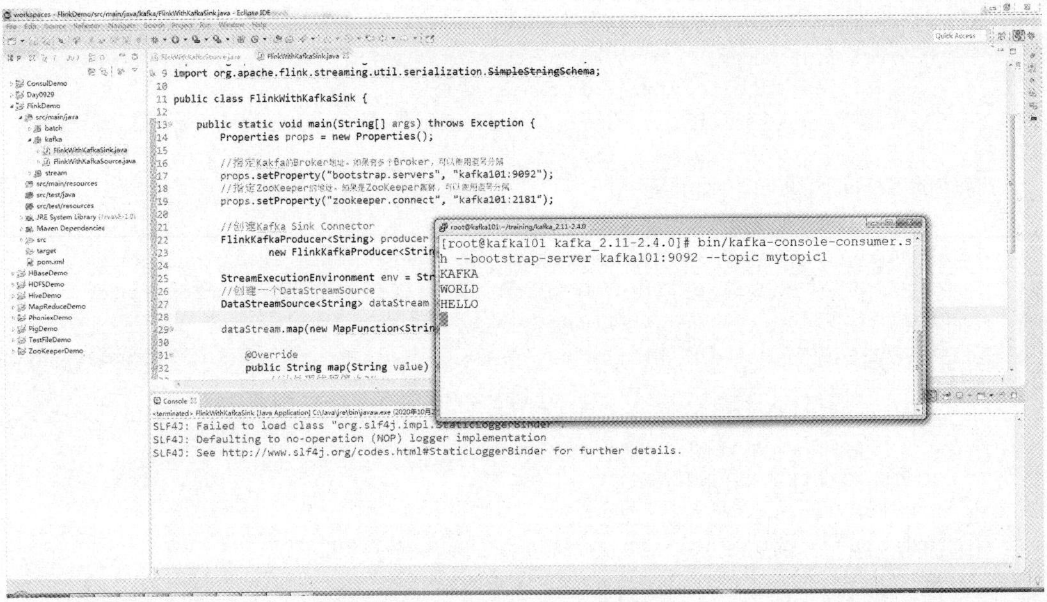

```
 9 import org.apache.flink.streaming.util.serialization.SimpleStringSchema;
10
11 public class FlinkWithKafkaSink {
12
13     public static void main(String[] args) throws Exception {
14         Properties props = new Properties();
15
16         //指定Kakfa的Broker地址，如果有多个Broker，可以使用逗号分隔
17         props.setProperty("bootstrap.servers", "kafka101:9092");
18         //指定ZooKeeper的地址，如果是ZooKeeper集群，可以使用逗号分隔
19         props.setProperty("zookeeper.connect", "kafka101:2181");
20
21         //创建Kafka Sink Connector
22         FlinkKafkaProducer<String> producer
23                 new FlinkKafkaProducer<Strin
24
25         StreamExecutionEnvironment env = Str
26         //创建一个DataStreamSource
27         DataStreamSource<String> dataStream
28
29         dataStream.map(new MapFunction<Strin
30
31             @Override
32             public String map(String value)
```

```
[root@kafka101 kafka_2.11-2.4.0]# bin/kafka-console-consumer.s
h --bootstrap-server kafka101:9092 --topic mytopic1
KAFKA
WORLD
HELLO
```

```
SLF4J: Failed to load class "org.slf4j.impl.StaticLoggerBinder".
SLF4J: Defaulting to no-operation (NOP) logger implementation
SLF4J: See http://www.slf4j.org/codes.html#StaticLoggerBinder for further details.
```

图 8.30　启动 Flink Sink Connector 应用程序

# 第 9 章　Kafka 与 Storm 集成

Storm 为分布式实时计算提供了一种实时数据处理方式，该方式可用于“流处理”中，实时处理消息并更新数据库，这是管理队列及工作者集群的另一种方式。Storm 也可用于连续计算（continuous computation），对数据流进行连续查询，在计算时就将结果以流的形式输送给用户。它还可用于“分布式 RPC”，以并行的方式进行昂贵的运算。

Storm 可以方便地在一个计算机集群中编写与扩展复杂的实时计算，Storm 用于实时处理，就像 Hadoop 用于批处理。Storm 保证每个消息都会得到处理，而且它处理消息的速度很快，在一个小集群中，每秒可以处理数以百万计的消息，并且可以使用任意编程语言进行开发。

图 9.1 是 Storm 的官方网站上提供的一张图片。从图中可以看到数据源（水龙头）在源源不断地产生数据，后续的 Storm 节点在源源不断地处理数据。通过后面的学习，可以开发 Storm 对应的组件，例如，开发 Spout 组件或 Bolt 组件来完成这样的处理过程。

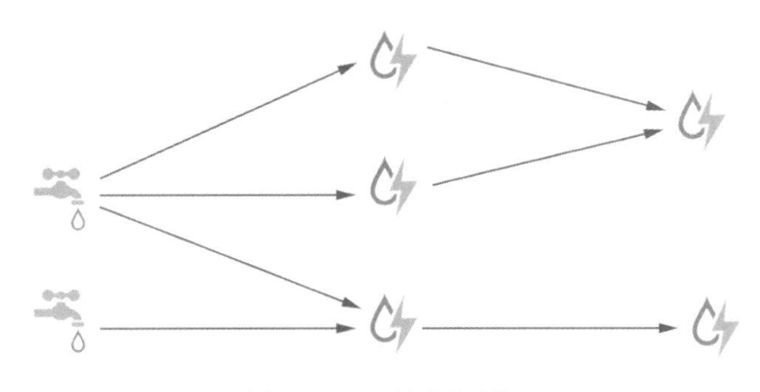

图 9.1　Storm 的流式计算

## 9.1　离线计算与流式计算

在大数据计算的领域中，有大数据的离线计算（也称批处理计算）和大数据的流式计算（也称实时计算）。表 9.1 对比了大数据离线计算和大数据流式计算的特征，帮助我们掌握它们各自的特点及区别。

表 9.1 大数据离线计算与流式计算

| | 离线计算 | 流式计算 |
|---|---|---|
| 特征本质 | 批量获取数据<br>批量传输数据<br>周期性批量计算数据<br>数据展示 | 数据实时产生<br>数据实时传输<br>数据实时计算<br>实时展示 |
| 数据的采集 | Sqoop 批量导入数据<br>Flume 批量采集离线数据 | Flume 实时获取数据 |
| 数据的存储 | HDFS 批量存储数据<br>HBase 批量存储数据 | Kafka 实时数据存储<br>Redis 实时结果存储 |
| 计算引擎 | MapReduce 批量计算<br>Spark Core 离线计算<br>Flink DataSet API | Storm<br>Spark Streaming<br>Flink DataStream API |

大数据离线计算的一个典型案例就是离线数据仓库的搭建。图 9.2 展示了一个典型的数据仓库搭建的过程。在这个过程中，数据源的类型多种多样，比如，有 Oracle 和 MySQL 中的数据，统称为 RDBMS（关系型数据库管理系统）；有文本类型的数据，如日志文件等，总之能够提供数据的都可以称为数据源。有了数据源的数据，我们就需要通过 ETL 进行数据的采集。ETL 可以使用已有的组件或工具完成，比如 Hadoop 中的 Sqoop 和 Flume；也可以采用第三方的组件完成，比如 Kettle。将采集到的数据存储在数据仓库中，可以使用 Oracle、MySQL、HDFS、HBase 等进行数据仓库的搭建，存储最原始的数据。原始数据经过分析处理过程后放到数据集中，最终进行结果的展示。这样就完成了一个典型离线数据仓库的搭建过程，当然这个过程也是实时的，也就是实时数据仓库，在实时数据仓库中进行的实时计算，也可以称为流式计算。那么什么是实时计算和流式计算呢？

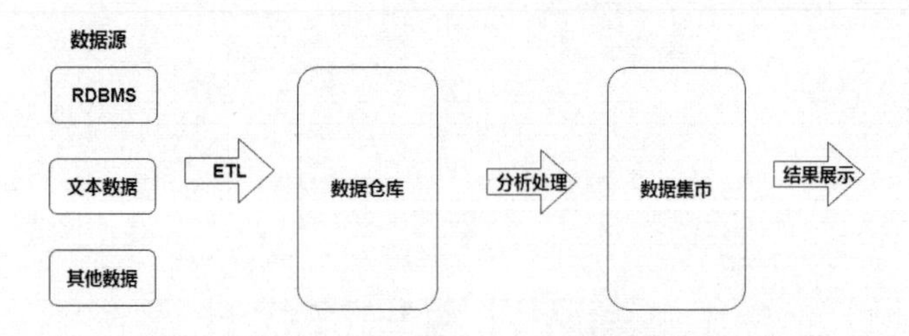


图 9.2 离线数据仓库的搭建过程

图 9.3 展示了一个典型的实时计算系统，这个实时计算系统与自来水厂处理自来水的过程类似。

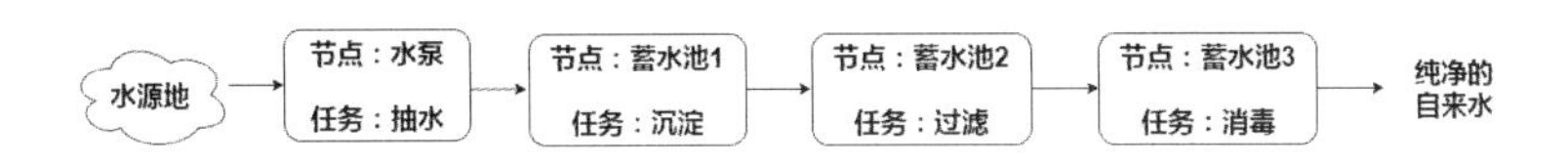


图 9.3　自来水厂处理自来水的过程

在这个过程中，水泵源源不断地采集水源，即采集数据；后续各个蓄水池中的水又可以被不同的任务源源不断地处理；最终提供干净的自来水，即处理完成的结果。这个过程只要不被人为停止，将永远运行。这就是实时计算与离线的批处理计算的最大差别。

## 9.2　Apache Storm 的体系架构

Apache Storm 是一个免费的、开源的分布式实时计算系统。在 Hadoop 体系中，我们执行的是典型的批处理过程。有了 Apache Storm 就不会使处理无边界的数据流变得冗余，并且处理的过程非常可靠。Apache Storm 非常简单，同时也支持多种编程的语言。Apache Storm 的体系架构也是一种主从式的架构，如图 9.4 所示。

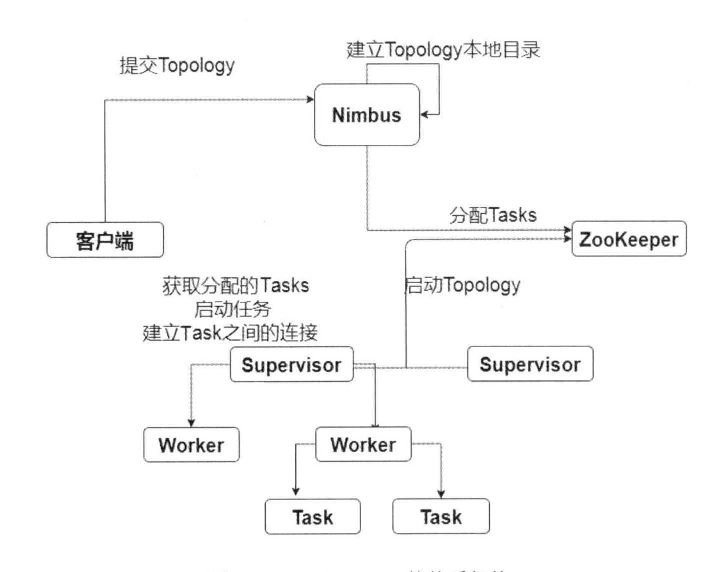


图 9.4　Apache Storm 的体系架构

其中，

- Nimbus：负责资源分配和任务调度，它是 Storm 集群的主节点。
- Supervisor：它是 Storm 集群从节点负责接收 Nimbus 分配的任务，启动和停止属于自己管理的 Worker 进程。通过配置文件设置当前 Supervisor 上启动多少个 Worker。
- Worker：运行具体处理组件逻辑的进程。Worker 运行的任务类型只有两种，一种是 Spout 任务，另一种是 Bolt 任务。

- Executor：在 Storm 0.8 之后，Executor 是 Worker 进程中的具体物理线程，同一个 Spout/Bolt 的 Task 可能会共享一个物理线程，一个 Executor 中只能运行隶属于同一个 Spout/Bolt 的 Task。
- Task：Worker 中每一个 Spout/Bolt 的线程称为一个 Task。在 Storm 0.8 之后，Task 不再与物理线程对应，不同 Spout/Bolt 的 Task 可能会共享一个物理线程，该线程称为 Executor。

图 9.5 展示了 Worker、Task 和 Executor 之间的关系。

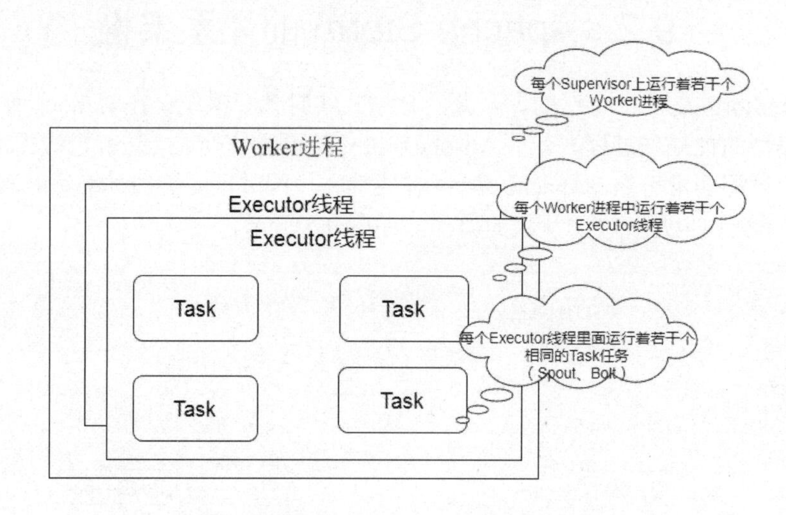


图 9.5　Worker、Task 和 Executor 之间的关系

在了解了 Apache Storm 的体系架构以后，Storm 处理数据的运行机制是什么？我们结合自来水厂处理自来水的过程进行说明。图 9.6 展示了 Storm 的处理机制。

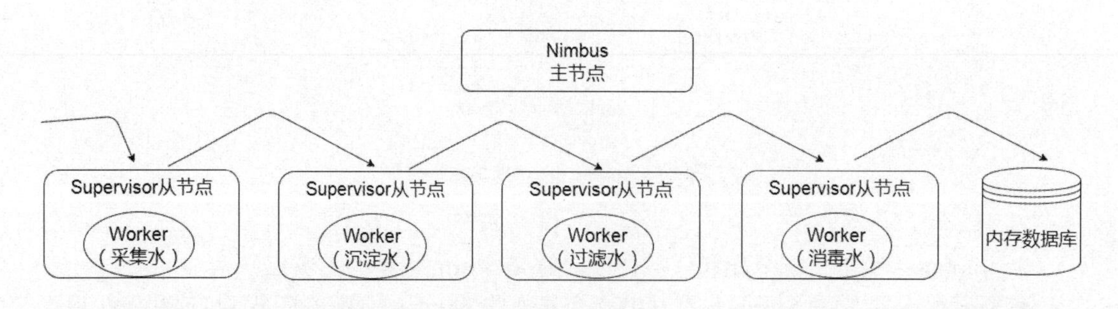


图 9.6　Storm 的处理机制

图 9.7 更详细地描述了这一过程。

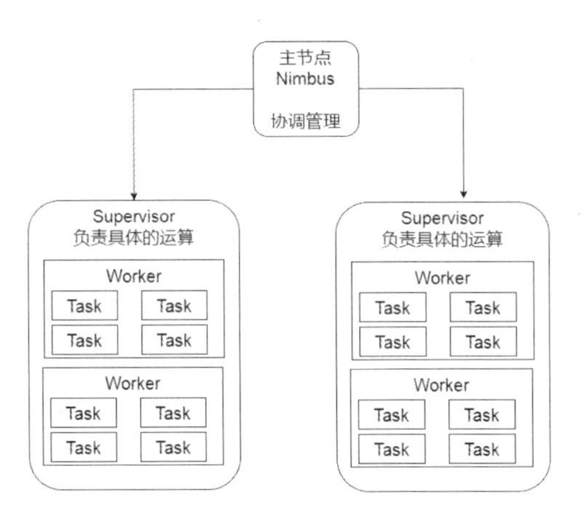


图 9.7　Storm 任务处理的流程

# 9.3　部署 Apache Storm

由于在 Storm 的体系架构中需要 ZooKeeper 的支持，可以在之前搭建好的 kafka101、kafka102 和 kafka103 的虚拟机上完成部署。我们搭建三种不同的模式，即 Storm 的伪分布模式、Storm 的全分布模式和 Storm HA 模式。

Storm 部署完成后的体系架构如图 9.8 所示。

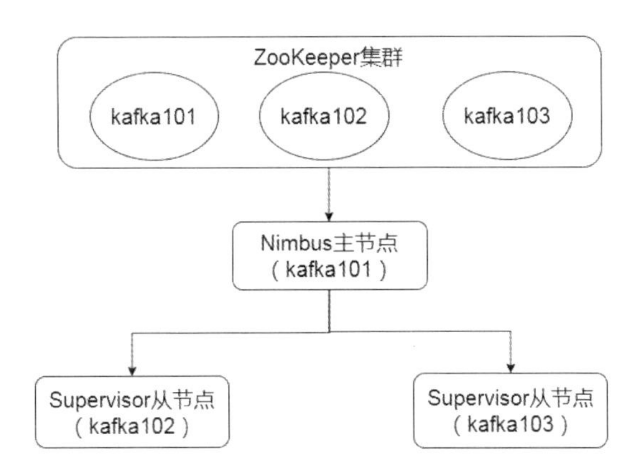


图 9.8　Storm 部署完成后的体系架构

在部署过程中，Apache Storm 的核心配置文件是 conf 目录下的 storm.yaml 文件，下面展示了这个文件的全部内容。

```
# Licensed to the Apache Software Foundation (ASF) under one
# or more contributor license agreements.  See the NOTICE file
# distributed with this work for additional information
```

```
# regarding copyright ownership.  The ASF licenses this file
# to you under the Apache License, Version 2.0 (the
# "License"); you may not use this file except in compliance
# with the License.  You may obtain a copy of the License at
#
# http://www.apache.org/licenses/LICENSE-2.0
#
# Unless required by applicable law or agreed to in writing, software
# distributed under the License is distributed on an "AS IS" BASIS,
# WITHOUT WARRANTIES OR CONDITIONS OF ANY KIND, either express or implied.
# See the License for the specific language governing permissions and
# limitations under the License.

########### These MUST be filled in for a storm configuration
# storm.zookeeper.servers:
#     - "server1"
#     - "server2"
#
# nimbus.seeds: ["host1", "host2", "host3"]
#
#
# ##### These may optionally be filled in:
#
## List of custom serializations
# topology.kryo.register:
#     - org.mycompany.MyType
#     - org.mycompany.MyType2: org.mycompany.MyType2Serializer
#
## List of custom kryo decorators
# topology.kryo.decorators:
#     - org.mycompany.MyDecorator
#
## Locations of the drpc servers
# drpc.servers:
#     - "server1"
#     - "server2"

## Metrics Consumers
# topology.metrics.consumer.register:
#   - class: "org.apache.storm.metric.LoggingMetricsConsumer"
#     parallelism.hint: 1
#   - class: "org.mycompany.MyMetricsConsumer"
#     parallelism.hint: 1
#     argument:
```

```
#       - endpoint: "metrics-collector.mycompany.org"
```

下面列举两个需要注意的参数说明。

- storm.zookeeper.servers：ZooKeeper 服务器列表。
- nimbus.seeds：Nimbus 服务器地址。

除了配置文件中列举的参数，Apache Storm 中还有一些隐含的配置参数。我们会在具体配置的过程中进行说明。

### 9.3.1　部署 Storm 的伪分布模式

可以在 kafka101 的虚拟机上部署一个伪分布模式的 Storm Standalone 模式，即只有一个 Nimbus 和一个 Supervisor。

（1）首先，将压缩包解压至/root/training 目录。

```
tar -zxvf apache-storm-1.0.3.tar.gz -C ~/training/
```

这里我们使用的是 apache-storm-1.0.3.tar.gz 的版本。

（2）为了方便执行 Storm 的命令脚本，编辑~/.bash_profile 文件进行环境变量的设置。

```
STORM_HOME=/root/training/apache-storm-1.0.3
export STORM_HOME

PATH=$STORM_HOME/bin:$PATH
export PATH
```

（3）生效 Storm 的环境变量。

```
source ~/.bash_profile
```

（4）修改 storm.yaml 文件，下面列举出修改后的文件内容。

```
########### These MUST be filled in for a storm configuration
storm.zookeeper.servers:
    - "kafka101"
    - “kafka102”
    - “kafka103”

nimbus.seeds: ["kafka101"]
storm.local.dir: "/root/training/apache-storm-1.0.3/tmp"
supervisor.slots.ports:
     - 6700
     - 6701
     - 6702
     - 6703
"topology.eventlogger.executors": 1

# ##### These may optionally be filled in:
#
## List of custom serializations
```

```
# topology.kryo.register:
#     - org.mycompany.MyType
#     - org.mycompany.MyType2: org.mycompany.MyType2Serializer
#
## List of custom kryo decorators
# topology.kryo.decorators:
#     - org.mycompany.MyDecorator
#
## Locations of the drpc servers
# drpc.servers:
#     - "server1"
#     - "server2"

## Metrics Consumers
# topology.metrics.consumer.register:
#   - class: "org.apache.storm.metric.LoggingMetricsConsumer"
#     parallelism.hint: 1
#   - class: "org.mycompany.MyMetricsConsumer"
#     parallelism.hint: 1
#     argument:
#       - endpoint: "metrics-collector.mycompany.org"
```

参数说明如下。

- storm.zookeeper.servers:

  - "kafka101"

  - "kafka102"

  - "kafka103"

这里我们指定 ZooKeeper 地址分别运行在 kafka101、kafka102 和 kafka103 的虚拟机上。

- nimbus.seeds: ["kafka101"]

指定将 Nimbus 的主节点运行在 kafka101 的虚拟机上。如果要搭建 Storm 的 HA 环境，只需要在这里指定多个 Nimbus 主节点即可。

- storm.local.dir: "/root/training/apache-storm-1.0.3/tmp":

storm 使用的本地文件系统目录（必须存在并且 storm 进程可读写）。

- supervisor.slots.ports:

  - 6700

  - 6701

  - 6702

  - 6703

Supervisor 上能够运行 Workers 的端口列表。每个 Worker 占用一个端口，且每个端口只运行一个 Worker。通过这项配置可以调整每台机器上运行的 Worker 数（调整 Slot 数/每机）。这里在每个 Supervisor 上运行了 4 个 Worker，即 4 个 Slot。

- "topology.eventlogger.executors": 1 通过查看 Storm UI 上每个组件的 Events 链接，可以查看 Storm 的每个组件（Spout、Blot）发送的消息。但 Storm 的 Event Logger 功能默认是禁用的，需要在配置文件中设置这个参数，具体说明如下。

"topology.eventlogger.executors": 0，默认，禁用。

"topology.eventlogger.executors": 1，一个 Topology 分配一个 Event Logger。

"topology.eventlogger.executors": nil，每个 Worker 分配一个 Event Logger。

（5）启动 Nimbus 的主节点。

```
storm nimbus &
```

成功启动的日志输出，如图 9.9 所示。

```
[root@kafka101 ~]# storm nimbus &
[1] 67696
[root@kafka101 ~]# Running: /root/training/jdk1.8.0_181/bin/java -server -Ddaemon.name=nimbus -Dstorm.options= -Dstorm.home=
/root/training/apache-storm-1.0.3 -Dstorm.log.dir=/root/training/apache-storm-1.0.3/logs -Djava.library.path=/usr/local/lib:
/opt/local/lib:/usr/lib -Dstorm.conf.file= -cp /root/training/apache-storm-1.0.3/lib/storm-core-1.0.3.jar:/root/training/apa
che-storm-1.0.3/lib/kryo-3.0.3.jar:/root/training/apache-storm-1.0.3/lib/reflectasm-1.10.1.jar:/root/training/apache-storm-1
.0.3/lib/asm-5.0.3.jar:/root/training/apache-storm-1.0.3/lib/minlog-1.3.0.jar:/root/training/apache-storm-1.0.3/lib/objenesi
s-2.1.jar:/root/training/apache-storm-1.0.3/lib/clojure-1.7.0.jar:/root/training/apache-storm-1.0.3/lib/disruptor-3.3.2.jar:
/root/training/apache-storm-1.0.3/lib/log4j-api-2.1.jar:/root/training/apache-storm-1.0.3/lib/log4j-core-2.1.jar:/root/train
ing/apache-storm-1.0.3/lib/log4j-slf4j-impl-2.1.jar:/root/training/apache-storm-1.0.3/lib/slf4j-api-1.7.7.jar:/root/training
/apache-storm-1.0.3/lib/log4j-over-slf4j-1.6.6.jar:/root/training/apache-storm-1.0.3/lib/servlet-api-2.5.jar:/root/training/
apache-storm-1.0.3/lib/storm-rename-hack-1.0.3.jar:/root/training/apache-storm-1.0.3/conf -Xmx1024m -Dlogfile.name=nimbus.lo
g -DLog4jContextSelector=org.apache.logging.log4j.core.async.AsyncLoggerContextSelector -Dlog4j.configurationFile=/root/trai
ning/apache-storm-1.0.3/log4j2/cluster.xml org.apache.storm.daemon.nimbus

[root@kafka101 ~]#
```

图 9.9　成功启动 Storm Nimbus 节点

（6）启动 Supervisor 的从节点。

```
storm supervisor &
```

成功启动的日志输出，如图 9.10 所示。

```
[root@kafka101 ~]# storm supervisor &
[2] 67815
[root@kafka101 ~]# Running: /root/training/jdk1.8.0_181/bin/java -server -Ddaemon.name=supervisor -Dstorm.options= -Dstorm.h
ome=/root/training/apache-storm-1.0.3 -Dstorm.log.dir=/root/training/apache-storm-1.0.3/logs -Djava.library.path=/usr/local/
lib:/opt/local/lib:/usr/lib -Dstorm.conf.file= -cp /root/training/apache-storm-1.0.3/lib/storm-core-1.0.3.jar:/root/training
/apache-storm-1.0.3/lib/kryo-3.0.3.jar:/root/training/apache-storm-1.0.3/lib/reflectasm-1.10.1.jar:/root/training/apache-sto
rm-1.0.3/lib/asm-5.0.3.jar:/root/training/apache-storm-1.0.3/lib/minlog-1.3.0.jar:/root/training/apache-storm-1.0.3/lib/obje
nesis-2.1.jar:/root/training/apache-storm-1.0.3/lib/clojure-1.7.0.jar:/root/training/apache-storm-1.0.3/lib/disruptor-3.3.2.
jar:/root/training/apache-storm-1.0.3/lib/log4j-api-2.1.jar:/root/training/apache-storm-1.0.3/lib/log4j-core-2.1.jar:/root/t
raining/apache-storm-1.0.3/lib/log4j-slf4j-impl-2.1.jar:/root/training/apache-storm-1.0.3/lib/slf4j-api-1.7.7.jar:/root/trai
ning/apache-storm-1.0.3/lib/log4j-over-slf4j-1.6.6.jar:/root/training/apache-storm-1.0.3/lib/servlet-api-2.5.jar:/root/train
ing/apache-storm-1.0.3/lib/storm-rename-hack-1.0.3.jar:/root/training/apache-storm-1.0.3/conf -Xmx256m -Dlogfile.name=superv
isor.log -DLog4jContextSelector=org.apache.logging.log4j.core.async.AsyncLoggerContextSelector -Dlog4j.configurationFile=/ro
ot/training/apache-storm-1.0.3/log4j2/cluster.xml org.apache.storm.daemon.supervisor.Supervisor

[root@kafka101 ~]#
```

图 9.10　成功启动 Storm Supervisor 节点

（7）启动 Storm UI 界面。

```
storm ui &
```

成功启动的日志输出，如图 9.11 所示。

```
[root@kafka101 ~]# storm ui &
[3] 67921
[root@kafka101 ~]# Running: /root/training/jdk1.8.0_181/bin/java -server -Ddaemon.name=ui -Dstorm.options= -Dstorm.home=/roo
t/training/apache-storm-1.0.3 -Dstorm.log.dir=/root/training/apache-storm-1.0.3/logs -Djava.library.path=/usr/local/lib:/opt
/local/lib:/usr/lib -Dstorm.conf.file= -cp /root/training/apache-storm-1.0.3/lib/storm-core-1.0.3.jar:/root/training/apache-
storm-1.0.3/lib/kryo-3.0.3.jar:/root/training/apache-storm-1.0.3/lib/reflectasm-1.10.1.jar:/root/training/apache-storm-1.0.3
/lib/asm-5.0.3.jar:/root/training/apache-storm-1.0.3/lib/minlog-1.3.0.jar:/root/training/apache-storm-1.0.3/lib/objenesis-2.
1.jar:/root/training/apache-storm-1.0.3/lib/clojure-1.7.0.jar:/root/training/apache-storm-1.0.3/lib/disruptor-3.3.2.jar:/roo
t/training/apache-storm-1.0.3/lib/log4j-api-2.1.jar:/root/training/apache-storm-1.0.3/lib/log4j-core-2.1.jar:/root/training/
apache-storm-1.0.3/lib/log4j-slf4j-impl-2.1.jar:/root/training/apache-storm-1.0.3/lib/slf4j-api-1.7.7.jar:/root/training/apa
che-storm-1.0.3/lib/log4j-over-slf4j-1.6.6.jar:/root/training/apache-storm-1.0.3/lib/servlet-api-2.5.jar:/root/training/apac
he-storm-1.0.3/lib/storm-rename-hack-1.0.3.jar:/root/training/apache-storm-1.0.3:/root/training/apache-storm-1.0.3/conf -Xmx
768m -Dlogfile.name=ui.log -DLog4jContextSelector=org.apache.logging.log4j.core.async.AsyncLoggerContextSelector -Dlog4j.con
figurationFile=/root/training/apache-storm-1.0.3/log4j2/cluster.xml org.apache.storm.ui.core

[root@kafka101 ~]#
```

图 9.11　成功启动 Storm UI

（8）启动 Storm LogViewer。

```
storm logviewer &
```

成功启动的日志输出，如图 9.12 所示。

```
[root@kafka101 ~]# storm logviewer &
[4] 68011
[root@kafka101 ~]# Running: /root/training/jdk1.8.0_181/bin/java -server -Ddaemon.name=logviewer -Dstorm.options= -Dstorm.ho
me=/root/training/apache-storm-1.0.3 -Dstorm.log.dir=/root/training/apache-storm-1.0.3/logs -Djava.library.path=/usr/local/l
ib:/opt/local/lib:/usr/lib -Dstorm.conf.file= -cp /root/training/apache-storm-1.0.3/lib/storm-core-1.0.3.jar:/root/training/
apache-storm-1.0.3/lib/kryo-3.0.3.jar:/root/training/apache-storm-1.0.3/lib/reflectasm-1.10.1.jar:/root/training/apache-stor
m-1.0.3/lib/asm-5.0.3.jar:/root/training/apache-storm-1.0.3/lib/minlog-1.3.0.jar:/root/training/apache-storm-1.0.3/lib/objen
esis-2.1.jar:/root/training/apache-storm-1.0.3/lib/clojure-1.7.0.jar:/root/training/apache-storm-1.0.3/lib/disruptor-3.3.2.j
ar:/root/training/apache-storm-1.0.3/lib/log4j-api-2.1.jar:/root/training/apache-storm-1.0.3/lib/log4j-core-2.1.jar:/root/tr
aining/apache-storm-1.0.3/lib/log4j-slf4j-impl-2.1.jar:/root/training/apache-storm-1.0.3/lib/slf4j-api-1.7.7.jar:/root/train
ing/apache-storm-1.0.3/lib/log4j-over-slf4j-1.6.6.jar:/root/training/apache-storm-1.0.3/lib/servlet-api-2.5.jar:/root/traini
ng/apache-storm-1.0.3/lib/storm-rename-hack-1.0.3.jar:/root/training/apache-storm-1.0.3:/root/training/apache-storm-1.0.3/co
nf -Xmx128m -Dlogfile.name=logviewer.log -DLog4jContextSelector=org.apache.logging.log4j.core.async.AsyncLoggerContextSelect
or -Dlog4j.configurationFile=/root/training/apache-storm-1.0.3/log4j2/cluster.xml org.apache.storm.daemon.logviewer

[root@kafka101 ~]#
```

图 9.12　成功启动 Storm LogViewer

（9）执行 jps 命令查看后台的进程信息。

```
[root@kafka101 ~]# jps
67921 core
67696 nimbus
67815 Supervisor
67643 QuorumPeerMain
68011 logviewer
68107 Jps
```

（10）通过浏览器访问 8080 端口，打开 Storm UI 界面，如图 9.13 所示。由于这里我们没有运行任何任务，所以看不到与任何任务相关的信息。但是通过 Web UI 可以了解到集群的信息，比如 Nimbus、Supervisor 和 Slot 信息等。

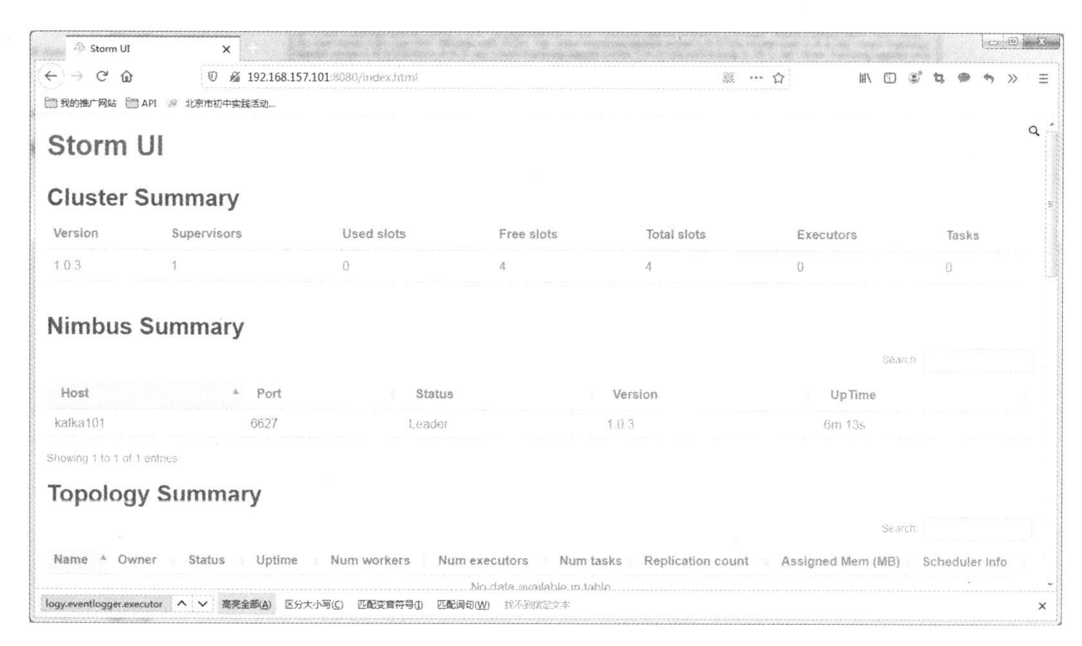


图 9.13　Storm UI 界面

## 9.3.2　部署 Storm 的全分布模式

在之前部署完成的单节点伪分布模式的基础上，进行下面的操作。

（1）在 kafka101 的虚拟机上，进入/root/training 目录。

```
cd /root/training/
```

（2）将配置好的 Storm 目录复制到 kafka102 和 kafka103 的虚拟机上。

```
scp -r apache-storm-1.0.3/ root@kafka102:/root/training
scp -r apache-storm-1.0.3/ root@kafka103:/root/training
```

（3）在 kafka102 和 kafka103 的虚拟机上设置 Storm 的环境变量，并生效环境变量。

```
编辑文件  vi ~/.bash_profile
输入以下内容：
STORM_HOME=/root/training/apache-storm-1.0.3
export STORM_HOME

PATH=$STORM_HOME/bin:$PATH
export PATH

生效环境变量：source ~/.bash_profile
```

（4）在三台虚拟机上启动 ZooKeeper 集群，如图 9.14 所示。

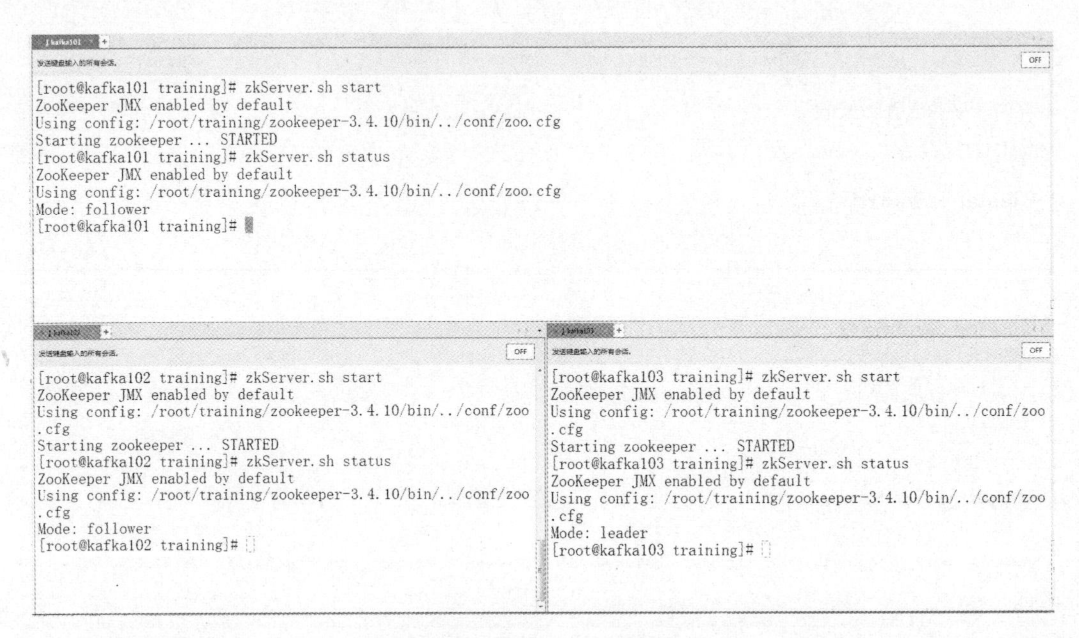


图 9.14　启动 ZooKeeper 集群

（5）在 kafka101 的虚拟机上启动 Nimbus、LogViewer 和 Storm UI，并执行 jps 命令查看后台的进程信息，如图 9.15 所示。

```
storm nimbus &
storm logviewer &
storm ui &
jps
```

```
[root@kafka101 training]# jps
70209 nimbus
70385 core
70311 logviewer
70137 QuorumPeerMain
70474 Jps
[root@kafka101 training]#
```

图 9.15　启动主节点进程

（6）在 kafka102 和 kafka103 的虚拟机上启动 Supervisor，并执行 jps 命令查看后台的进程信息，如图 9.16 所示

```
storm supervisor &
jps
```

图 9.16　启动 Supervisor

（7）访问 Storm UI 界面，看到有两个 Supervisor 的从节点信息，如图 9.17 所示。

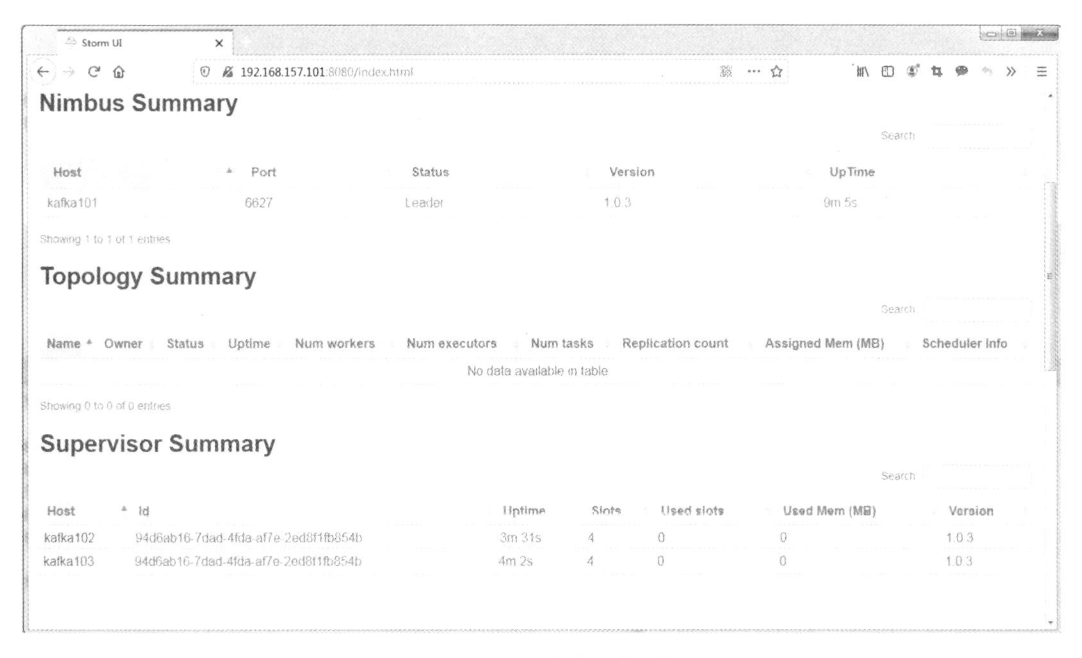


图 9.17　Storm 的全分布环境

### 9.3.3 Storm HA 模式

从 Storm 的体系架构中可以看到 ZooKeeper 和 Supervisor 都是多节点，任意一个 ZooKeeper 节点宕机或 Supervisor 节点宕机均不会对系统整体运行造成影响，但 Nimbus 和 UI 都是单节点，UI 的单节点对系统的稳定运行没有影响，仅提供 Storm ui 界面展示统计信息。Nimbus 承载了集群的许多工作，如果 Nimbus 单节点宕机，将会对系统整体的稳定运行造成极大风险。因此解决 Nimbus 的单节点问题，需要更加完善 Storm 集群的稳定性。

在引入了 ZooKeeper 以后，我们启动了两个主节点 Nimbus，它们都向 ZooKeeper 去注册。在同一个时刻，这两个主节点的状态是不一样的，只有一个主节点是 Leader 的状态；而另一个主节点是 Not Leader 的状态。如果其中 Leader 的主节点挂了，这时候它所注册的节点将被 ZooKeeper 自动删除， ZooKeeper 会自动感知节点的变化，然后再次发出选举，将另一个 Not Leader 的主节点切换成 Leader 的状态，以替代之前的主节点，图 9.18 描述了这一过程。

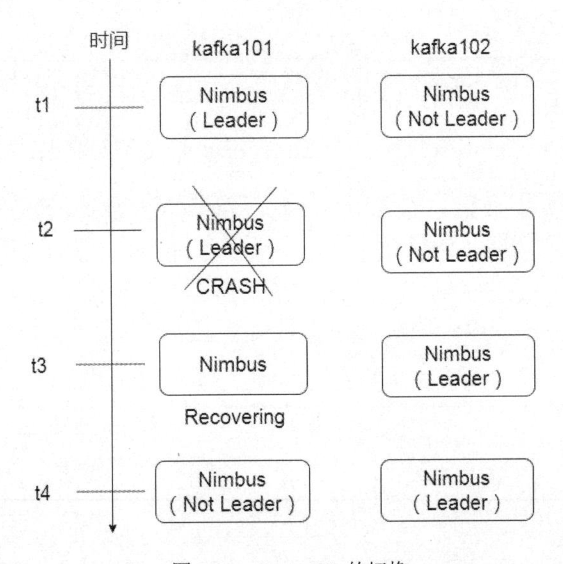


图 9.18　Storm HA 的切换

通过图 9.18 可以看到，在 kafka101 和 kafka102 的虚拟机上各有一个 Nimbus 的主节点，其中，kafka101 的虚拟机上的 Nimbus 是 Leader；kafka102 的虚拟机上的 Nimbus 是 Not Leader。在 t2 时，kafka101 的虚拟机上的 Nimbus 出现宕机，通过 ZooKeeper 的选举将 kafka102 的虚拟机上的 Nimbus 选举为 Leader 后，如果 kafka101 的虚拟机上的 Nimbus 又恢复运行，就只能是 Not Leader 的状态。

可以通过以下步骤来搭建 Storm HA 架构，最终将在 kafka101 和 kafka102 的虚拟机上各部署一个 Nimbus 的主节点。

（1）修改 kafka101、kafka102 和 kafka103 的虚拟机上 Storm 的配置文件，即 storm.yaml 文件，增加一个新的 Nimbus 节点信息。

```
nimbus.seeds: ["kafka101", "kafka102"]
```

配置 Storm HA 的 Nimbus 节点，如图 9.19 所示。

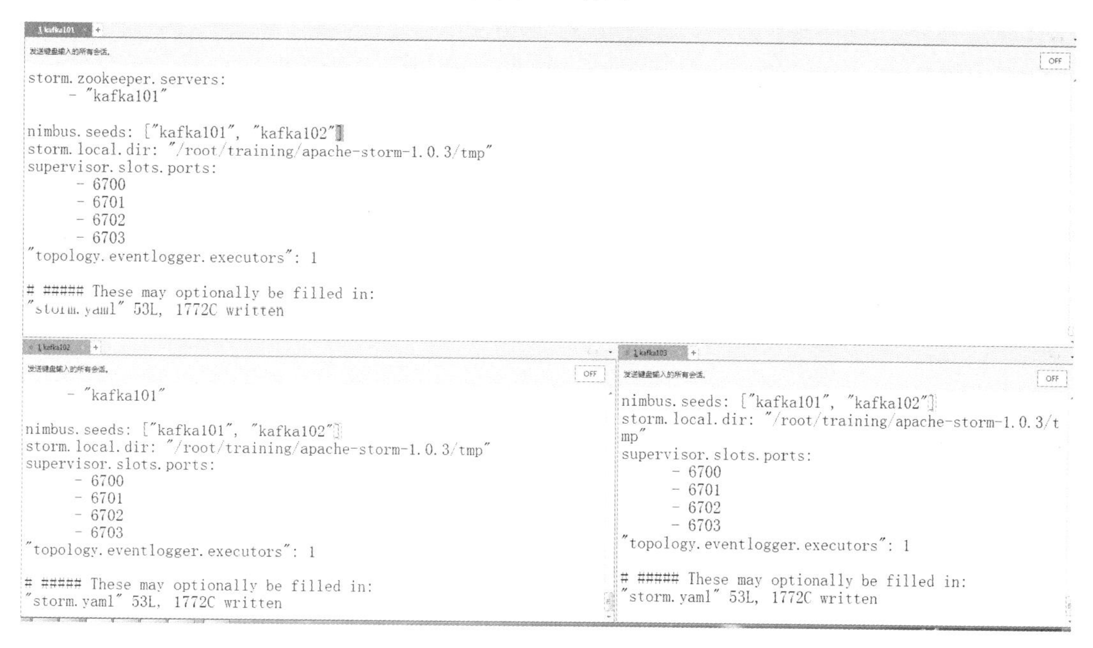


图 9.19　配置 Storm HA 的 Nimbus 节点

（2）在 kafka101、kafka102 和 kafka103 的虚拟机上启动 ZooKeeper 集群。

（3）在 kafka101 的虚拟机上启动 Nimbus、LogViewer 和 Storm UI。

```
storm nimbus &
storm logviewer &
storm ui &
```

（4）在 kafka102 的虚拟机上启动 Supervisor 和另一个 Nimbus。

```
storm nimbus &
storm supervisor &
```

（5）在 kafka103 的虚拟机上启动 Supervisor。

```
storm supervisor &
```

（6）在 kafka101、kafka102 和 kafka103 的虚拟机上执行 jps 命令，可以看到在 kafka101 和 kafka102 的虚拟机上各有一个 Nimbus 的主节点，如图 9.20 所示。

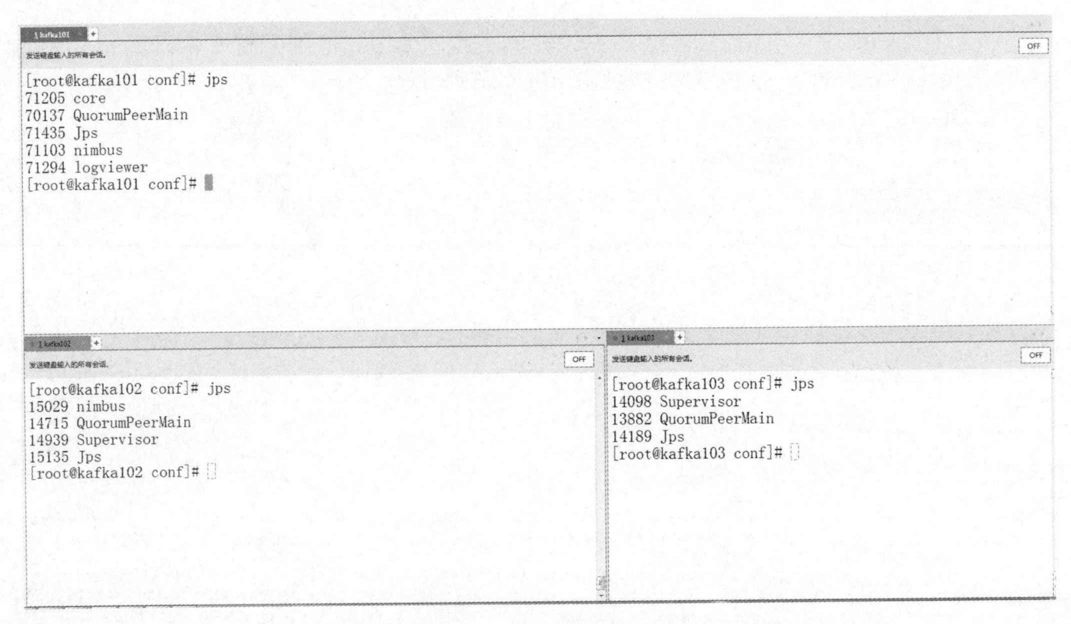


图 9.20 Storm HA 的进程信息

（7）访问 kafka101 的 Storm UI 界面，可以看到有两个 Nimbus，其中一个状态是 Leader；而另一个状态是 Not a Leader，如图 9.21 所示。

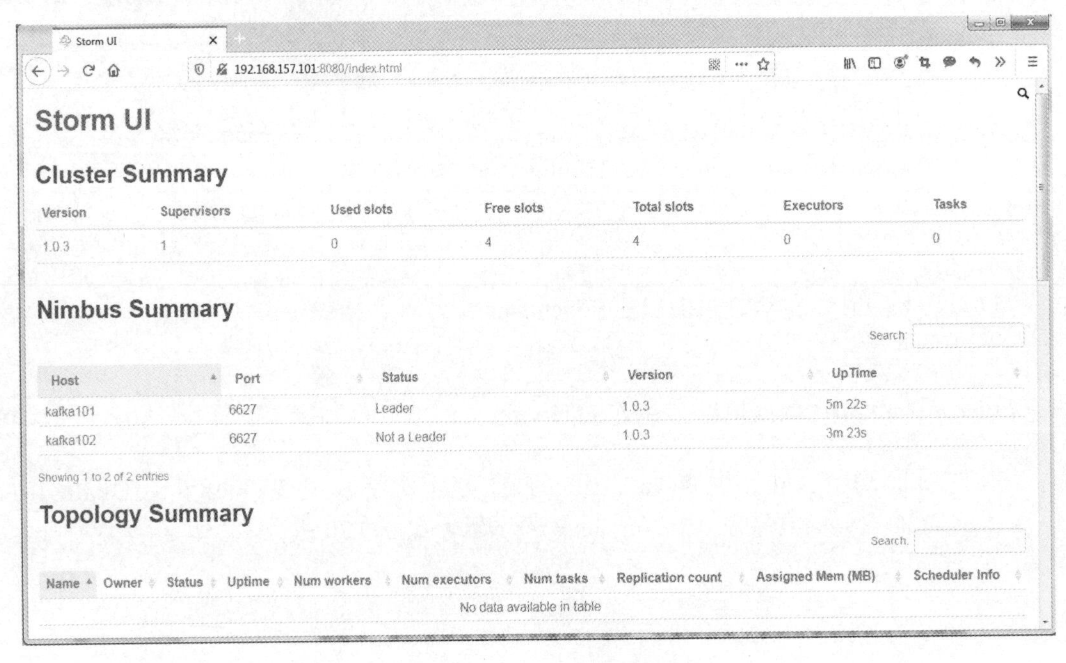


图 9.21 Storm HA 的 UI 界面

（8）下面来进行 HA 的测试。在 kafka101 的虚拟机上杀掉 Nimbus 的进程信息。通过 jps 命令可以看到 Nimbus 的进程号是 71103，如图 9.22 所示。

```
jps
kill -9 71103
```

```
[root@kafka101 conf]# jps
71205 core
70137 QuorumPeerMain
71103 nimbus
71503 Jps
71294 logviewer
[root@kafka101 conf]# kill -9 71103
[root@kafka101 conf]#
```

图 9.22　杀掉 kafka101 上的 Nimbus 进程

（9）刷新 Storm UI 界面，可以看到 kafka102 的虚拟机上的 Nimbus 变成了 Leader 的状态，如图 9.23 所示。

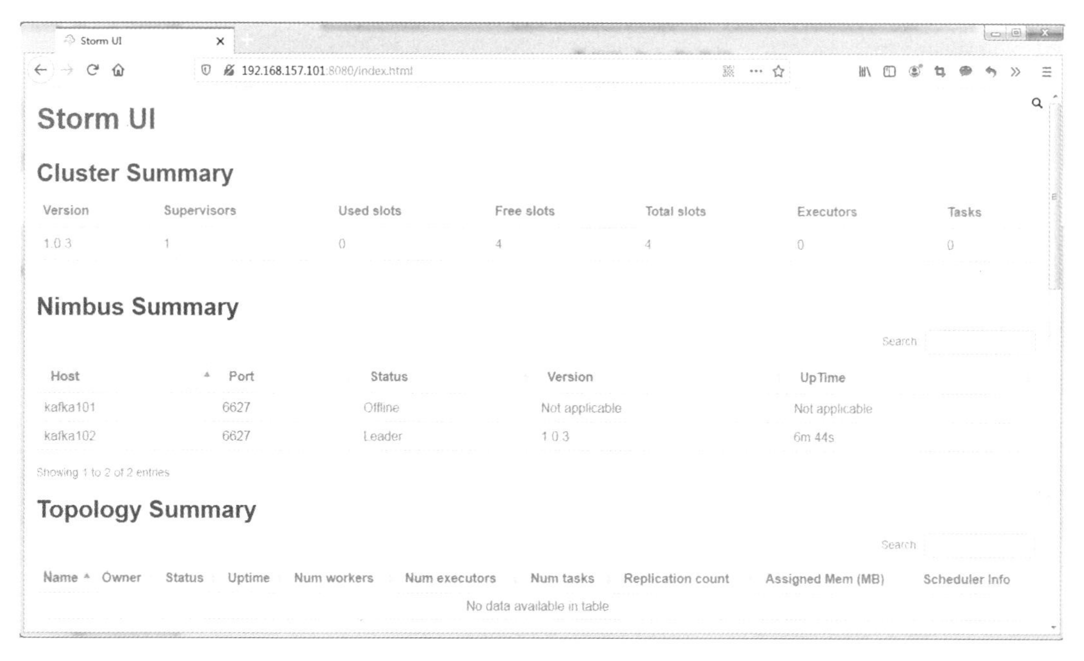


图 9.23　Storm HA 的切换

# 9.4 执行 Apache Storm 任务

在部署完成 Storm 的环境后，我们就可以执行 Storm 的实时计算任务。在 Storm 中，一个计算任务称为 Topology。在安装部署完成的环境中，官方为我们提供了一个 Example 的示例程序，直接将其提交到集群上运行。我们可以在下面的目录中找到这个示例的 Example。

```
/root/training/apache-storm-1.0.3/examples/storm-starter/storm-starter-topologies-1.0.3.jar
```

下面在之前搭建好的单节点环境上演示说明。

## 9.4.1 执行 WordCountTopology

使用 storm jar 命令可以将一个 Topology 任务的 jar 提交到集群上运行。命令的格式说明如下所示。

```
storm jar 【jar路径】【拓扑包名.拓扑类名】【拓扑名称】
```

下面将 WordCountTopology 任务提交到集群上运行，执行下面的命令。

```
cd /root/training/apache-storm-1.0.3/examples/storm-starter
storm jar storm-starter-topologies-1.0.3.jar org.apache.storm.starter.WordCountTopology MyWC1
```

任务成功提交后，将打印如下输出信息，如图 9.24 所示。

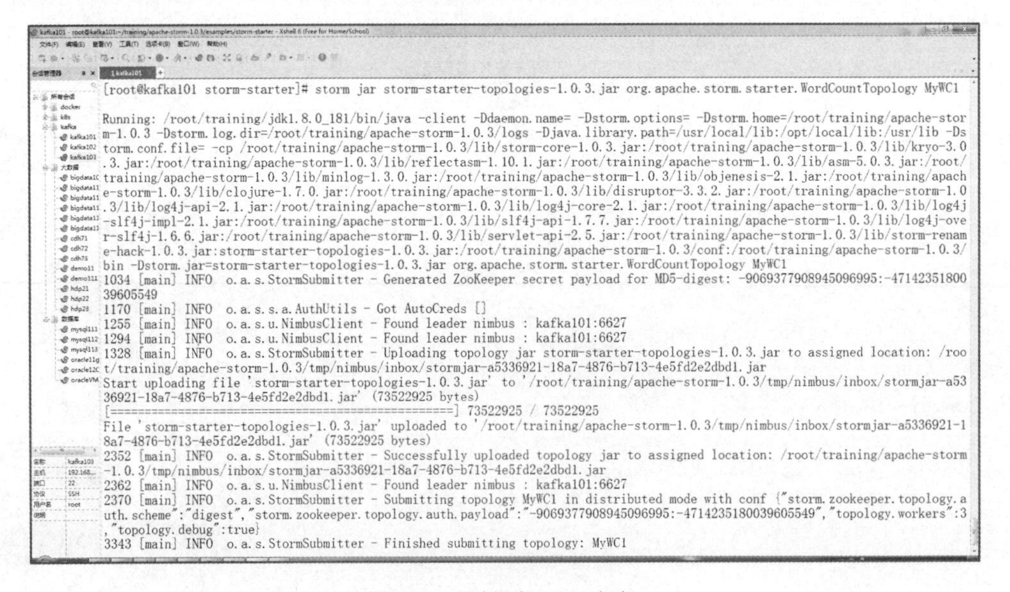


图 9.24 成功提交 Storm 任务

刷新 Storm UI 界面，可以看到 Topology 任务运行的信息，如图 9.25 所示。

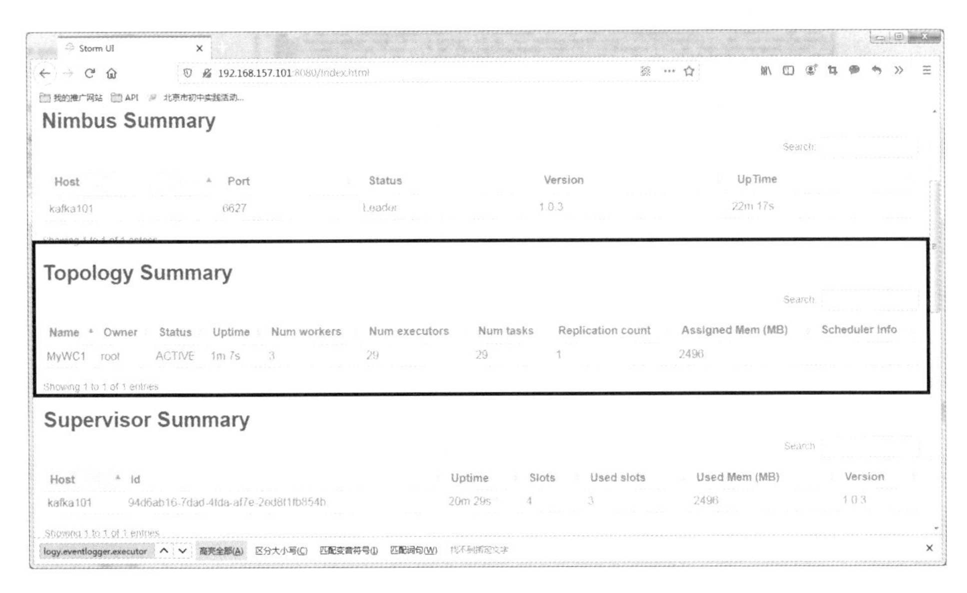


图 9.25　在 Storm UI 上监控 Topology 任务

可以看到当前的 Topology 任务 MyWC1 是 ACTIVE 的状态。它占用了 3 个 Workers 和 29 个 Executors。接下来，可以进入 Topology 的 Debug 模式，查看 MyWC1 的任务处理的数据是什么。单击任务名称“MyWC1”，进入“Topology summary”界面，单击该界面上的“Debug”按钮，如图 9.26 所示。

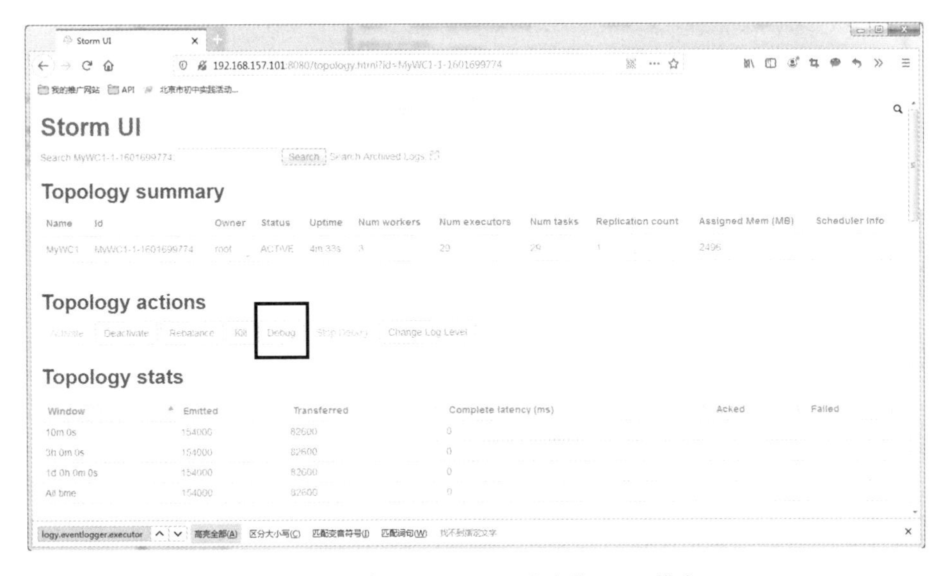


图 9.26　开启 Storm Topology 任务的 Debug 模式

接下来需要设置采样率是多少，保持默认的 10%即可，即每处理 100 条数据，采样 10

条数据进行观察，单击“确定”按钮，如图 9.27 所示。

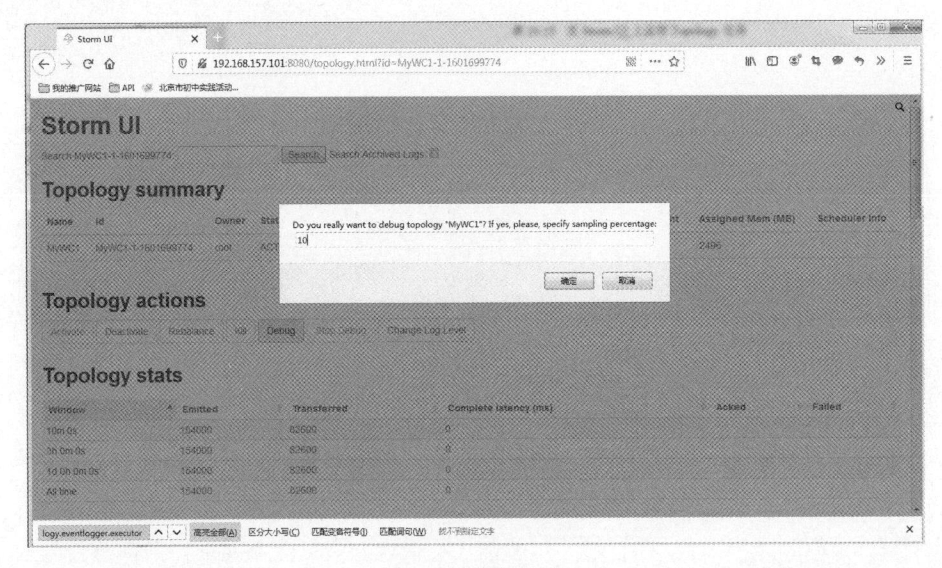


图 9.27　设置采样率

在“Topology summary”界面上可以看到单击 Spouts 组件和 Bolts 组件的名称，如图 9.28 所示。

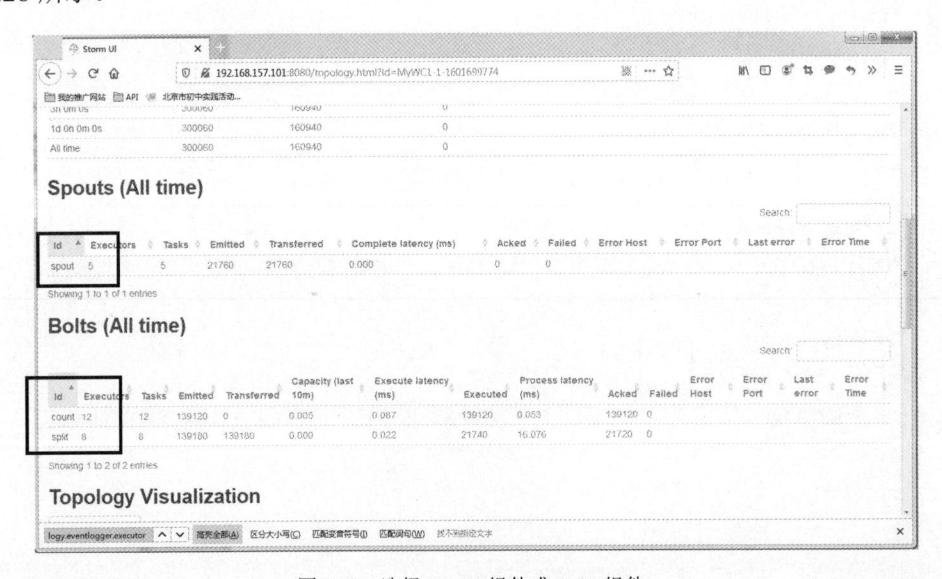


图 9.28　选择 Spouts 组件或 Bolts 组件

在图 9.28 的界面上，单击 Spouts 组件或 Bolts 组件的名称，即可切换到“Component summary”界面。在“Component summary”界面单击“events”即可观察到 Storm 中实时

处理的数据，如图 9.29 所示。

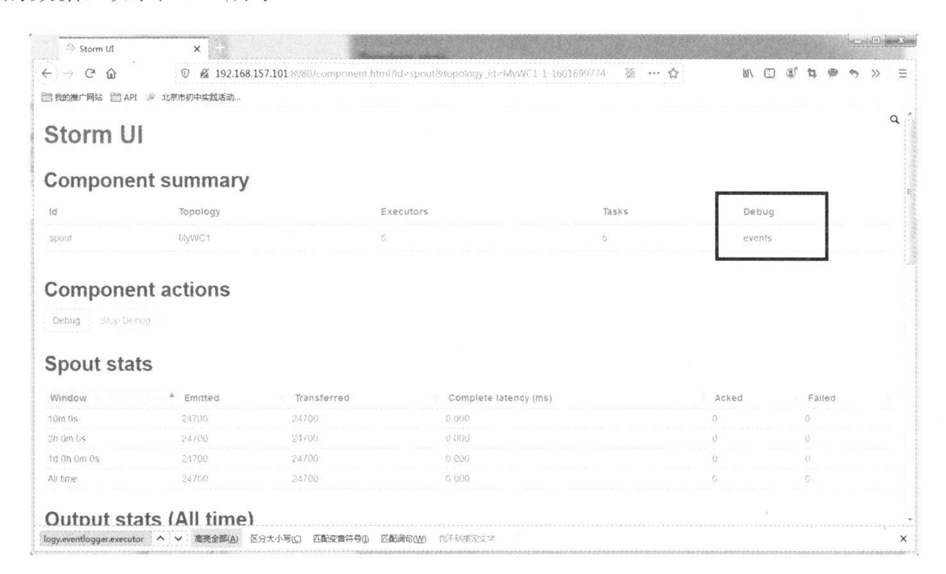


图 9.29　单击“events”

图 9.30 展示了提交的 WordCountTopology 任务中实时处理的数据。每次刷新页面时，统计出来的结果将会更新。

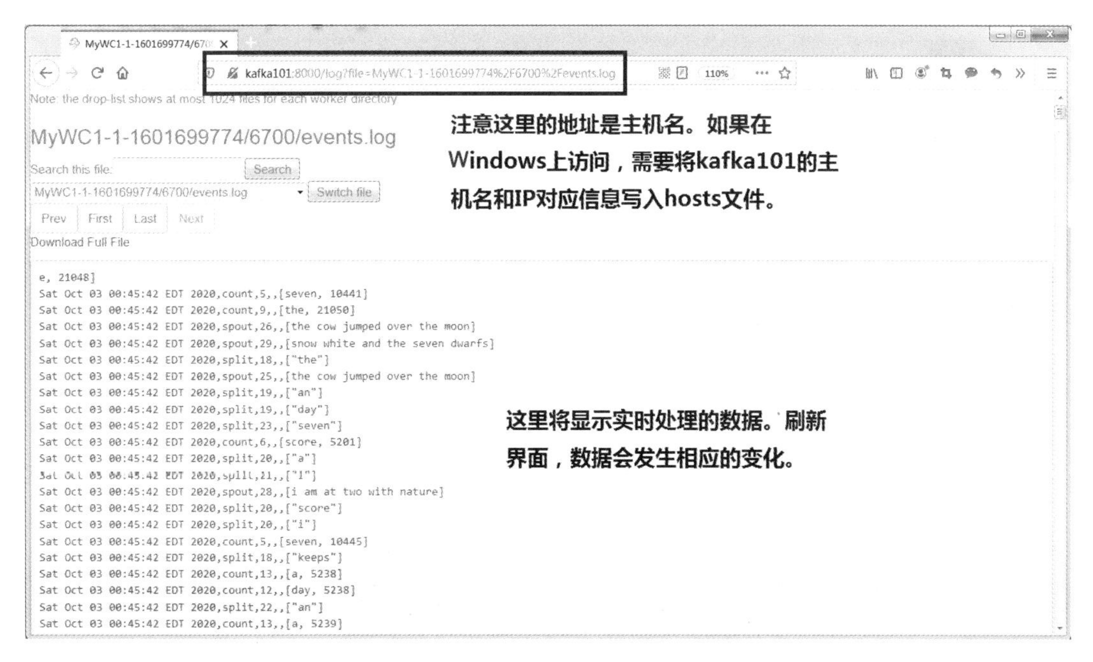


图 9.30　WordCountTopology 处理的结果

### 9.4.2 Storm 的其他管理命令

（1）杀死任务命令格式：storm kill【拓扑名称】-w 10。

执行 kill 命令时可以通过-w [等待秒数]指定拓扑停用以后的等待时间。例如，

```
storm kill topology-name -w 10
```

（2）停用任务命令格式：storm deactivte【拓扑名称】。

```
storm deactivte topology-name
```

（3）启用任务命令格式：storm activate【拓扑名称】。

```
storm activate topology-name
```

（4）重新部署任务命令格式：storm rebalance【拓扑名称】。

```
storm rebalance topology-name
```

再平衡命令能够重分配集群任务。这是个很强大的命令。比如，向一个运行中的集群增加了节点。再平衡命令将会停用拓扑，然后在相应超时时间之后重新分配工人，并重启拓扑。

## 9.5 开发自己的 Storm 任务

### 9.5.1 Storm Topology 任务处理的数据模型

在前面的章节中，我们提交并执行了 Storm 官方的 Example 示例程序 WordCountTopology，并且知道了 Storm 的 Topology 任务由 Spout 任务和 Bolt 任务组成。现在以 WordCountTopology 任务为例，分析 Storm 数据处理的模型，并以此为基础开发自己的 WordCountTopology 程序。

#### 1. Storm 的数据分组策略

在讨论 Storm 任务处理的数据模型之前，首先来了解一下 Storm 的数据分组策略。数据分组策略为每一个 Bolt 定义怎么接收上一级组件发送来的数据。Storm 自带 8 种数据分组策略，也可以通过实现 CustomStreamingGrouping 接口来自定义分组策略。

- Shuffle grouping：随机处理。
  - 保证每个 task 处理的数据量是相同的。数据会被随机分发到下游 Bolt 的 task 中。
- Fields grouping：按字段分组。
  - 相同字段放到一个 task 中处理，可能发生数据倾斜。数据流会根据指定的字段分区，比如按照 userId 字段分区，相同的 userId 会分到相同的 task 中处理。
- Partial Key grouping。
  - 数据流会按照指定的字段分组，和 Fields grouping 类似，但是它能在下游 Bolt 中负载均衡，当输入数据发生数据倾斜的时候能让资源充分利用。

- All grouping。
  - ➢ 每个数据都会发到 Bolt 所有 task 中去。Bolt 的每个 task 收到的是相同的、完整的数据，如数字累加。设置 Bolt 有 3 个 task。spout 发 1，每个 task 都能收到 1，每个 task 累加后的值都是 1。spout 再发 2，每个 task 都能收到 2，每个 task 累加后的值都是 3。
- Global grouping。
  - ➢ 所有数据会发到同一个 task 中处理。
- None grouping。
  - ➢ 这种分组策略并不关心怎么分组，目前和 Shuffle grouping 一样。
- Direct grouping。
  - ➢ 由数据发送者直接指定哪一个 task 接收数据，只能在声明了 direct 的流中使用，要用 emitDirect 发送。
- Local or shuffle grouping。
  - ➢ 当下游 Bolt 中有 task 与上游的 task 运行在同样的 worker 中时，那么上游的 task 发出的数据由下游同进程的 task 处理。否则跟 shuffle grouping 一样。

### 2．WordCountTopology 数据处理的模型

在了解 Storm 数据分组的策略后，我们还需要知道 Storm 组件之间传递的数据格式为 Tuple，它相当于一张“表”。当在一级的组件中完成了数据的处理后，需要定义相应的 Tuple 格式 Schema（相当于表结构），然后根据 Schema 的格式将数据发送给下一级组件处理。

图 9.31 展示了 WordCountTopology 的数据处理的模型。

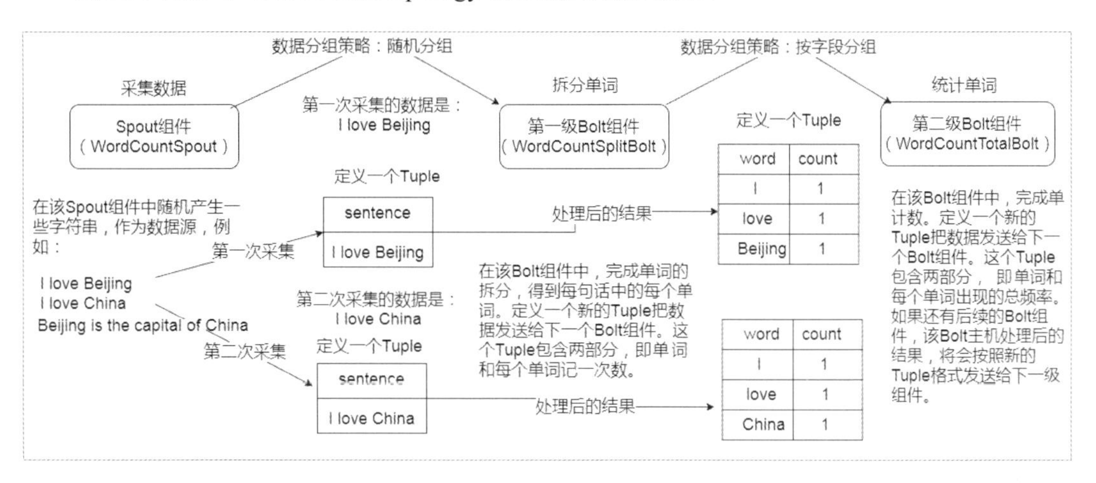


图 9.31　WordCountTopology 处理的数据模型

图 9.31 详细描述了 WordCountTopology 数据处理的模型。假设第一次由

WordCountSpout 组件采集的数据是“I love Beijing”，数据将按照 Tuple 的格式送入下一级组件 WordCountSplitBolt 进行处理。这里定义的 Tuple 中只包含一个字段，即 sentence，它代表采集的数据是什么，或者可以将其看成表中的列。当 WordCountSplitBolt 收到采集的数据后，完成分词的操作，得到这句话中的每个单词，再按照新定义的 Tuple 格式发送给 WordCountTotalBolt 进行单词的计数。这个新定义的 Tuple 格式包含两个字段，即<单词，每个单词记一次数>。

同时，需要考虑组件之间的分组策略。这里，WordCountSpout 与 WordCountSplitBolt 之间的分组策略为随机分组；而 WordCountSplitBolt 和 WordCountTotalBolt 之间的分组策略是字段分组。需要在创建 Topology 任务的主程序指定对应的分组策略。

在完成了整个处理过程后，如果还要进行后续的处理逻辑，可以定义新的 Bolt 组件完成对应的业务逻辑，例如，可以把处理完成的结果输出到 Redis 或其他一些持久化存储的介质中。

## 9.5.2 开发自己的 WordCountTopology 任务

基于前面的数据处理模型，我们接下来开发自己的 WordCountTopology 任务。该任务由一个 Spout 组件和两个 Bolt 组件组成。在开发 Storm 应用程序时，需要添加以下依赖信息。

```
01  <dependency>
02      <groupId>org.apache.storm</groupId>
03      <artifactId>storm-core</artifactId>
04      <version>1.0.3</version>
05      <scope>provided</scope>
06  </dependency>
```

具体的代码如下。

### 1. WordCountSpout 组件

```
01  import java.util.Map;
02  import java.util.Random;
03
04  import org.apache.storm.spout.SpoutOutputCollector;
05  import org.apache.storm.task.TopologyContext;
06  import org.apache.storm.topology.OutputFieldsDeclarer;
07  import org.apache.storm.topology.base.BaseRichSpout;
08  import org.apache.storm.tuple.Fields;
09  import org.apache.storm.tuple.Values;
10  import org.apache.storm.utils.Utils;
11
12  //数据采集
13  public class WordCountSpout extends BaseRichSpout{
```

```
14
15      private SpoutOutputCollector collector;
16
17      //模拟一些数据
18      private String[] data = {"I love Beijing",
19                               "I love China",
20                                   "Beijing is the capital of China"};
21
22      @Override
23      public void nextTuple() {
24          Utils.sleep(2000); //每两秒采集一次数据
25
26          //每次采集到数据后，如何处理
27
28          //随机采集一条数据
29          int random = (new Random()).nextInt(3);
30
31          //发送给下一级组件
32          this.collector.emit(new Values(data[random]));
33
34          System.out.println("采集的数据是：" + data[random]);
35      }
36
37      @Override
38      public void open(Map arg0, TopologyContext arg1,
39                      SpoutOutputCollector collector) {
40          //初始化的方法
41          //SpoutOutputCollector：代表 spout 的输出流
42          this.collector = collector;
43      }
44
45      @Override
46      public void declareOutputFields(OutputFieldsDeclarer declare) {
47          //申明输出的 Tuple 的格式
48          declare.declare(new Fields("sentence"));
49      }
50  }
```

2．WordCountSplitBolt 组件

```
01  import java.util.Map;
02
03  import org.apache.storm.task.OutputCollector;
04  import org.apache.storm.task.TopologyContext;
05  import org.apache.storm.topology.OutputFieldsDeclarer;
```

```
06  import org.apache.storm.topology.base.BaseRichBolt;
07  import org.apache.storm.tuple.Fields;
08  import org.apache.storm.tuple.Tuple;
09  import org.apache.storm.tuple.Values;
10
11  //进行单词拆分
12  public class WordCountSplitBolt extends BaseRichBolt{
13
14      private OutputCollector collector;
15
16      @Override
17      public void execute(Tuple tuple) {
18          //如何处理上一级组件发来的 Tuple
19          //取出数据: I love Beijing
20          String data = tuple.getStringByField("sentence");
21          String[] words = data.split(" ");
22
23          for(String w:words) {
24              //这里输出的是：每个单词记一次数
25               this.collector.emit(new Values(w,1));
26          }
27      }
28
29      @Override
30      public void prepare(Map arg0, TopologyContext arg1,
31                      OutputCollector collector) {
32          //OutputCollector 该级组件的输出流
33          this.collector = collector;
34      }
35
36      @Override
37      public void declareOutputFields(OutputFieldsDeclarer declare) {
38          declare.declare(new Fields("word","count"));
39      }
40  }
```

### 3. WordCountTotalBolt 组件

```
01  import java.util.HashMap;
02  import java.util.Map;
03
04  import org.apache.storm.task.OutputCollector;
05  import org.apache.storm.task.TopologyContext;
06  import org.apache.storm.topology.OutputFieldsDeclarer;
07  import org.apache.storm.topology.base.BaseRichBolt;
```

```
08  import org.apache.storm.tuple.Fields;
09  import org.apache.storm.tuple.Tuple;
10  import org.apache.storm.tuple.Values;
11
12  //单词的计数：累加
13  //注意Storm本身没有状态的管理
14  public class WordCountTotalBolt extends BaseRichBolt{
15
16      private OutputCollector collector;
17
18      //定义一个集合，用于存储每次统计的结果，相当于Redis、MySQL
19      private Map<String, Integer> result = new HashMap<String, Integer>();
20
21      @Override
22      public void execute(Tuple tuple) {
23          String word = tuple.getStringByField("word");
24          Integer count = tuple.getIntegerByField("count");
25
26          //累加
27          if(result.containsKey(word)) {
28              //已经存在，累加
29              int total = result.get(word);
30              result.put(word, total + count);
31          }else {
32              //第一次出现
33              result.put(word, count);
34          }
35
36          System.out.println("结果是: " + result);
37
38          //把结果发送给下一级组件
39          this.collector.emit(new Values(word,result.get(word)));
40      }
41
42      @Override
43      public void prepare(Map arg0, TopologyContext arg1,
44                      OutputCollector collector) {
45          this.collector = collector;
46      }
47
48      @Override
49      public void declareOutputFields(OutputFieldsDeclarer declare) {
50          //申明输出的Tuple的格式
51          declare.declare(new Fields("word","total"));
```

```
52      }
53  }
```

### 4. WordCountTopology 主程序

在开发主程序时，可以将其运行在本地模式上，也可以将其运行在集群模式上。本地模式是指在 IDE 开发环境中直接运行，这种模式和运行一个普通的 Java 程序是一样的；而集群模式需要将程序打包成一个 jar 包，通过 storm jar 命令提交到集群上执行。

下面的代码展示了本地模式下的 Topology 任务。

```
01  import org.apache.storm.Config;
02  import org.apache.storm.LocalCluster;
03  import org.apache.storm.StormSubmitter;
04  import org.apache.storm.generated.StormTopology;
05  import org.apache.storm.topology.TopologyBuilder;
06  import org.apache.storm.tuple.Fields;
07
08  public class WordCountTopology {
09
10      public static void main(String[] args) {
11          //创建一个 Topology
12          TopologyBuilder builder = new TopologyBuilder();
13
14          //指定任务的 spout 组件
15          builder.setSpout("myspout", new WordCountSpout());
16
17          //指定拆分单词的 bolt，是随机分组
18          builder.setBolt("mysplit",
19                      new WordCountSplitBolt())
20                          .shuffleGrouping("myspout");
21
22          //指定单词计数的 bolt
23          builder.setBolt("mytotal",
24                      new WordCountTotalBolt())
25                          .fieldsGrouping("mysplit", new Fields("word"));
26
27          //创建任务
28          StormTopology topology = builder.createTopology();
29
30          //配置参数
31          Config conf = new Config();
32
33          //本地模式
34          LocalCluster cluster = new LocalCluster();
35          cluster.submitTopology("MyWC", conf, topology);
```

```
36      }
37  }
```

只需要在上面代码的基础上，稍进行修改就可以将其改成集群模式。下面代码展示了集群模式下的 Topology 代码。

```
01  import org.apache.storm.Config;
02  import org.apache.storm.LocalCluster;
03  import org.apache.storm.StormSubmitter;
04  import org.apache.storm.generated.StormTopology;
05  import org.apache.storm.topology.TopologyBuilder;
06  import org.apache.storm.tuple.Fields;
07
08  public class WordCountTopology {
09
10      public static void main(String[] args) {
11          //创建一个 Topology
12          TopologyBuilder builder = new TopologyBuilder();
13
14          //指定任务的 spout 组件
15          builder.setSpout("myspout", new WordCountSpout());
16
17          //指定拆分单词的 bolt，是随机分组
18          builder.setBolt("mysplit",
19                          new WordCountSplitBolt())
20                              .shuffleGrouping("myspout");
21
22          //指定单词计数的 bolt
23          builder.setBolt("mytotal",
24                          new WordCountTotalBolt())
25                              .fieldsGrouping("mysplit", new Fields("word"));
26
27          //创建任务
28          StormTopology topology = builder.createTopology();
29
30          //配置参数
31          Config conf = new Config();
32
33          //集群模式
34          try {
35              StormSubmitter.submitTopology(args[0], conf, topology);
36          } catch (Exception e) {
37              e.printStackTrace();
38          }
39      }
40  }
```

开发集群模式后，可以直接在 IDE 环境中导出 jar 包，如图 9.32 所示。

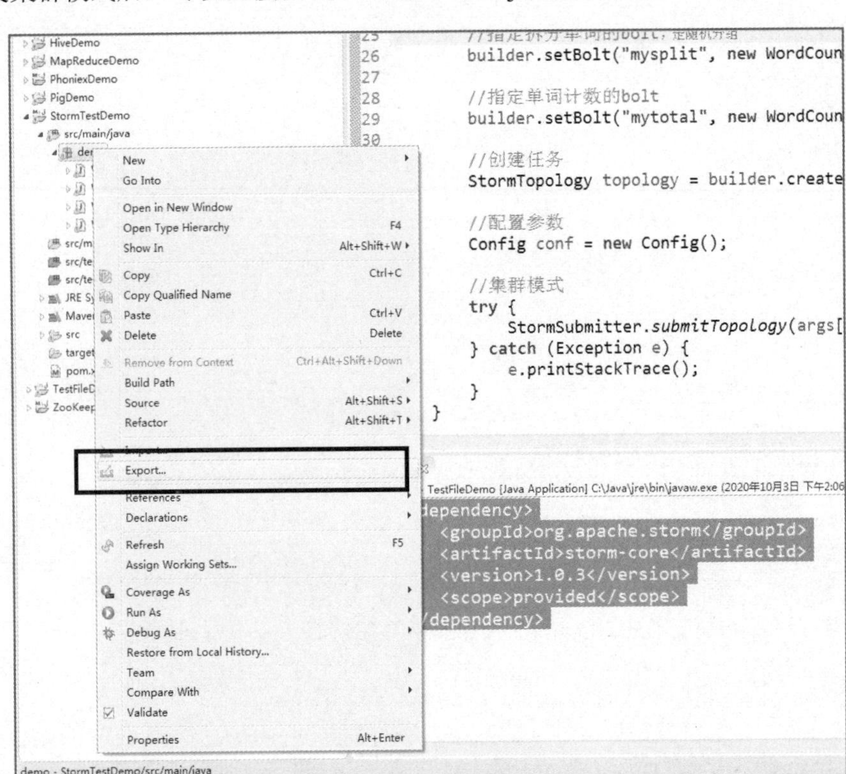

图 9.32 导出 Topology 的 jar 包

## 9.6 集成 Kafka 与 Storm

截至目前，我们已经知道 Storm 的作用主要是进行流式计算的，对于源源不断的均匀数据流的流入处理是非常有效的，而现实生活中大部分场景并不是均匀的数据流，而是时而多时而少的数据流入，在这种情况下用批量处理显然是不合适的，如果使用 Storm 进行实时计算可能因为数据拥堵导致服务器挂掉，应对这种情况，使用 Kafka 作为消息队列是非常合适的选择，Kafka 可以将不均匀的数据转换成均匀的消息流，从而和 Storm 进行完善的结合，这样才可以实现稳定的流式计算，那么接下来开发一个简单的案例来现 Storm 和 Kafka 的集成。

集成 Storm 和 Kafka 就是数据先由 Kafka 生产者发送至 Kafka 集群进行存储，然后 Storm 作为 Kafka 消费者进行消息的处理，最后将 Storm 处理的结果输出或保存到文件、数据库、分布式存储等，具体框架如图 9.33 所示。

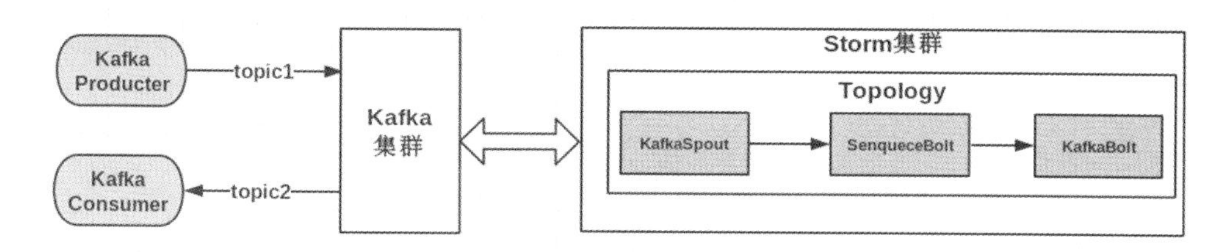


图 9.33　集成 Kafka 与 Storm 的框架

## 9.6.1　Storm 从 Kafka 中接收数据

为了实现 Kafka 与 Storm 的集成，需要添加以下依赖信息。

```
01  <dependency>
02      <groupId>org.apache.kafka</groupId>
03      <artifactId>kafka_2.9.2</artifactId>
04      <version>0.8.2.2</version>
05      <exclusions>
06          <exclusion>
07              <groupId>org.apache.zookeeper</groupId>
08              <artifactId>zookeeper</artifactId>
09          </exclusion>
10          <exclusion>
11              <groupId>log4j</groupId>
12              <artifactId>log4j</artifactId>
13          </exclusion>
14      </exclusions>
15  </dependency>
16  <dependency>
17      <groupId>org.apache.storm</groupId>
18      <artifactId>storm-kafka</artifactId>
19      <version>1.0.3</version>
20  </dependency>
```

修改之前的 WordCountTopology 主程序，创建一个新的 KafkaSpout 组件，并将其作为 WordCountTopology 任务的输入。KafkaSpout 将从 Kafka 中接收消息，作为 Kafka 消息的消费者使用。下面是创建这个 KafkaSpout 组件的核心代码程序。

```
01  //支持从 Kafka 消息系统中读取数据
02  private static KafkaSpout createKafkaSpout() {
03      //指定 ZooKeeper 的地址信息
04      BrokerHosts brokerHosts = new ZkHosts("kafka101:2181");
05  
06      //创建 KafkaSpout 的配置信息
07      SpoutConfig spoutConfig = new SpoutConfig(brokerHosts,
08                                      "mytopic1", "/mytopic1",
```

```
09                                             UUID.randomUUID().toString());
10      spoutConfig.scheme = new SchemeAsMultiScheme(new StringScheme());
11      spoutConfig.startOffsetTime = kafka.api.OffsetRequest.LatestTime();
12      //返回一个 KafkaSpout
13      return new KafkaSpout(spoutConfig);
14  }
```

需要注意的是，Storm 默认从头开始读取 Kafka 的消息，如果要从 Kafka 最新的地址开始读取数据，需要添加如下代码。

```
conf.startOffsetTime =  kafka.api.OffsetRequest.LatestTime();
```

另一方面，由于应用程序将从 Kafka 中接收消息，Kafka 发送来的消息中并没有“sentence”字段，所以需要修改 WordCountSplitBolt 的代码。

```
01  //进行单词拆分
02  public class WordCountSplitBolt extends BaseRichBolt{
03
04      private OutputCollector collector;
05
06      @Override
07      public void execute(Tuple tuple) {
08          //如何处理上一级组件发来的 Tuple
09          //取出数据，如：I love Beijing
10          //数据取出后赋值给下面的 data 变量
11          String data = tuple.getString(0);
12
13          String[] words = data.split(" ");
14
15          for(String w:words) {           //k2  v2
16              this.collector.emit(new Values(w,1));
17          }
18      }
19
20      @Override
21      public  void  prepare(Map  arg0,  TopologyContext  arg1,  OutputCollector
collector) {
22          //OutputCollector 该级组件的输出流
23          this.collector = collector;
24      }
25
26      @Override
27      public void declareOutputFields(OutputFieldsDeclarer declare) {
28          declare.declare(new Fields("word","count"));
```

```
29        }
30    }
```

需要注意的是，这里的第 10 行代码和第 11 行代码的区别。

为了测试方便，将 WordCountTopology 任务运行在本地模式下。下面展示了完整的代码程序。

```
01    import org.apache.storm.Config;
02    import org.apache.storm.LocalCluster;
03    import org.apache.storm.StormSubmitter;
04    import org.apache.storm.generated.AlreadyAliveException;
05    import org.apache.storm.generated.AuthorizationException;
06    import org.apache.storm.generated.InvalidTopologyException;
07    import org.apache.storm.generated.StormTopology;
08    import org.apache.storm.kafka.BrokerHosts;
09    import org.apache.storm.kafka.KafkaSpout;
10    import org.apache.storm.kafka.SpoutConfig;
11    import org.apache.storm.kafka.StringScheme;
12    import org.apache.storm.kafka.ZkHosts;
13    import org.apache.storm.spout.SchemeAsMultiScheme;
14    import org.apache.storm.topology.IRichBolt;
15    import org.apache.storm.topology.IRichSpout;
16    import org.apache.storm.topology.TopologyBuilder;
17    import org.apache.storm.tuple.Fields;
18    import org.apache.storm.tuple.ITuple;
19
20    public class WordCountTopology {
21
22        public static void main(String[] args) {
23            //创建一个 Topology
24            TopologyBuilder builder = new TopologyBuilder();
25
26            //指定任务的 spout 组件
27            //builder.setSpout("myspout", new WordCountSpout());
28            builder.setSpout("myspout", createKafkaSpout());
29
30            //指定拆分单词的 bolt，是随机分组
31            builder.setBolt("mysplit",
32                            new WordCountSplitBolt())
33                         .shuffleGrouping("myspout");
34
35            //指定单词计数的 bolt
36            builder.setBolt("mytotal",
37                            new WordCountTotalBolt())
```

```
38                    .fieldsGrouping("mysplit",
39                                   new Fields("word"));
40
41         //创建任务
42         StormTopology topology = builder.createTopology();
43
44         //配置参数
45         Config conf = new Config();
46
47         //本地模式
48         LocalCluster cluster = new LocalCluster();
49         cluster.submitTopology("MyWC", conf, topology);
50     }
51
52     private static IRichSpout createKafkaSpout() {
53
54         BrokerHosts zkHost = new ZkHosts("kafka101:2181");
55         SpoutConfig conf = new SpoutConfig(zkHost,
56                                      "mytopic1",
57                                           "/mytopic1",
58                                           "mygroupID");
59
60         //指定反序列化机制
61         conf.scheme = new SchemeAsMultiScheme(new StringScheme());
62
63         conf.startOffsetTime = kafka.api.OffsetRequest.LatestTime();
64
65         return new KafkaSpout(conf);
66     }
67 }
```

## 9.6.2 测试 Kafka 与 Storm 的集成

（1）在 kafka101、kafka102 和 kafka103 上启动 ZooKeeper 集群，执行下面的命令。

```
zkServer.sh start
```

（2）启动 Kafka 集群。

（3）执行下面的命令，启动 Kafka Producer Console。

```
bin/kafka-console-producer.sh --broker-list kafka101:9092 --topic mytopic1
```

（4）启动 Storm 应用程序，如图 9.34 所示。

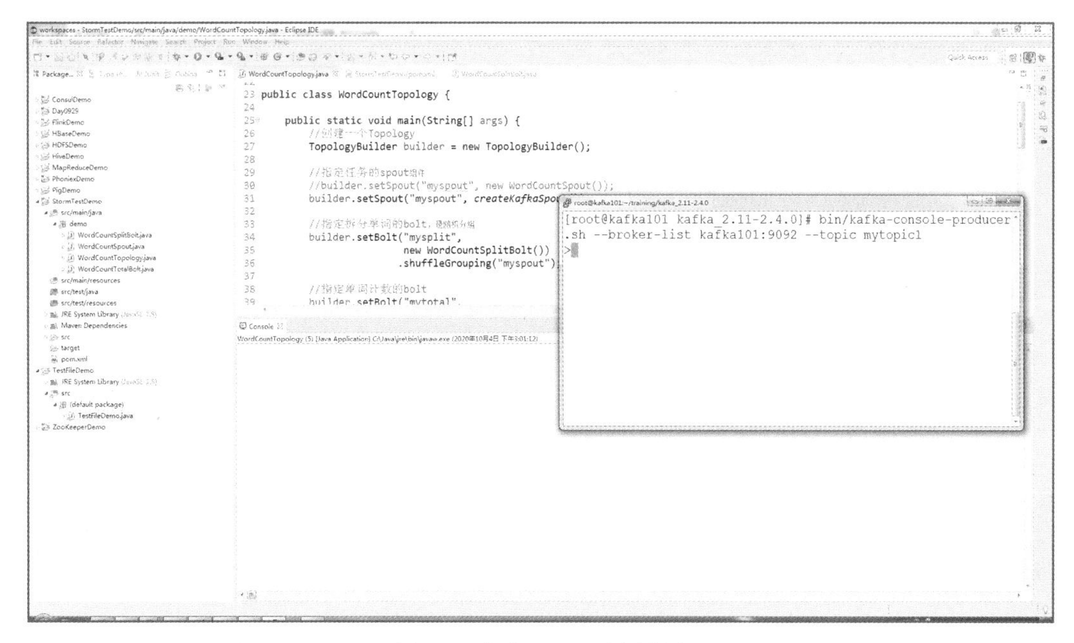


图 9.34 启动 Storm 应用程序

（5）在 Kafka Producer Console 中输入一些字符串，并按 Enter 键发送。观察 Storm 应用程序的结果。可以看到当在 Kafka Producer Console 中输入了数据时，在 Storm 的应用程序中将实时接收消息并处理消息，如图 9.35 所示。

图 9.35　Storm 应用程序接收 Kafka 的消息

## 9.6.3 Storm 将数据输出到 Kafka

在前面的案例中，我们集成了 Storm 和 Kafka。将 Kafka 作为 Storm 的 Spout，Storm 将从 Kafka 中接收数据并处理数据。其实还有另一种情况，就是 Storm 处理完成后，也可以将数据输出到 Kafka 中，图 9.36 描述了这一过程。

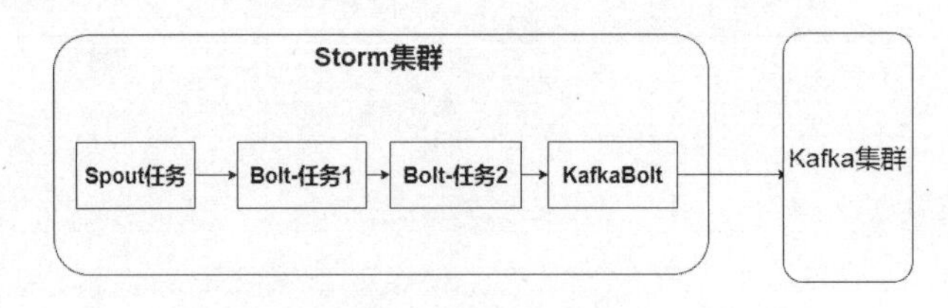


图 9.36 Storm 应用程序将消息发送到 Kafka

在了解了上面的过程后，可以改造一下之前的 Topology 主程序的代码，创建一个新的 Bolt 任务，将任务处理完成的数据输出到 Kafka 中，下面列出了改造后的代码程序。

```
01  private static IRichBolt createKafkaBolt() {
02      Properties props = new Properties();
03      //broker 的地址
04      props.put("bootstrap.servers", "kafka101:9092");
05      //说明了使用何种序列化方式将用户提供的 key 和 value 值序列化成字节
06      props.put("key.serializer",
07              "org.apache.kafka.common.serialization.StringSerializer");
08      props.put("value.serializer",
09              "org.apache.kafka.common.serialization.StringSerializer");
10
11      props.put("acks", "1");
12
13      KafkaBolt<String, String> bolt = new KafkaBolt<String, String>();
14
15      //指定 Kafka 的配置信息
16      bolt.withProducerProperties(props);
17
18      //指定 Topic 的名字
19      bolt.withTopicSelector(new DefaultTopicSelector("mytopic1"));
20
21      //指定上一级 Bolt 处理完成后的 Key 和 Value
22      //KafkaBolt 将会按照这里指定的 Key 和 Value 将数据发送到 Kafka
23      bolt.withTupleToKafkaMapper(new FieldNameBasedTupleToKafkaMapper
24                              <String, String>("word","total"));
25
26      return bolt;
27  }
```

创建好新的 Bolt 组件后，可以将其添加到 Topology 任务中，下面列出了改造后的主程序代码。

```
01  public static void main(String[] args) {
02      //创建一个 Topology
03      TopologyBuilder builder = new TopologyBuilder();
04
05      //指定任务的 Spout 组件
06      //builder.setSpout("myspout", new WordCountSpout());
07      builder.setSpout("myspout", createKafkaSpout());
08
09      //指定拆分单词的 Bolt，是随机分组
10      builder.setBolt("mysplit",
11                  new WordCountSplitBolt())
12                     .shuffleGrouping("myspout");
13
14      //指定单词计数的 Bolt
15      builder.setBolt("mytotal",
16                  new WordCountTotalBolt())
17                     .fieldsGrouping("mysplit",
18                                 new Fields("word"));
19
20      //创建 KafkaBolt，将结果发送到 Kafka 中
21      builder.setBolt("kafkabolt", createKafkaBolt())
22             .shuffleGrouping("mytotal");
23
24      //创建任务
25      StormTopology topology = builder.createTopology();
26
27      //配置参数
28      Config conf = new Config();
29
30      //本地模式
31      LocalCluster cluster = new LocalCluster();
32      cluster.submitTopology("MyWC", conf, topology);
33  }
```

注意这里的第21行代码和第22行代码，我们使用了createKafkaBolt方法，将KafkaBolt任务添加到 Topology 任务中。

# 第 10 章 Kafka 与 Spark 集成

Spark 是一种快速、通用、可扩展的大数据分析引擎，2009 年诞生于加州大学伯克利分校 AMPLab，2010 年开源，2013 年 6 月成为 Apache 孵化项目，2014 年 2 月成为 Apache 顶级项目。目前，Spark 生态系统已经发展成为一个包含多个子项目的集合，其中包含 SparkSQL、Spark Streaming、GraphX、MLlib 等。Spark 是基于内存计算的大数据并行计算框架。网站的首页对 Spark 有相应的描述，如图 10.1 所示。

图 10.1　Spark 官网的描述

从这里的描述可以看出 Spark 是一个通用的、用于处理大规模数据集的分析引擎。利用 Spark Core 能够完成大数据的离线计算，类似 Hadoop 中的 MapReduce 或 Flink 中的 DataSet API；利用 Spark Streaming 能够完成大数据的流式计算，类似 Apache Storm 或 Flink 中的 DataStream API。

## 10.1 Spark 基础

Spark 基于内存计算的大数据计算框架，提高了在大数据环境下数据处理的实时性，同时保证了高容错性和高可伸缩性，允许用户将 Spark 部署在大量廉价硬件之上，形成集群。众多大数据公司中的产品都支持 Spark，其中包括 Intel、京东、Hortonworks、携程、优酷、土豆、百度、阿里巴巴、腾讯、IBM 等。那么 Spark 为什么如此受欢迎，接下来我们了解一下 Spark 的特点。

## 10.1.1　Spark 的特点

### 1. 快

与 Hadoop 的 MapReduce 相比，Spark 基于内存的运算速度要快 100 倍以上，Spark 基于硬盘的运算也要快 10 倍。Spark 实现了高效的 DAG 执行引擎，从而可以通过内存来高效处理数据流。

### 2. 易用性

Spark 的易用性主要体现在支持多种编程语言。Spark 支持 Java、Python 和 Scala 的 API，还支持 80 多种高级算法，使用户可以快速构建不同的应用。Spark 支持交互式的 Python 和 Scala 的 shell，可以非常方便地在这些 shell 中使用 Spark 集群来验证解决问题的方法。

### 3. 兼容性

Spark 可以非常方便地与其他开源产品进行融合。比如，Spark 可以使用 Hadoop 的 YARN 和 Apache Mesos 作为其资源管理和调度器，并且可以处理所有 Hadoop 支持的数据，包括 HDFS、HBase 和 Cassandra 等。这对于已经部署 Hadoop 集群的用户特别重要，不需要做任何数据迁移就可以使用 Spark 的强大处理能力。Spark 也可以不依赖于第三方的资源管理和调度器，它实现了 Standalone 作为其内置的资源管理和调度框架，进一步降低了 Spark 的使用门槛，使所有人都可以非常容易地部署和使用 Spark。此外，Spark 还提供了在 EC2 上部署 Standalone 的 Spark 集群的工具。

### 4. 通用性

Spark 提供了统一的解决方案。Spark 可以用于批处理、交互式查询、实时流处理、机器学习和图计算。这些不同类型的处理都可以在同一个应用中无缝使用。Spark 统一的解决方案非常具有吸引力，毕竟任何公司都想用统一的平台去处理遇到的问题，减少开发和维护的人力成本及部署平台的物力成本。另外，Spark 还可以很好地融入 Hadoop 的体系结构中直接操作 HDFS，并提供 Hive on Spark、Pig on Spark 的框架集成 Hadoop。

图 10.2 展示了 Spark 的生态圈体系中的组件。

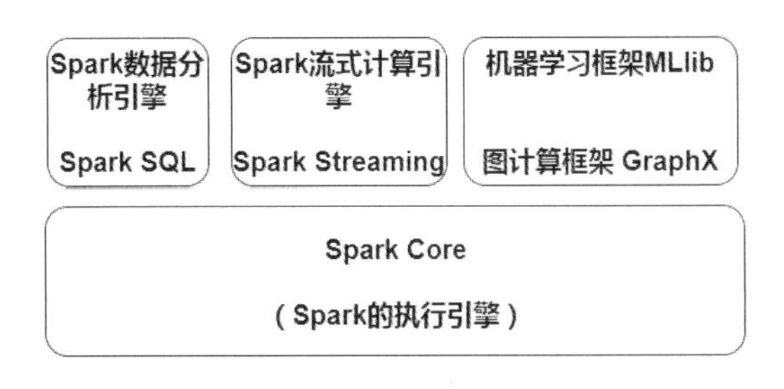


图 10.2　Spark 的生态圈组件

其中，

- Spark Core 用于大数据的批处理离线计算，它是整个 Spark 中的核心部分，也是 Spark 的执行引擎。
- Spark SQL 用于交互式查询，是 Spark 提供的数据分析引擎，在 Spark SQL 中可以使用标准的 SQL 语句处理大数据。
- Spark Streaming 是 Spark 中用于实时流处理计算的框架。
- Spark MLlib 是机器学习的框架，其本质是一系列的算法。
- GraphX 用于 Spark 的图计算框架，类似 Hadoop 中 GraphX 或 Flink 中的 Gelly。

## 10.1.2 Spark 的体系架构

Spark 的体系架构分为客户端和服务器端，即 Client-Server 结构。图 10.3 描述了 Spark 体系架构。

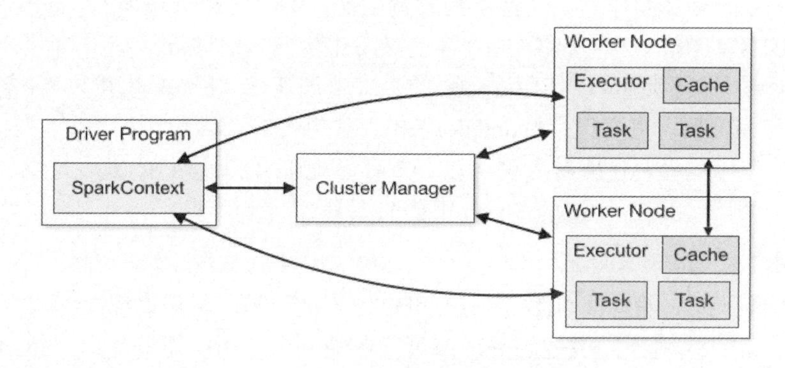


图 10.3 Spark 的体系架构

Spark 应用程序在 Spark 集群上作为独立的应用程序运行。我们知道 Spark Core 是整个 Spark 中的核心部分，想要访问 Spark Core，就需要通过它的核心对象 SparkContext。这就决定了 SparkContext 是整个 Spark 中非常重要的对象，也是非常重要的编程接口。可以把 SparkContext 对象理解成客户端程序，称为 Driver Program。

成功创建 SparkContext 对象后，可以连接几种类型的集群管理器，例如，YARN 和 Mesos 可以在应用程序之间分配资源；当然 Spark 也可以独立运行，即 Spark 的 Standalone 模式。连接后，Spark 会在集群中的节点上获取执行程序，这些节点是运行计算。这个运算过程中的核心数据模型是 Spark RDD（弹性分布式数据集，这个概念在后续章节中介绍）。

这里需要注意以下几点。

（1）Spark 的服务器端是主从式架构，Cluster Manager 是整个 Spark 的主节点，如果我们把 Spark 部署在 Standalone 模式下，它就是 Spark 的 Master 节点；任务最终由 Spark

的从节点 Worker 来执行。

（2）Spark 的服务器端既然是一种主从式架构，就存在单点故障的问题；要解决这样的问题，就需要使用 ZooKeeper 实现 Spark 的 HA 特性。

（3）每个 Spark 应用都有自己的执行程序进程，这些进程在整个应用程序期间保持不变，并在多个线程中运行任务。但是，如果不将数据写入外部存储系统，就无法实现在不同的 Spark 应用程序之间共享数据。

（4）驱动程序（Driver Program）在其整个生命周期中必须侦听并接受其执行程序的传入连接。因此，驱动程序必须是可从工作程序节点访问的网络。由于驱动程序在集群上调度任务，并最终由 Worker 从节点执行，最好把驱动程序和 Worker 运行在同一局域网上。

（5）如果想将 Spark 任务请求远程发送到集群，最好在驱动程序打开 RPC，并让它在集群附近提交任务请求，而不是在远离 Spark Worker 工作节点的地方运行驱动程序。

图 10.4 从 Spark 任务调度的角度对 Spark 集群进行了描述。

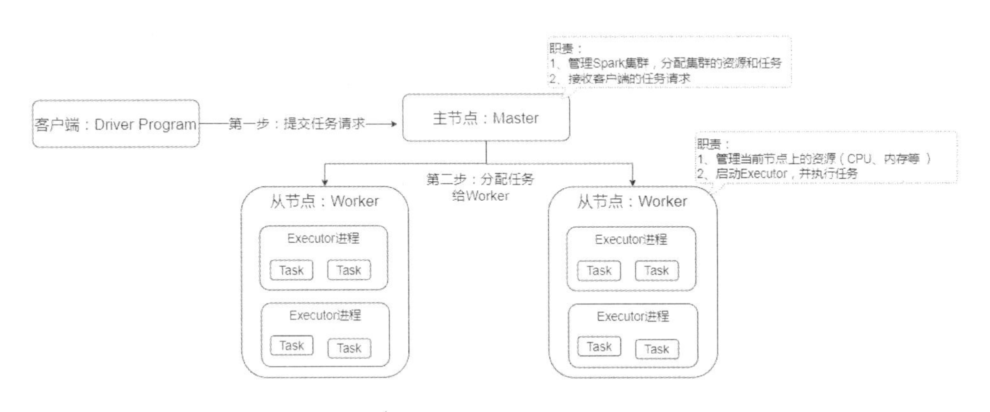


图 10.4　Spark 的任务分配

这里需要说明的是 Spark 客户端，即 Driver Program。当完成 Spark 的安装部署后，Spark 提供了两个不同的客户端工具，帮助应用开发人员将 Spark 任务提交到 Spark 集群运行。这个客户端工具分别是 spark-submit 和 spark-shell。它们的具体使用方法会在部署完成 Spark 集群后，再进行详细介绍。

## 10.2　安装部署 Spark 环境

这里我们使用的版本是 spark-3.0.0-bin-hadoop3.2.tgz。Spark 可以部署到多种资源管理平台上，主要有以下几种。

- Standalone 是 Spark 的独立运行模式，也是一种典型的 Mater-Slave 模式，Master 节点存在单点故障的问题，需要使用 ZooKeeper 实现 Spark HA。

- YARN 将 Spark 集群运行在 YARN 资源管理器框架之上，由 YARN 负责资源管理，Spark 负责任务调度和计算。从 Hadoop 2.x 版本后，Hadoop 开始提供 YARN 的资源管理容器，它是 Hadoop 的一个资源管理系统。
- Mesos 将 Spark 集群运行在 Mesos 资源管理器框架上，由 Mesos 负责资源管理，Spark 负责任务调度和计算。
- Cloud 将 Spark 部署在云环境下，如 Docker、Kubernetes 或 AWS 的 EC2，使用这个模式能很方便地访问云模式下的存储介质，如 HDFS 和 S3 等。

接下来，以 Spark Standalone 的模式介绍 Spark 的部署，在之前配置好的虚拟主机上安装部署 Spark 集群，重点介绍以下两种环境的安装部署。

- 伪分布模式的单节点环境部署。
- 全分布模式的环境安装部署。

下面列出了 Spark 的核心配置文件——spark-env.sh 的文件内容。需要注意的是，在默认情况下，这个配置文件是没有的，需要手动生成这个文件。

```
01  #!/usr/bin/env bash
02
03  #
04  # Licensed to the Apache Software Foundation (ASF) under one or more
05  # contributor license agreements.  See the NOTICE file distributed with
06  # this work for additional information regarding copyright ownership.
07  # The ASF licenses this file to You under the Apache License, Version 2.0
08  # (the "License"); you may not use this file except in compliance with
09  # the License.  You may obtain a copy of the License at
10  #
11  #    http://www.apache.org/licenses/LICENSE-2.0
12  #
13  # Unless required by applicable law or agreed to in writing, software
14  # distributed under the License is distributed on an "AS IS" BASIS,
15  # WITHOUT WARRANTIES OR CONDITIONS OF ANY KIND, either express or implied.
16  # See the License for the specific language governing permissions and
17  # limitations under the License.
18  #
19
20  # This file is sourced when running various Spark programs.
21  # Copy it as spark-env.sh and edit that to configure Spark for your site.
22
23  # Options read when launching programs locally with
24  # ./bin/run-example or ./bin/spark-submit
25  # - HADOOP_CONF_DIR, to point Spark towards Hadoop configuration files
26  # - SPARK_LOCAL_IP, to set the IP address Spark binds to on this node
27  # - SPARK_PUBLIC_DNS, to set the public dns name of the driver program
28
```

```
29  # Options read by executors and drivers running inside the cluster
30  # - SPARK_LOCAL_IP, to set the IP address Spark binds to on this node
31  # - SPARK_PUBLIC_DNS, to set the public DNS name of the driver program
32  # - SPARK_LOCAL_DIRS, storage directories to use on this node for shuffle
and RDD data
33  # - MESOS_NATIVE_JAVA_LIBRARY, to point to your libmesos.so if you use Mesos
34
35  # Options read in YARN client/cluster mode
36  # - SPARK_CONF_DIR, Alternate conf dir. (Default: ${SPARK_HOME}/conf)
37  # - HADOOP_CONF_DIR, to point Spark towards Hadoop configuration files
38  # - YARN_CONF_DIR, to point Spark towards YARN configuration files when you
use YARN
39  # - SPARK_EXECUTOR_CORES, Number of cores for the executors (Default: 1).
40  # - SPARK_EXECUTOR_MEMORY, Memory per Executor (e.g. 1000M, 2G) (Default: 1G)
41  # - SPARK_DRIVER_MEMORY, Memory for Driver (e.g. 1000M, 2G) (Default: 1G)
42
43  # Options for the daemons used in the standalone deploy mode
44  # - SPARK_MASTER_HOST, to bind the master to a different IP address or
hostname
45  # - SPARK_MASTER_PORT / SPARK_MASTER_WEBUI_PORT, to use non-default ports
for the master
46  # - SPARK_MASTER_OPTS, to set config properties only for the master (e.g. "-
Dx=y")
47  # - SPARK_WORKER_CORES, to set the number of cores to use on this machine
48  # - SPARK_WORKER_MEMORY, to set how much total memory workers have to give
executors (e.g. 1000m, 2g)
49  # - SPARK_WORKER_PORT / SPARK_WORKER_WEBUI_PORT, to use non-default ports
for the worker
50  # - SPARK_WORKER_DIR, to set the working directory of worker processes
51  # - SPARK_WORKER_OPTS, to set config properties only for the worker (e.g. "-
Dx=y")
52  # - SPARK_DAEMON_MEMORY, to allocate to the master, worker and history server
themselves (default: 1g).
53  # - SPARK_HISTORY_OPTS, to set config properties only for the history server
(e.g. "-Dx=y")
54  # - SPARK_SHUFFLE_OPTS, to set config properties only for the external shuffle
service (e.g. "-Dx=y")
55  # - SPARK_DAEMON_JAVA_OPTS, to set config properties for all daemons (e.g.
"-Dx=y")
56  # - SPARK_DAEMON_CLASSPATH, to set the classpath for all daemons
57  # - SPARK_PUBLIC_DNS, to set the public dns name of the master or workers
58
59  # Options for launcher
```

```
60  # - SPARK_LAUNCHER_OPTS, to set config properties and Java options for the launcher (e.g. "-Dx=y")
61
62  # Generic options for the daemons used in the standalone deploy mode
63  # - SPARK_CONF_DIR     Alternate conf dir. (Default: ${SPARK_HOME}/conf)
64  # - SPARK_LOG_DIR         Where log files are stored.  (Default: ${SPARK_HOME}/logs)
65  # - SPARK_PID_DIR      Where the pid file is stored. (Default: /tmp)
66  # - SPARK_IDENT_STRING  A string representing this instance of spark. (Default: $USER)
67  # - SPARK_NICENESS     The scheduling priority for daemons. (Default: 0)
68  # - SPARK_NO_DAEMONIZE  Run the proposed command in the foreground. It will not output a PID file.
69  # Options for native BLAS, like Intel MKL, OpenBLAS, and so on.
70  # You might get better performance to enable these options if using native BLAS (see SPARK-21305).
71  # - MKL_NUM_THREADS=1       Disable multi-threading of Intel MKL
72  # - OPENBLAS_NUM_THREADS=1   Disable multi-threading of OpenBLAS
```

在部署 Spark 集群时，可以根据实际部署的模式选择需要配置的参数，不需要对上面的每个参数进行配置。例如，如果部署 Spark on YARN 只需要配置以下参数即可。

```
01  # Options read in YARN client/cluster mode
02  # - SPARK_CONF_DIR, Alternate conf dir. (Default: ${SPARK_HOME}/conf)
03  # - HADOOP_CONF_DIR, to point Spark towards Hadoop configuration files
04  # - YARN_CONF_DIR, to point Spark towards YARN configuration files when you use YARN
05  # - SPARK_EXECUTOR_CORES, Number of cores for the executors (Default: 1).
06  # - SPARK_EXECUTOR_MEMORY, Memory per Executor (e.g. 1000M, 2G) (Default: 1G)
07  # - SPARK_DRIVER_MEMORY, Memory for Driver (e.g. 1000M, 2G) (Default: 1G)
```

## 10.2.1 伪分布模式的单节点环境部署

在这种模式下，只存在一个 Master 节点和一个 Worker 节点。直接在 kafka101 的主机上进行部署即可。

（1）将 Spark 安装包解压到/root/training/目录下。

```
tar -zxvf spark-3.0.0-bin-hadoop3.2.tgz -C /root/training/
```

（2）切换到 Spark 的 conf 目录，并生成 spark-env.sh 文件。

```
cd /root/training/spark-3.0.0-bin-hadoop3.2/conf/
cp spark-env.sh.template spark-env.sh
```

（3）最后在 spark-env.sh 文件中增加以下内容。

```
export JAVA_HOME=/root/training/jdk1.8.0_181
export SPARK_MASTER_HOST=kafka101
export SPARK_MASTER_PORT=7077
```

其中，

- SPARK_MASTER_HOST 为 Spark Master 节点的地址信息。
- SPARK_MASTER_PORT 为客户端 Driver Program 连接集群的端口。

（4）启动 Spark 集群。由于没有设置 Spark 的环境变量，所以需要进入 Spark 的安装目录执行下面的语句命令。

```
cd /root/training/spark-3.0.0-bin-hadoop3.2/
sbin/start-all.sh
```

这里需要注意的是，需要配置主机的免密码登录。

图 10.5 展示了伪分布模式 Spark 集群的启动过程。

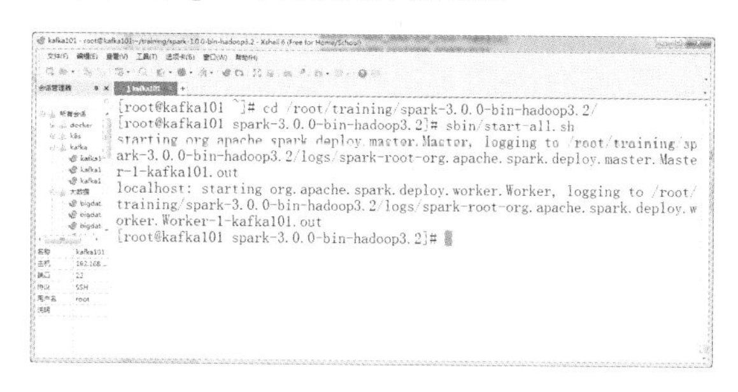


图 10.5　伪分布模式 Spark 集群的启动过程

（5）执行 jps 命令查看 Spark 的后台进程信息。

```
jps
```

可以看到有一个 Master 进程和一个 Worker 进程在运行，即 Spark 的主节点和从节点。

（6）通过浏览器访问 Spark 的 Web Console，端口号为 8080，如图 10.6 所示。

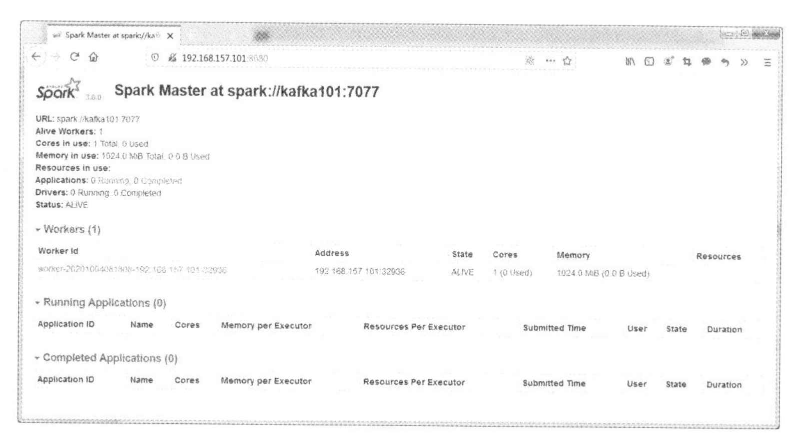


图 10.6　伪分布模式 Spark 的 Web Console

在 Spark 的 Web Console 上可以看到只有一个 Worker 的从节点信息，并且没有任何任务在运行。稍后我们会通过 spark-submit 和 spark-shell 连接到集群上，并将 Spark

任务提交到集群。当任务成功提交并运行后，通过 Spark Web Console 可以监控任务的运行状态。

## 10.2.2 全分布模式的环境安装部署

在 Spark 伪分布环境的基础上，可以进一步搭建 Spark 的全分布环境，这里需要三台主机进行部署。图 10.7 展示了全分布模式 Spark 集群架构。

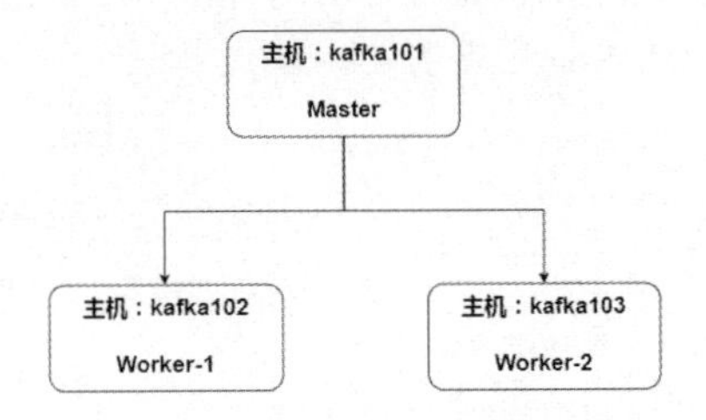


图 10.7 全分布模式 Spark 集群架构

（1）在 kafka101 主机上，停止之前部署好的 Spark 伪分布环境，执行下面的语句。

```
sbin/stop-all.sh
```

（2）进入 Spark 的 conf 目录，生成 slaves 文件。

```
cd /root/training/spark-3.0.0-bin-hadoop3.2/conf/
cp slaves.template slaves
```

（3）修改 slaves 文件的内容如下。

```
# A Spark Worker will be started on each of the machines listed below.
kafka102
kafka103
```

（4）将 kafka101 主机上的 Spark 目录复制到 kafka102 主机和 kafka103 主机上。

```
scp -r /root/training/spark-3.0.0-bin-hadoop3.2/ root@kafka102:/root/training
scp -r /root/training/spark-3.0.0-bin-hadoop3.2/ root@kafka103:/root/training
```

（5）在 kafka101 主机上启动 Spark 集群。

```
cd /root/training/spark-3.0.0-bin-hadoop3.2/
sbin/start-all.sh
```

图 10.8 展示了全分布模式 Spark 集群的启动过程。

```
kafka103: stopping org.apache.spark.deploy.worker.Worker
stopping org.apache.spark.deploy.master.Master
[root@kafka101 spark-3.0.0-bin-hadoop3.2]#
[root@kafka101 spark-3.0.0-bin-hadoop3.2]# sbin/start-all.sh
starting org.apache.spark.deploy.master.Master, logging to /root/training/spark-3.0.0-bin-hadoop3.2/logs/spark-root-org.apache.spark.deploy.master.Master-1-kafka101.out
kafka102: starting org.apache.spark.deploy.worker.Worker, logging to /root/training/spark-3.0.0-bin-hadoop3.2/logs/spark-root-org.apache.spark.deploy.worker.Worker-1-kafka102.out
kafka103: starting org.apache.spark.deploy.worker.Worker, logging to /root/training/spark-3.0.0-bin-hadoop3.2/logs/spark-root-org.apache.spark.deploy.worker.Worker-1-kafka103.out
[root@kafka101 spark-3.0.0-bin-hadoop3.2]#
```

图 10.8 全分布模式 Spark 集群的启动过程

（6）在 kafka101 主机、kafka102 主机和 kafka103 主机上执行 jps 命令查看后台的进程信息，如图 10.9 所示。在 kafka101 主机上启动了 Master，在 kafka102 主机和 kafka103 主机上各启动了一个 Worker。

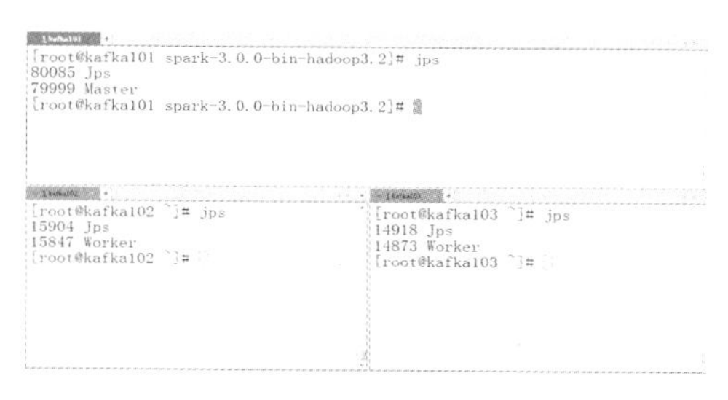


图 10.9　全分布模式 Spark 集群的进程信息

（7）访问 kafka101 主机上的 Spark Web Console 界面，如图 10.10 所示，可以看到有两个 Worker 的从节点，分别运行在 192.168.157.102 主机和 192.168.157.103 主机上，即 kafka102 主机和 kafka103 主机。

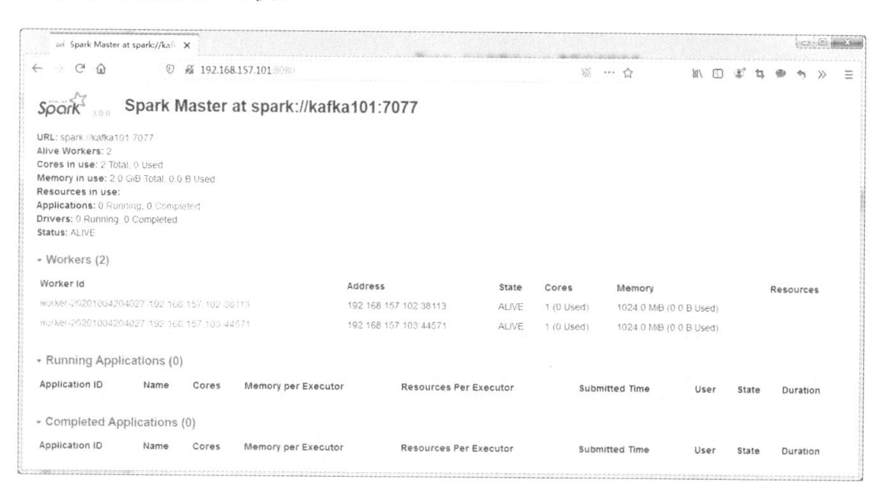


图 10.10　全分布模式的 Spark Web Console 界面

至此，我们已经成功部署了 Spark 的两种运行环境，环境部署完成后，就可以使用 Spark 提供 Driver Program 客户端工具将任务提交到集群运行了。

# 10.3　执行 Spark 任务

## 10.3.1　使用 spark-submit 提交任务

部署好 Spark 环境后，可以将 Spark 程序打包成 jar 包，使用 bin/spark-submit 将应用程序提交到 Spark 集群运行。这个脚本负责设置 Spark 使用的 classpath 和依赖，支持不同类型的集群管理器和发布模式。

下面列出了这个脚本的命令基本用法和参数说明。

```
01  ./bin/spark-submit \
02  --class <main-class>
03  --master <master-url> \
04  --deploy-mode <deploy-mode> \
05  --conf <key>=<value> \
06  ... # other options
07  <application-jar> \
08  [application-arguments]
```

其中，

- --class：应用的启动类，例如，org.apache.spark.examples.SparkPi。
- --master：集群的 master URL，例如，spark://kafka101:7077 可以是本地，也可以是 YARN。
- --deploy-mode：是否将驱动发布到 worker 节点（cluster）或作为一个本地客户端（client）。
- --conf：任意的 Spark 配置属性，其格式为 key=value。如果值包含空格，则可以加引号“key=value”，默认 Spark 配置。
- application-jar：打包好的应用 jar，包含依赖。这个 URL 在集群中全局可见。比如 hdfs:// 共享存储系统，如果是 file:// path，那么所有节点的 path 都包含同样的 jar。
- application-arguments：传给 main()方法的参数。

现在以官方提供的示例程序蒙特·卡罗算法求 PI，演示 spark-submit 的用法。进入 Spark 的安装目录执行下面的语句。

```
cd /root/training/spark-3.0.0-bin-hadoop3.2/

bin/spark-submit \
--class org.apache.spark.examples.SparkPi \
--master spark://kafka101:7077 \
--executor-memory 1G \
--total-executor-cores 2 \
examples/jars/spark-examples_2.12-3.0.0.jar \
100
```

其中，

--master spark://kafka101:7077 指定 Master 的地址。

--executor-memory 1G 指定每个 executor 可用内存为 1G。

--total-executor-cores 2 指定每个 executor 使用的 CPU 核数为 2。

图 10.11 展示了执行蒙特•卡罗算法求 PI。

```
20/10/04 20:08:52 INFO TaskSetManager: Finished task 96.0 in stage 0.0 (TID 96) in 40 ms on 192.168.157.101 (executor 0) (97/100)
20/10/04 20:08:52 INFO TaskSetManager: Starting task 98.0 in stage 0.0 (TID 98, 192.168.157.101, executor 0, partition 98, PROCESS_LOCAL, 7397 bytes)
20/10/04 20:08:52 INFO TaskSetManager: Finished task 97.0 in stage 0.0 (TID 97) in 47 ms on 192.168.157.101 (executor 0) (98/100)
20/10/04 20:08:52 INFO TaskSetManager: Starting task 99.0 in stage 0.0 (TID 99, 192.168.157.101, executor 0, partition 99, PROCESS_LOCAL, 7397 bytes)
20/10/04 20:08:52 INFO TaskSetManager: Finished task 98.0 in stage 0.0 (TID 98) in 26 ms on 192.168.157.101 (executor 0) (99/100)
20/10/04 20:08:52 INFO TaskSetManager: Finished task 99.0 in stage 0.0 (TID 99) in 26 ms on 192.168.157.101 (executor 0) (100/100)
20/10/04 20:08:52 INFO DAGScheduler: ResultStage 0 (reduce at SparkPi.scala:38) finished in 11.816 s
20/10/04 20:08:52 INFO DAGScheduler: Job 0 is finished. Cancelling potential speculative or zombie tasks for this job
20/10/04 20:08:52 INFO TaskSchedulerImpl: Removed TaskSet 0.0, whose tasks have all completed, from pool
20/10/04 20:08:52 INFO TaskSchedulerImpl: Killing all running tasks in stage 0: Stage finished
20/10/04 20:08:52 INFO DAGScheduler: Job 0 finished: reduce at SparkPi.scala:38, took 12.137306 s
Pi is roughly 3.1412643141264316
20/10/04 20:08:52 INFO SparkUI: Stopped Spark web UI at http://kafka101:4040
20/10/04 20:08:52 INFO StandaloneSchedulerBackend: Shutting down all executors
20/10/04 20:08:52 INFO CoarseGrainedSchedulerBackend$DriverEndpoint: Asking each executor to shut down
20/10/04 20:08:52 INFO MapOutputTrackerMasterEndpoint: MapOutputTrackerMasterEndpoint stopped!
20/10/04 20:08:52 INFO MemoryStore: MemoryStore cleared
20/10/04 20:08:52 INFO BlockManager: BlockManager stopped
20/10/04 20:08:52 INFO BlockManagerMaster: BlockManagerMaster stopped
20/10/04 20:08:52 INFO OutputCommitCoordinator$OutputCommitCoordinatorEndpoint: OutputCommitCoordinator stopped!
20/10/04 20:08:52 INFO SparkContext: Successfully stopped SparkContext
20/10/04 20:08:52 INFO ShutdownHookManager: Shutdown hook called
20/10/04 20:08:52 INFO ShutdownHookManager: Deleting directory /tmp/spark-e9a21a68-3c00-4873-ac54-7275e9fe70e4
20/10/04 20:08:52 INFO ShutdownHookManager: Deleting directory /tmp/spark-89011dc9-d18a-44b0-ae53-75123e17a573
[root@kafka101 spark-3.0.0-bin-hadoop3.2]#
```

图 10.11 执行蒙特·卡罗算法求 PI

通过打印输出的日志，可以看到在循环迭代 100 次的条件下，计算处理的圆周率是 3.1412643141264316。

打开浏览器访问 Spark 的 Web Console，可以监控任务的执行过程，如图 10.12 所示。

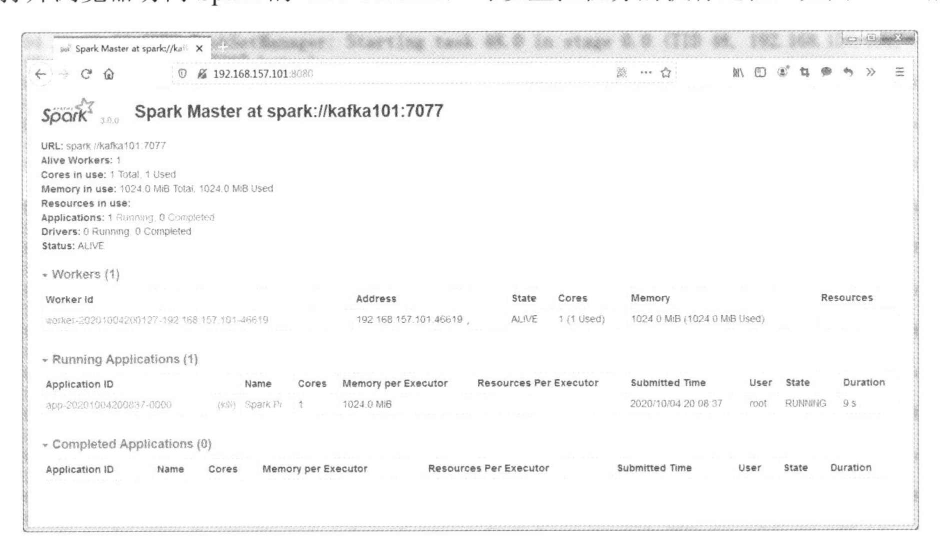


图 10.12 使用 Spark Web Console 监控任务

## 10.3.2 交互式命令行工具 spark-shell

spark-shell 是 Spark 自带的交互式命令行工具，类似 Scala 语言的 REPL 命令行，方便用户进行交互式编程，用户可以在该命令行下用 scala 编写 Spark 程序并执行程序。spark-shell 有两种运行模式，即 local 模式和 cluster 模式。

### 1. local 模式

local 模式是 spark-shell 的本地模式。在这种模式下，客户端并没有真正连接到 Spark 集群，应用程序也是在本地执行的，与在本地执行一个普通的 Java 程序或 Scala 程序一样。通过下面的方式启动 spark-shell 的本地模式。

```
bin/spark-shell
```

图 10.13 展示了 spark-shell 本地模式的启动过程。

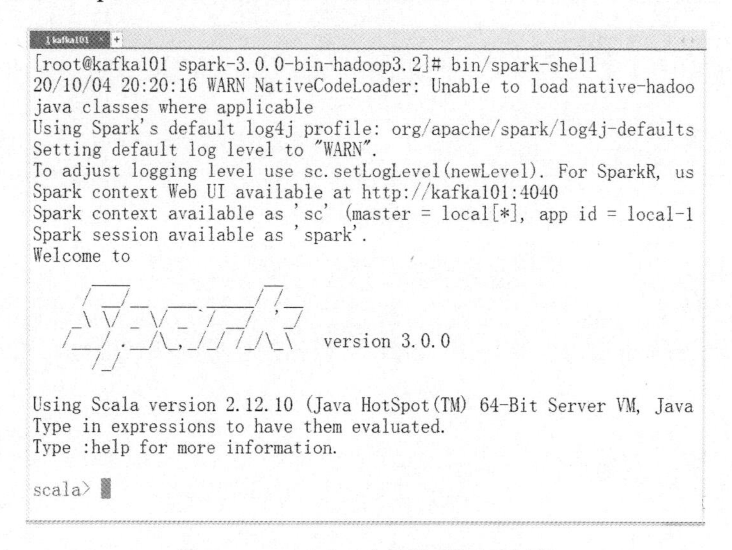


图 10.13 spark-shell 本地模式的启动过程

命令脚本启动后，会生成一个 SparkContext 的上下文对象 sc。可以看到这个 SparkContext 对象的 master 地址指向的是 local。利用 SparkContext 对象可以访问 Spark Core，执行批处理的离线计算。同时，spark-shell 还会创建一个 SparkSession 的对象 spark。SparkSession 是 Spark 2.x 以后提供的一个统一访问接口，利用 SparkSession 可以访问 Spark 的各个功能模块，如 Spark Core、Spark SQL 和 Spark Streaming。

### 2. cluster 模式

cluster 模式是 spark-shell 的集群模式。在这种模式下，客户端需要 Master 的地址，需要真正连接到 Spark 集群，执行的应用程序将提交给大集群执行。通过下面的方式启动 spark-shell 的集群模式。

```
bin/spark-shell --master spark://kafka101:7077
```

图 10.14 展示了 spark-shell 本地模式的启动过程。可以看到创建 SparkContext 对象的 master 地址指向的是集群地址。

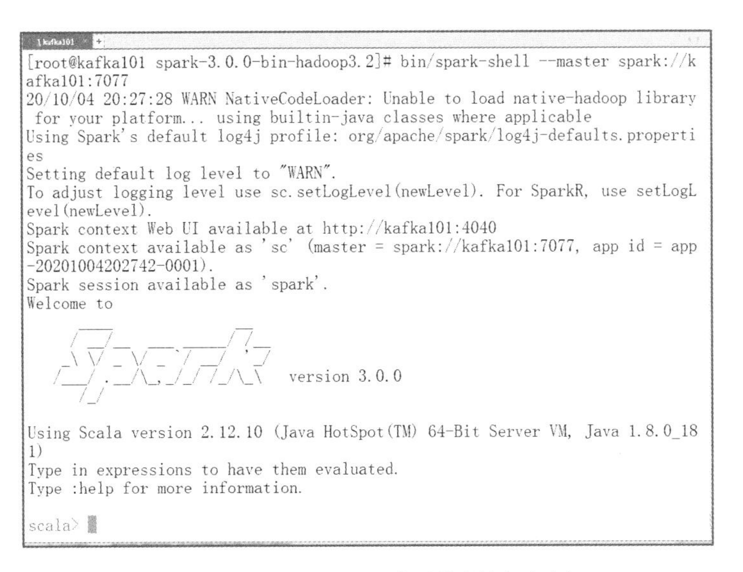


图 10.14　启动 spark-shell 集群模式的启动过程

打开浏览器访问 Spark 的 Web Console 可以查看 spark-shell 的运行状态。只要不退出 spark-shell，它的状态将一直是 RUNNING，如图 10.15 所示。

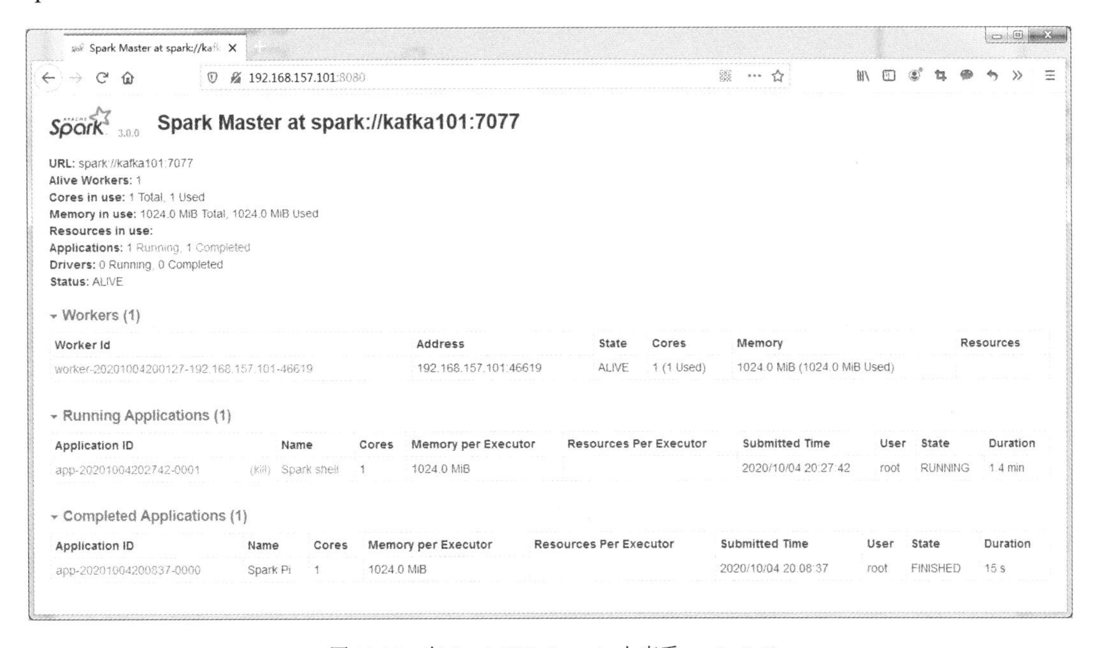


图 10.15　在 Spark Web Console 上查看 spark-shell

下面以 WordCount 程序为例，介绍如何使用 spark-shell 开发 Spark 应用程序。由于这里没有 Hadoop 的 HDFS，程序将处理本地文件系统上的文件。

（1）创建测试数据文件，执行下面的语句。

```
cd /root
mkdir -p temp/input
vi data.txt
```

（2）输入以下的文本，保存并退出。

```
I love Beijing
I love China
Beijing is the capital of China
```

（3）在 spark-shell 中输入以下的程序代码，并按 Enter 键，如图 10.16 所示。

```
sc.textFile("/root/temp/input/data.txt").flatMap(_.split("
")).map((_,1)).reduceByKey(_+_).collect
```

```
scala> sc.textFile("/root/temp/input/data.txt").flatMap(_.split(" ")).map((_,1))
.reduceByKey(_+_).collect
```

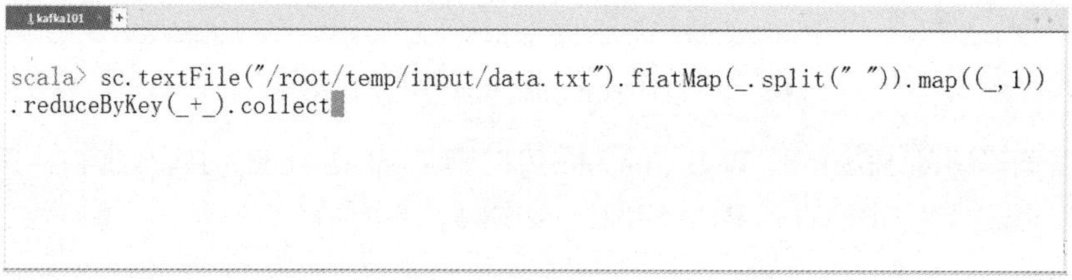

图 10.16　在 spark-shell 中书写 Scala 程序

（4）查看程序输出的结果，如图 10.17 所示。

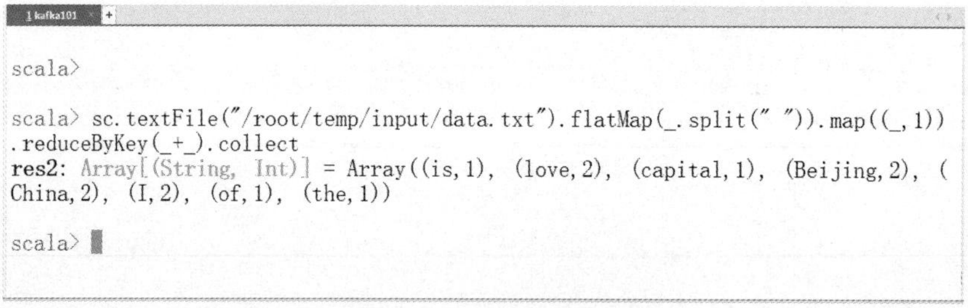

```
scala>

scala> sc.textFile("/root/temp/input/data.txt").flatMap(_.split(" ")).map((_,1))
.reduceByKey(_+_).collect
res2: Array[(String, Int)] = Array((is,1), (love,2), (capital,1), (Beijing,2), (
China,2), (I,2), (of,1), (the,1))

scala>
```

图 10.17　spark-shell 的输出结果

（5）打开浏览器访问 Spark 的 Web Console。可以查看到刚才执行程序的详细执行过程。单击界面“RUNNING APPLICATIONS”中的“Spark Shell”链接，进入“Spark Jobs”界面，如图 10.18 所示。

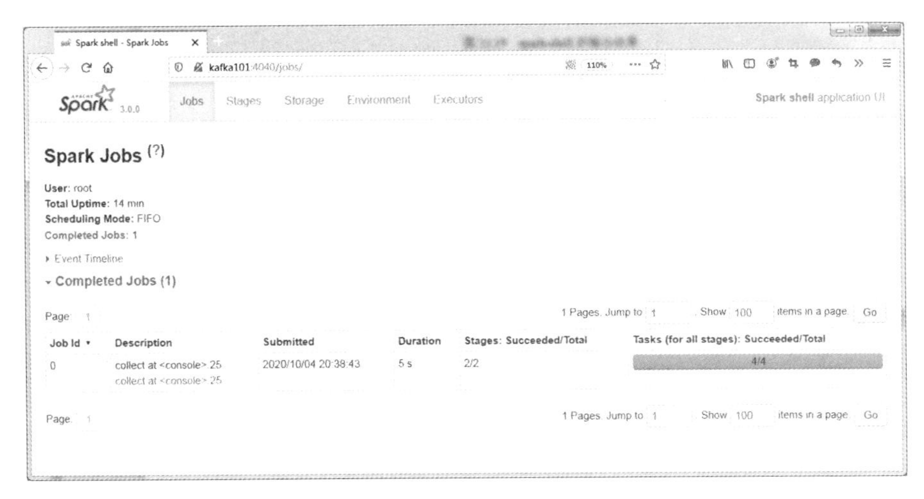


图 10.18　Spark Jobs 界面

（6）单击链接“collect at <console>:25”，进入任务的详细界面；然后单击“Details for Job 0”链接，就可以看到 Spark 为应用程序构建的 DAG 图了，如图 10.19 所示。

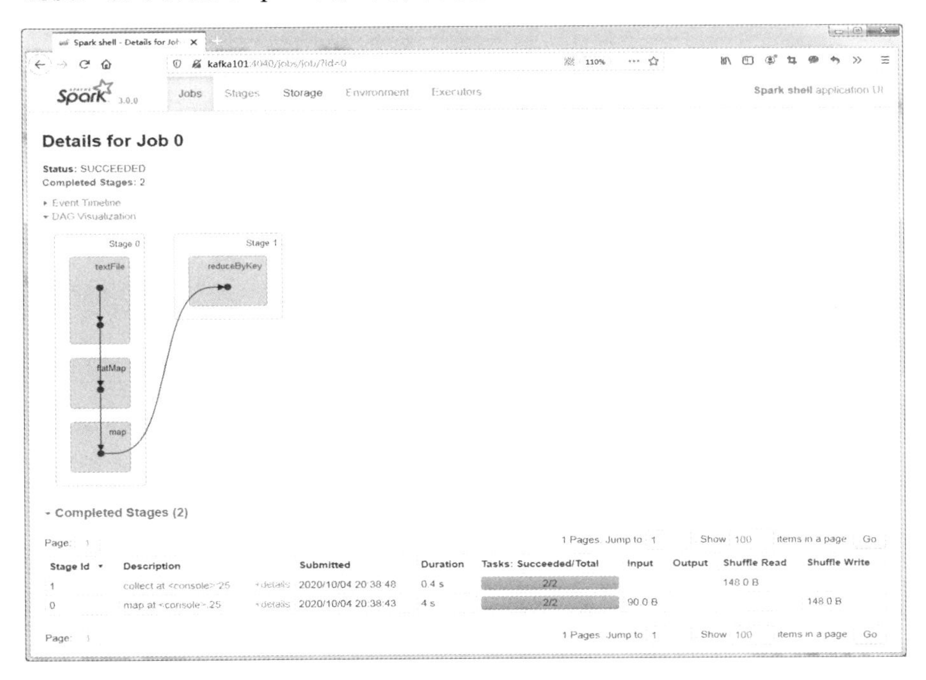


图 10.19　任务的 DAG 图

（7）在 spark-shell 中执行下面的语句退出 spark-shell。

```
:quit
```

# 10.4 Spark 的核心编程模型

## 10.4.1 什么是 RDD

RDD（Resilient Distributed Dataset）称为弹性分布式数据集，是 Spark 中最基本的数据抽象之一，它代表一个不可变、可分区中的元素可并行计算的集合。RDD 具有数据流模型的特点：自动容错、位置感知性调度和可伸缩性。RDD 允许用户在执行多个查询时显式地将工作集缓存在内存中，后续的查询能够重用工作集，极大地提升了查询速度。

在 Spark 的源代码中，对 RDD 的特性进行如下说明。

```
01   - A list of partitions
02   - A function for computing each split
03   - A list of dependencies on other RDDs
04   - Optionally, a Partitioner for key-value RDDs (e.g. to say that the RDD is
hash-partitioned)
05   - Optionally, a list of preferred locations to compute each split on (e.g.
block locations for an HDFS file)
```

下面给出相应的解释。

### 1．一组分区（Partition）

分区即数据集的基本组成单位。对 RDD 来说，每个分区都会被一个计算任务处理，并决定并行计算的粒度。用户可以在创建 RDD 时指定 RDD 的分区个数，如果没有指定，那么就会采用默认值。默认值就是程序所分配到的 CPU Core 的数目。图 10.20 说明了 RDD 分区和从节点 Worker 的关系。

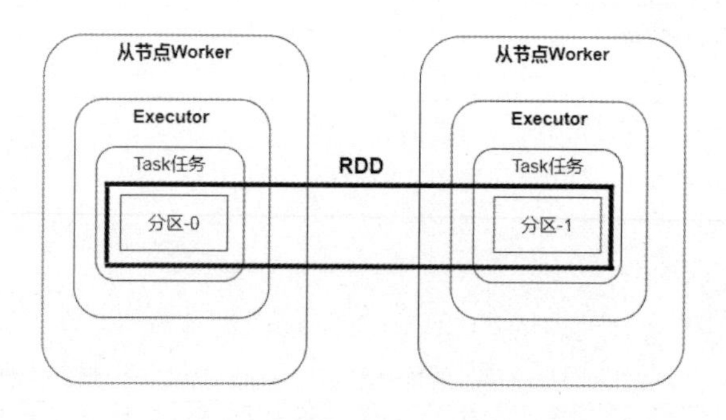


图 10.20　RDD 分区和从节点 Worker 的关系

图 10.20 中的黑色加粗的方框代表 RDD。RDD 由两个分区组成，每个分区又运行在不同的计算节点上，从而支持分布式计算。

### 2．一个计算每个分区中数据的函数

Spark 中 RDD 的计算是以分区为单位的，每个 RDD 都会实现 compute 函数以达到这

个目的。compute 函数会对迭代器进行复合，不需要保存每次计算的结果。

#### 3．一个存储 RDD 依赖关系的列表

RDD 的每次转换都会生成一个新的 RDD，所以 RDD 之间就会形成类似流水线一样的前后依赖关系。在部分分区数据丢失时，Spark 可以通过这个依赖关系重新计算丢失的分区数据，而不是对 RDD 的所有分区进行重新计算。

#### 4．一个 Partitioner 的分区函数

当前 Spark 中实现了两种类型的分区函数，一种是基于哈希的 HashPartitioner，另一种是基于范围的 RangePartitioner。

#### 5．一个存储存取每个 Partition 的优先位置的列表

对一个 HDFS 文件来说，这个列表保存的就是每个 Partition 所在的数据块位置。Spark 在进行任务调度的时候，会尽可能地将计算任务分配到其所要处理数据块的存储位置进行计算。

### 10.4.2　RDD 的算子

Spark RDD 的算子，即函数或方法。通过这些算子可以操作处理 RDD 集合中的数据。算子分为两种不同的类型：Transformation 和 Action。

RDD 中的所有 Transformation 算子都是延迟加载的，也就是说，它们并不会直接计算结果。相反的，它们只是记住这些应用到基础数据集（如一个文件）上的转换动作。只有当发生一个要求将结果返回给 Driver 的动作时，这些转换才会真正运行。这种设计让 Spark 更加有效率的运行。

Action 算子会立即执行当前算子及之前懒执行的算子。

#### 1．Transformation 算子

表 10.1 列出了常见的 Transformation 算子。

表 10.1　Transformation 算子

| Transformation 算子 | 算子的作用 |
| --- | --- |
| map(func) | 返回一个新的 RDD，该 RDD 由每一个输入元素经过 func 函数转换后组成 |
| filter(func) | 返回一个新的 RDD，该 RDD 由经过 func 函数计算后返回值为 true 的输入元素组成 |

续表

| Transformation 算子 | 算子的作用 |
| --- | --- |
| flatMap(func) | 类似于 map,但是每一个输入元素可以被映射为 0 或多个输出元素(所以 func 应该返回一个序列，而不是单一元素) |
| mapPartitions(func) | 类似于 map，但独立地在 RDD 的每一个分区上运行，因此在类型为 T 的 RDD 上运行时，func 的函数类型必须是 Iterator[T] => Iterator[U] |
| mapPartitionsWithIndex(func) | 类似于 mapPartitions，func 带有一个整数参数表示分区的索引值，因此在类型为 T 的 RDD 上运行时,func 的函数类型必须是(Int, Interator[T]) => Iterator[U] |
| sample(withReplacement, fraction, seed) | 根据 fraction 指定的比例对数据进行采样，可以选择是否使用随机数进行替换，seed 用于指定随机数生成器种子 |
| union(otherDataset) | 对源 RDD 和参数 RDD 求并集后返回一个新的 RDD |
| intersection(otherDataset) | 对源 RDD 和参数 RDD 求交集后返回一个新的 RDD |
| distinct([numTasks])) | 对源 RDD 去重后返回一个新的 RDD |
| groupByKey([numTasks]) | 在一个(K,V)的 RDD 上调用，返回一个(K, Iterator[V])的 RDD |
| reduceByKey(func, [numTasks]) | 在一个(K,V)的 RDD 上调用，返回一个(K,V)的 RDD，使用指定的 reduce 函数，将相同的 key 值聚合到一起，与 groupByKey 类似，reduce 任务的个数可以通过第二个可选的参数来设置 |
| sortByKey([ascending], [numTasks]) | 在一个(K,V)的 RDD 上调用，K 必须实现 Ordered 接口，返回一个按照 key 进行排序的(K,V)的 RDD |
| sortBy(func,[ascending], [numTasks]) | 与 sortByKey 类似，但是更灵活 |
| join(otherDataset, [numTasks]) | 在类型为(K,V)和(K,W)的 RDD 上调用，返回一个相同 key 对应的所有元素对在一起的(K,(V,W))的 RDD |
| cogroup(otherDataset, [numTasks]) | 在类型为(K,V)和(K,W)的 RDD 上调用，返回一个(K,(Iterable<V>,Iterable<W>))类型的 RDD |
| cartesian(otherDataset) | 生成两个集合的笛卡儿积 |

## 2. Action 算子

表 10.2 列出了常见的 Action 算子。

表 10.2 Action 算子

| Transformation 算子 | 算子的作用 |
| --- | --- |
| reduce(func) | 通过 func 函数聚集 RDD 中的所有元素，这个功能必须是可交换且可并联的 |

续表

| Transformation 算子 | 算子的作用 |
| --- | --- |
| collect() | 在驱动程序中，以数组的形式返回数据集中的所有元素 |
| count() | 返回 RDD 的元素个数 |
| first() | 返回 RDD 的第一个元素（类似于 take(1)） |
| take(n) | 返回一个由数据集的前 n 个元素组成的数组 |
| takeSample(withReplacement,num, [seed]) | 返回一个数组，该数组由从数据集中随机采样的 num 个元素组成，可以选择是否用随机数替换不足的部分，seed 用于指定随机数生成器种子 |
| saveAsTextFile(path) | 将数据集中的元素以 textfile 的形式保存到 HDFS 文件系统或其他支持的文件系统，对于每个元素，Spark 将会调用 toString 方法，将它转换为文件中的文本 |
| saveAsSequenceFile(path) | 将数据集中的元素以 Hadoop sequencefile 的格式保存到指定的目录下，可以使用 HDFS 或其他 Hadoop 支持的文件系统 |
| countByKey() | 针对(K,V)类型的 RDD，返回一个(K,Int)的 map，表示每一个 key 对应的元素个数 |

### 3．算子示例 Transformation 与 Action

（1）创建 RDD。

```
val rdd1 = sc.parallelize(Array(1,2,3,4,5,6,7,8))
```

（2）map（func）算子：将输入的每个元素重写组合成一个元组。

```
将 rdd1 中的元素乘以 10
val rdd2 = rdd1.map((_ * 10))
```

（3）filter(func)：返回一个新的 RDD，该 RDD 是经过 func 运算后返回 true 的元素。

```
找出 rdd1 中大于 5 的元素
val rdd3 = rdd1.filter(_ > 5)
```

（4）flatMap(func)压平操作。

```
val books = sc.parallelize(List("Hadoop","Hive","HDFS"))
books.flatMap(_.toList).collect
结果：res18: Array[Char] = Array(H, a, d, o, o, p, H, i, v, e, H, D, F, S)
```

（5）union(otherDataset)：并集运算，这里需要注意的是，参与运算的集合类型要一致。

```
val rdd4 = sc.parallelize(List(5,6,4,7))
val rdd5 = sc.parallelize(List(1,2,3,4))
val rdd6 = rdd4.union(rdd5)
```

（6）intersection(otherDataset)：交集运算，这里需要注意的是，参与运算的集合类型要一致。

```
val rdd7 = rdd5.intersection(rdd4)
```

（7）distinct([numTasks]))：去掉重复数据。

```
val rdd8 = sc.parallelize(List(5,6,4,7,5,5,5))
rdd8.distinct.collect
```

（8）groupByKey([numTasks])：对于一个<k,v>的 RDD，按照 key 进行分组。

```
val rdd = sc.parallelize(Array(("I",1),("love",2),("I",3)))
rdd.groupByKey.collect
结果: res38:  Array[(String,  Iterable[Int])]  =  Array((love,CompactBuffer(2)),
(I,CompactBuffer(1, 3)))
```

（9）groupByKey 复杂的例子。

```
val sen = sc.parallelize(List("I love Beijing","I love China","Beijing is the
capital of China"))
sen.flatMap(_.split(" ")).map((_,1)).groupByKey.collect
```

（10）reduceByKey(func, [numTasks])：类似于 groupByKey，二者的区别是 reduceByKey 会有一个 combiner 的过程对每个分区上的数据先进行一次合并。

```
val a = sc.parallelize(List("dog", "cat", "owl", "gnu", "ant"), 2)
val b = a.map(x => (x.length, x))
b.reduceByKey(_ + _).collect
```

（11）cartesian 笛卡儿积。

```
val rdd1 = sc.parallelize(List("tom", "jerry"))
val rdd2 = sc.parallelize(List("tom", "kitty", "shuke"))
val rdd3 = rdd1.cartesian(rdd2)
```

（12）reduce：将集合中的元素进行累计运算，例如，累加求和。

```
val rdd1 = sc.parallelize(List(1,2,3,4,5), 2)
rdd1.reduce(_+_)
```

（13）count：求 RDD 中元素的个数。

```
val rdd1 = sc.parallelize(List(1,2,3,4,5), 2)
rdd1.count
```

（14）top：求 RDD 集合中元素最大的某几个元素。

```
val rdd1 = sc.parallelize(List(1,2,3,4,5), 2)
rdd1.top(2)
```

（15）take：按顺序取出 RDD 集合中的某几个元素。

```
val rdd1 = sc.parallelize(List(1,2,3,4,5), 2)
rdd1.take(2)
```

（16）first：类似于 take(1)的操作。

```
val rdd1 = sc.parallelize(List(1,2,3,4,5), 2)
rdd1.first
```

### 10.4.3 开发自己的 WordCount 程序

在了解了 Spark RDD 的基本概念和常用的算子后，就可以开发自己的应用程序了。这里我们通过 Scala 语言和 Java 语言开发之前执行过的 WordCount 程序，下面列出了完整的代码。

### 1. Scala 版本的 WordCount 程序

开发 Scala 版本的 Spark 程序需要在工程中添加以下依赖。

```
01  <properties>
02      <spark.version>2.1.0</spark.version>
03      <scala.version>2.11</scala.version>
04  </properties>
05
06  <dependencies>
07      <dependency>
08          <groupId>org.apache.spark</groupId>
09          <artifactId>spark-core_${scala.version}</artifactId>
10          <version>${spark.version}</version>
11      </dependency>
12  </dependencies>
```

需要注意的是，由于 Kafka 与 Spark Streaming 有两种集成方式，这里我们使用的是一个旧版本的 Spark。

下面给出了完整的 Scala 代码。

```
01  import org.apache.spark.SparkContext
02  import org.apache.spark.SparkConf
03  import org.apache.log4j.Logger
04  import org.apache.log4j.Level
05
06  object WordCountDemo {
07    def main(args: Array[String]): Unit = {
08      //运行时，不输出日志
09      Logger.getLogger("org.apache.spark").setLevel(Level.ERROR)
10      Logger.getLogger("org.eclipse.jetty.server").setLevel(Level.OFF)
11
12      //本地模式
13      val conf = new SparkConf().setAppName("WordCountDemo")
14                            .setMaster("local")
15
16      //创建 SparkContext
17      val sc = new SparkContext(conf)
18
19      //创建 RDD
20      val data = sc.parallelize(List("I love Beijing",
21                                     "I love China",
22                                     "Beijing is the capital of China"))
23
24      val result = data.flatMap(_.split(" "))
25                   .map((_,1))
26                   .reduceByKey(_+_)
```

```
27
28      //输出到屏幕
29      result.collect.foreach(println)
30      sc.stop
31    }
32  }
```

程序运行的结果，如图 10.21 所示。

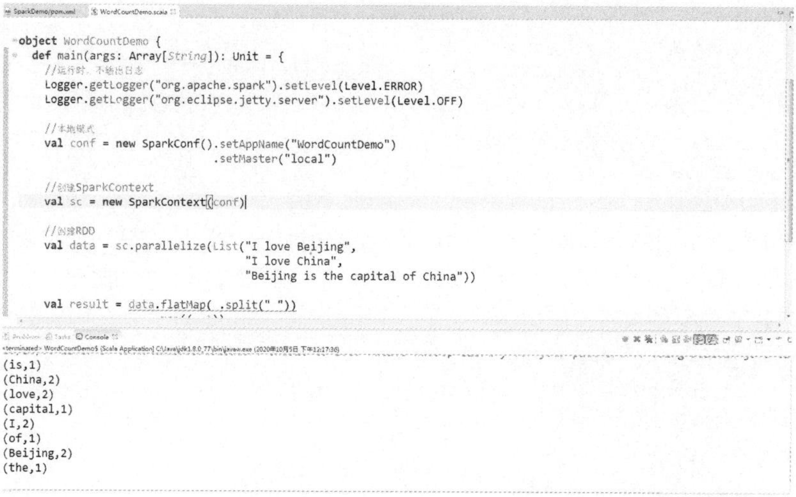

图 10.21 运行 Spark 的 Scala 程序

### 2. Java 版本的 WordCount 程序

开发 Java 版本的 Spark 程序需要在创建的 Java Maven 工程中添加以下依赖。

```
01  <dependency>
02      <groupId>org.apache.spark</groupId>
03      <artifactId>spark-core_2.12</artifactId>
04      <version>3.0.0</version>
05  </dependency>
```

下面给出了完整的 Java 代码。

```
01  import org.apache.spark.SparkConf;
02  import org.apache.spark.api.java.JavaPairRDD;
03  import org.apache.spark.api.java.JavaRDD;
04  import org.apache.spark.api.java.JavaSparkContext;
05  import org.apache.spark.api.java.function.FlatMapFunction;
06  import org.apache.spark.api.java.function.Function2;
07  import org.apache.spark.api.java.function.PairFunction;
08  import scala.Tuple2;
09  import java.util.ArrayList;
10  import java.util.Arrays;
```

```
11  import java.util.Iterator;
12  import java.util.List;
13  import java.util.regex.Pattern;
14
15  public final class JavaWordCount {
16      private static final Pattern SPACE = Pattern.compile(" ");
17
18      public static void main(String[] args) throws Exception {
19          //创建测试数据集合
20          List<String> list = new ArrayList<String>();
21          list.add("I love Beijing");
22          list.add("I love China");
23          list.add("Beijing is the capital of China");
24
25          //创建 SparkContext 对象
26          SparkConf sparkConf = new SparkConf()
27                                .setAppName("JavaWordCount")
28                                     .setMaster("local");
29          JavaSparkContext ctx = new JavaSparkContext(sparkConf);
30
31          //读入数据
32          JavaRDD<String> lines = ctx.parallelize(list);
33
34          //执行分词操作
35          JavaRDD<String> words = lines
36                  .flatMap(new FlatMapFunction<String, String>() {
37
38                      @Override
39                      public Iterator<String> call(String s) throws Exception
{
40                          return Arrays.asList(SPACE.split(s)).iterator();
41                      }
42
43                  });
44
45          //每个单词记一次数
46          JavaPairRDD<String, Integer> ones = words
47                  .mapToPair(new PairFunction<String, String, Integer>() {
48                      @Override
49                      public Tuple2<String, Integer> call(String s) {
50                          return new Tuple2<String, Integer>(s, 1);
51                      }
52                  });
53
```

```
54            //统计每个单词的频率
55            JavaPairRDD<String, Integer> counts = ones
56                    .reduceByKey(new Function2<Integer, Integer, Integer>() {
57                        @Override
58                        public Integer call(Integer i1, Integer i2) {
59                            return i1 + i2;
60                        }
61                    });
62
63            //输出结果
64            List<Tuple2<String, Integer>> output = counts.collect();
65            for (Tuple2<?, ?> tuple : output) {
66                System.out.println(tuple._1() + ": " + tuple._2());
67            }
68            ctx.stop();
69        }
70    }
```

程序运行的结果，如图 10.22 所示。

```
35
36         //执行分词操作
37         JavaRDD<String> words = lines
38                 .flatMap(new FlatMapFunction<String, String>() {
39
40                     @Override
41                     public Iterator<String> call(String s) throws Exception {
42                         return Arrays.asList(SPACE.split(s)).iterator();
43                     }
44
45                 });
46
47         //每个单词记一次数
48         JavaPairRDD<String, Integer> ones = words
49                 .mapToPair(new PairFunction<String, String, Integer>() {
50                     @Override
51                     public Tuple2<String, Integer> call(String s) {

Console
<terminated> JavaWordCount (2) [Java Application] C:\Java\jre\bin\javaw.exe (2020年10月5日 下午12:10:48)
20/10/05 12:10:54 INFO DAGScheduler: Job 0 is finished. Cancelling potential speculative or zombie tasks for this job
20/10/05 12:10:54 INFO TaskSchedulerImpl: Killing all running tasks in stage 1: Stage finished
is: 1
China: 2
love: 2
capital: 1
I: 2
of: 1
Beijing: 2
the: 1
20/10/05 12:10:54 INFO DAGScheduler: Job 0 finished: collect at JavaWordCount.java:66, took 1.146982 s
20/10/05 12:10:54 INFO SparkUI: Stopped Spark web UI at http://lenovo-PC:4040
20/10/05 12:10:54 INFO MapOutputTrackerMasterEndpoint: MapOutputTrackerMasterEndpoint stopped!
20/10/05 12:10:54 INFO MemoryStore: MemoryStore cleared
20/10/05 12:10:54 INFO BlockManager: BlockManager stopped
20/10/05 12:10:54 INFO BlockManagerMaster: BlockManagerMaster stopped
20/10/05 12:10:54 INFO OutputCommitCoordinator$OutputCommitCoordinatorEndpoint: OutputCommitCoordinator stopped!
20/10/05 12:10:54 INFO SparkContext: Successfully stopped SparkContext
```

图 10.22　运行 Spark 的 Java 程序

# 10.5　流式计算引擎 Spark Streaming

## 10.5.1　什么是 Spark Streaming

Spark Streaming 是核心 Spark API 的扩展，能够实现可扩展、高吞吐量、可容错的实时数据流处理。数据可以从诸如 Kafka、Flume、Kinesis 或 TCP 套接字等众多来源获取，并且可以使用由高级函数（如 map、reduce、join 和 window）开发的复杂算法进行流数据处理。最后，处理后的数据可以被推送到文件系统、数据库和实时仪表板，还可以在数据流上应用 Spark 提供的机器学习和图处理算法。

Spark Streaming 具备如下特点。

- 易用，Spark Streaming 支持多种编程语言，如 Java、Scala、Python。
- 高容错性，由于 Spark Streaming 底层依赖的 Spark RDD，而 RDD 具备很好的容错机制，因此 Spark Streaming 也继承了这一特点。
- 容易整合到 Spark 体系中，Spark Streaming 运行在 Spark 集群上，可以允许开发人员很方便地编写代码进行批处理技术，并且可以将历史的离线数据和流式数据整合在一起。同时在 Spark Streaming 中也可以使用 Spark SQL，从而使用 SQL 语句来处理流式数据。

通过官方提供的一个示例程序 NetworkWordCount 了解 Spark Streaming。

（1）在 kafka101 上启动 Spark 集群。

```
cd /root/training/spark-3.0.0-bin-hadoop3.2/
sbin/start-all.sh
```

（2）在 kafka101 上新开启一个命令终端并启动 netcat，在 1234 的端口上监听。

```
nc -l -p 1234
```

（3）启动 NetworkWordCount 应用程序，并从 kafka101 的 1234 端口上接收数据。

```
bin/run-example streaming.NetworkWordCount localhost 1234
```

（4）在 netcat 中发生数据，观察 NetworkWordCount 处理的结果，如图 10.23 所示。

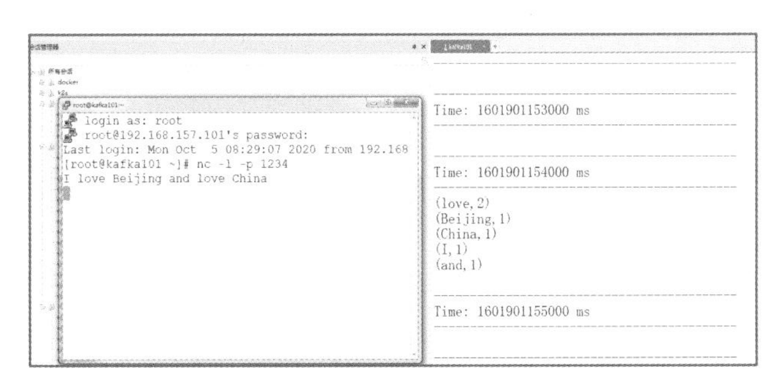


图 10.23　运行 NetworkWordCount 示例程序

这里可以看到，netcat 中发生的数据（“I love Beijing and love China”）在 Spark Streaming 中被实时处理了，处理的时间间隔是 1s。

## 10.5.2　离散流

执行的 NetworkWordCount 程序中每隔 1s 处理一次数据，也就是说在 Spark Streaming

中会存在一个时间采样的间隔，从而把连续的数据流变成不连续的数据集合 RDD。这就形成了 Spark Streaming 中的离散流 DStream。

Discretized Stream 或 DStream 是 Spark Streaming 对流式数据的基本抽象。它表示连续的数据流，这些连续的数据流可以是从数据源接收的输入数据流，也可以是通过对输入数据流执行转换操作而生成的经处理的数据流。在 Spark Streaming 内部，DStream 由一系列连续的 RDD 表示，如图 10.24 所示。

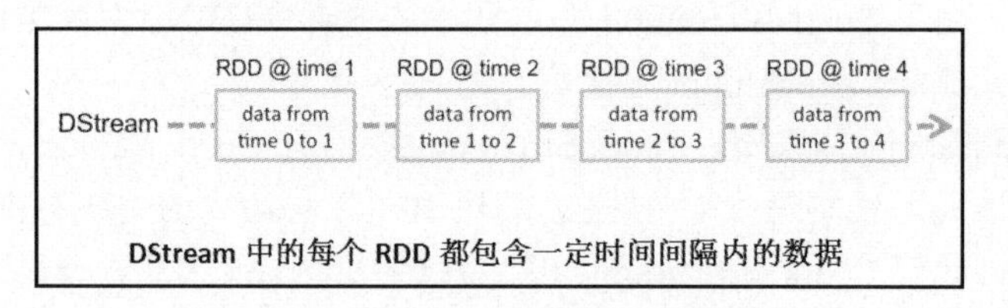


图 10.24　运行 DStream 与 RDD

在图 10.24 中可以看到，time 0 到 time 1 这段连续时间上的数据流被包含在第一个 RDD 的数据集合中；time 1 到 time 2 这段连续时间上的数据流被包含在第二个 RDD 的数据集合中；以此类推。通过设置的时间采样间隔，就把连续的数据流变成了不连续的 RDD，而在 Spark Streaming 中处理的正是这些不连续的 RDD 组成的集合。

从本质上看，Spark Streaming 依然是一个小批量的离线计算，其底层依然是前面介绍过的 Spark Core。

## 10.5.3　开发自己的 Spark Streaming 程序

要开发 Spark Streaming 的应用程序，首先就需要了解 Spark Streaming 的接收器和对应的输出操作；由于 Spark Streaming 依赖 Spark Core，所以 Spark Streaming 中的算子操作就类似之前介绍过的 Spark Core 的算子。

### 1. Spark Streaming 的接收器

接收器表示从数据源获取输入数据流的 DStream。在 NetworkWordCount 例子中，它代表从 netcat 服务器获取的数据流。每一个输入流 DStream 和一个 Receiver 对象相关联，这个 Receiver 从源中获取数据，并将数据存入内存中用于处理。

输入 DStream 表示从数据源获取的原始数据流。Spark Streaming 包含两类数据源。

- 基本源（Basic sources）：这些源在 StreamingContext API 中直接可用。例如，文件系统、套接字连接、Akka 的 actor 等。
- 高级源（Advanced sources）：这些源包括 Kafka、Flume、Kinesis、Twitter 等。

后续章节将介绍如何集成 Kafka 与 Spark Streaming。

### 2. Spark Streaming 的输出操作

输出操作允许 DStream 的操作推到如数据库、文件系统等外部系统中。因为输出操作实际上允许外部系统消费转换后的数据，它们触发的实际操作是 DStream 转换。目前，Spark Streaming 定义了下面几种输出操作，如表 10.3 所示。

表 10.3 Spark Streaming 的输出操作

| 输出操作 | 含义 |
| --- | --- |
| print() | 在 DStream 的每个批数据中打印前 10 条元素，这个操作在开发和调试中都非常有用。在 Python API 中调用 pprint()。 |
| saveAsObjectFiles(prefix, [suffix]) | 保存 DStream 的内容为一个序列化的文件 SequenceFile。每一个批间隔的文件，其文件名基于 prefix 和 suffix 生成。"prefix-TIME_IN_MS[.suffix]"，在 Python API 中不可用 |
| saveAsTextFiles(prefix, [suffix]) | 将 DStream 的内容保存为一个文本文件。每一个批间隔的文件，其文件名基于 prefix 和 suffix 生成。"prefix-TIME_IN_MS[.suffix]" |
| saveAsHadoopFiles(prefix, [suffix]) | 将 DStream 的内容保存为一个 hadoop 文件。每一个批间隔的文件，其文件名基于 prefix 和 suffix 生成。"prefix-TIME_IN_MS[.suffix]"，在 Python API 中不可用 |
| foreachRDD(func) | 从流中生成的每个 RDD 上应用函数 func 是十分通用的输出操作。这个函数应该将每个 RDD 的数据推送到外部系统，例如，保存 RDD 到文件通过网络写到数据库中。需要注意的是，func 函数在驱动程序中执行，并且通常都有 RDD action 在里面推动 RDD 流的计算 |

### 3. 开发自己的 Spark Streaming 程序

要开发自己的 Spark Streaming 程序，可以在之前的 Scala 工程上添加以下依赖。

```
01  <dependency>
02      <groupId>org.apache.spark</groupId>
03      <artifactId>spark-streaming_${scala.version}</artifactId>
04      <version>${spark.version}</version>
05  </dependency>
```

现在，我们来实现之前运行过的 NetworkWordCoun 程序，完整的 Scala 代码如下所示。

```
01  import org.apache.log4j.Logger
02  import org.apache.log4j.Level
03  import org.apache.spark.SparkConf
04  import org.apache.spark.storage.StorageLevel
05  import org.apache.spark.streaming.Seconds
06  import org.apache.spark.streaming.StreamingContext
07
08  object NetworkWordCoun {
```

```
09      def main(args: Array[String]) {
10        Logger.getLogger("org.apache.spark").setLevel(Level.ERROR)
11        Logger.getLogger("org.eclipse.jetty.server").setLevel(Level.OFF)
12
13         //创建一个本地的 StreamingContext，并设置两个工作线程，批处理时间间隔为 3 秒
14        val sparkConf = new SparkConf().setAppName("NetworkWordCount")
15                                    .setMaster("local[2]")
16
17        val ssc = new StreamingContext(sparkConf, Seconds(3))
18
19        //创建一个 DStream 对象，并连接到 netcat 的服务器端
20        val lines = ssc.socketTextStream("kafka101", 1234,
21                                     StorageLevel.MEMORY_AND_DISK_SER)
22
23        //采集数据并处理
24        val words = lines.flatMap(_.split(" "))
25        val wordCounts = words.map(x => (x, 1)).reduceByKey(_ + _)
26
27        //打印结果
28        wordCounts.print()
29
30        //启动 StreamingContext，开始计算
31        ssc.start()
32
33        //等待计算结束
34        ssc.awaitTermination()
35      }
36    }
```

启动 netcat 和应用程序进行测试，如图 10.25 所示。

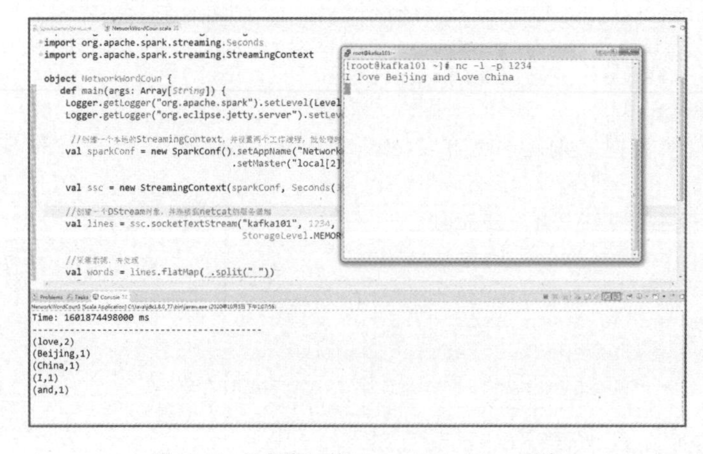

图 10.25　运行自己的 NetworkWordCoun 程序

可以看到，在 netcat 中发送的数据被 Spark Streaming 程序正确接收，并实时地计算出来，得到每个单词的频率。这里我们设置的采样时间间隔是 3 秒，即每隔 3 秒采样一次，生成一个 RDD 的数据集合。

## 10.6　集成 Kafka 与 Spark Streaming

Spark Streaming 是基于小批处理的流式计算引擎，通常利用 Spark Core 或与 Spark SQL 一起处理数据。在实际的流式实时处理架构中，通常将 Spark Streaming 和 Kafka 集成作为整个大数据处理架构的核心环节之一。针对不同的 Spark 和 Kafka 版本，集成处理数据的方式分为两种：基于 Receiver 的方式和直接读取的方式。

需要将以下依赖加入之前创建好的 Scala 工程中。

```
01   <dependency>
02       <groupId>org.apache.spark</groupId>
03       <artifactId>spark-streaming-kafka-0-8_2.11</artifactId>
04       <version>2.1.1</version>
05   </dependency>
```

### 10.6.1　基于 Receiver 的方式

Receiver 是使用 Kafka 高级 Consumer API 实现的。与所有接收器一样，Kafka 通过 Receiver 接收的数据存储在 Spark Executor 的内存中，然后由 Spark Streaming 启动的 job 来处理数据。然而在默认配置下，这种方式可能会因为底层的失败而丢失数据。如果要启用高可靠机制，确保零数据丢失，要启用 Spark Streaming 的预写日志机制（Write Ahead Log）。该机制会将接收到的 Kafka 数据保存到分布式文件系统（比如 HDFS）的预写日志中，以便底层节点在发生故障时也可以使用预写日志中的数据进行恢复。

图 10.26 展示了基于 Receiver 方式的工作原理。

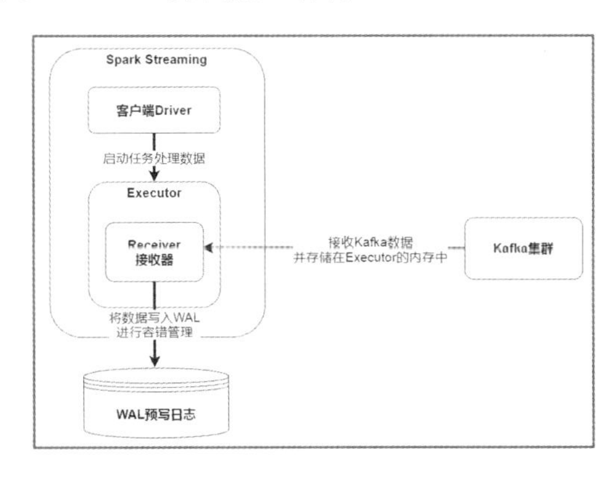


图 10.26　基于 Receiver 的方式接收 Kafka 的消息

完整的代码如下所示。

```
01  import org.apache.spark.SparkConf
02  import org.apache.spark.streaming.kafka.KafkaUtils
03  import org.apache.spark.streaming.Seconds
04  import org.apache.spark.streaming.StreamingContext
05  import org.apache.log4j.Logger
06  import org.apache.log4j.Level
07
08  object KafkaWordCount {
09    def main(args: Array[String]) {
10      Logger.getLogger("org.apache.spark").setLevel(Level.ERROR)
11      Logger.getLogger("org.eclipse.jetty.server").setLevel(Level.OFF)
12
13      val conf = new SparkConf()
14                    .setAppName("SparkFlumeNGWordCount")
15                    .setMaster("local[2]")
16
17      val ssc = new StreamingContext(conf, Seconds(10))
18
19      //创建topic名称，1表示一次从这个topic中获取一条记录
20      val topics = Map("mytopic1" ->1)
21
22      //创建Kafka的输入流，指定ZooKeeper的地址
23      val kafkaStream = KafkaUtils.createStream(ssc,
24                                        "kafka101:2181",
25                                        "mygroup",
26                                        topics)
27
28      //处理每次接收到的数据
29      val lineDStream = kafkaStream.map(e => {
30        new String(e.toString())
31      })
32
33      //直接输出结果
34      lineDStream.print()
35
36      ssc.start()
37      ssc.awaitTermination();
38    }
39  }
```

在这段程序代码中，使用 Spark Streaming 中的 Receiver 接收 Kafka 发送来的数据，并直接输出到屏幕上，如图 10.27 所示。

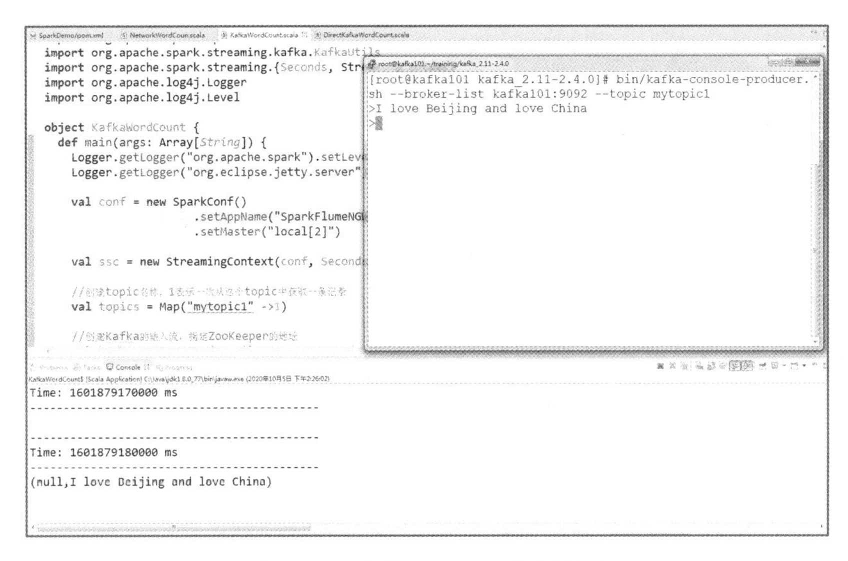


图 10.27　运行基于 Receiver 的应用程序

## 10.6.2　直接读取的方式

和基于 Receiver 的方式接收数据不一样，这种方式定期从 Kafka 的 topic+partition 中查询最新的偏移量，再根据定义的偏移量范围在每个 batch 中处理数据。当需要处理的数据来临时，Spark 通过调用 Kafka 的简单消费者 API 读取一定范围的数据。

图 10.28 展示了直接读取方式的工作原理。

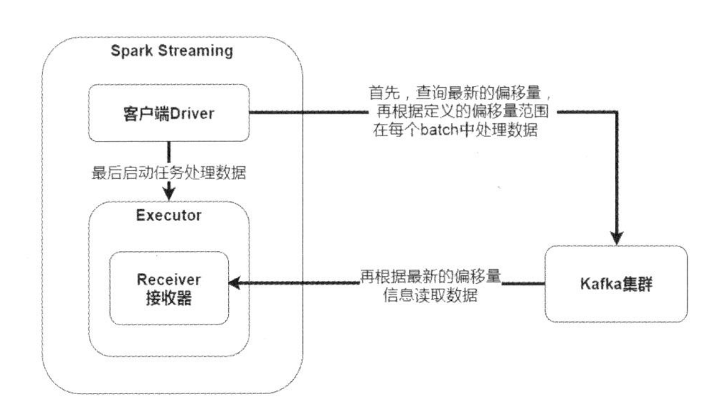


图 10.28　基于直接读取方式接收 Kafka 的消息

完整的代码如下所示。

```
01   import kafka.serializer.StringDecoder
02   import org.apache.spark.SparkConf
03   import org.apache.spark.streaming.Seconds
```

```
04  import org.apache.spark.streaming.StreamingContext
05  import org.apache.spark.streaming.kafka.KafkaUtils
06  import org.apache.log4j.Logger
07  import org.apache.log4j.Level
08
09  object DirectKafkaWordCount {
10    def main(args: Array[String]) {
11      Logger.getLogger("org.apache.spark").setLevel(Level.ERROR)
12      Logger.getLogger("org.eclipse.jetty.server").setLevel(Level.OFF)
13
14      val conf = new SparkConf()
15                    .setAppName("SparkFlumeNGWordCount")
16                    .setMaster("local[2]")
17      val ssc = new StreamingContext(conf, Seconds(10))
18
19      //创建 topic 名称，1 表示一次从这个 topic 中获取一条记录
20      val topics = Set("mytopic1")
21      //指定 Kafka 的 broker 地址
22      val kafkaParams = Map[String, String](
23                      "metadata.broker.list" -> "kafka101:9092")
24
25      //创建 DStream，接收 Kafka 的数据
26      val kafkaStream =
27          KafkaUtils
28        .createDirectStream[String, String, StringDecoder, StringDecoder](ssc,
29                                                                      kafkaParams,
30                                                                      topics)
31
32
33      //处理每次接收到的数据
34      val lineDStream = kafkaStream.map(e => {
35        new String(e.toString())
36      })
37      //输出结果
38      lineDStream.print()
39
40      ssc.start()
41      ssc.awaitTermination();
42    }
43  }
```

在这段程序代码中，使用 Spark Streaming 直接读取的方式接收 Kafka 发送来的数据，并直接将其输出到屏幕上，如图 10.29 所示。

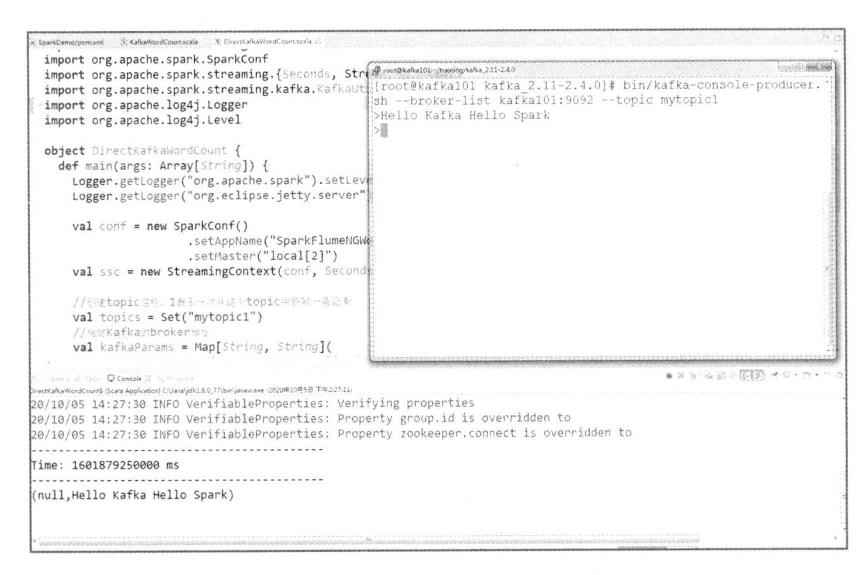


图 10.29　运行基于直接读取方式的应用程序

直接读取方式的优点在于没有接收器，不需要预先写入日志，所以其执行的效率会更高。因此在实际的大规模数据处理过程中，推荐使用直接读取的方式来消费 Kafka 中的数据。

# 第 11 章 Kafka 与 Flume 集成

Apache Flume 是一个高可用的、高可靠的、分布式的海量日志采集、聚合和传输的系统，Flume 支持在日志系统中定制各类数据发送方，用于收集数据；同时，Flume 提供对数据进行简单处理并写到各种数据接收方（可定制）的能力。

在前面的章节中，都是将 Flink、Storm 和 Spark Streaming 作为 Kafka 消息的消费者，从 Kafka 中接收数据；而在一个典型的实时计算系统中，通过 Flume 作为 ETL 工具实时采集数据源的数据，并将 Flume 作为 Kafka 消息的生产者，将采集的数据发送到 Kafka 中，图 11.1 说明了这一过程。

图 11.1 实时计算的典型架构

## 11.1 Apache Flume 基础

Flume 支持采集各类数据发送方产生的日志信息，并且可以将采集到的日志信息写到各种数据接收方，其核心是把数据从数据源（Source）收集过来，再将收集到的数据发送到指定的目的地（Sink）。为了保证输送过程的成功，在将数据发送到目的地之前，先缓存数据（Channel），待数据真正到达目的地后，Flume 再删除自己缓存的数据。

### 11.1.1 Apache Flume 的体系架构

Flume 分布式系统中的核心角色是 Agent。Agent 本身是一个 Java 进程，一般运行在日志收集节点。Flume 采集系统就是由多个 Agent 连接起来形成的。一个 Agent 相当于一个数据传递员，其内部有三个组件。

- Source：用于和数据源对接，以获取数据。

- Sink：采集数据的传送目的地，用于往下一级 Agent 传递数据或往最终存储系统传递数据。
- Channel：Agent 内部的数据传输通道，用于从数据源将数据传递到目的地。

在整个数据传输的过程中，流动的是 Event，它是 Flume 内部数据传输的基本单元。Event 将传输的数据进行封装。如果是文本文件，通常是一行记录，Event 也是事务的基本单位。Event 从 Source 流向 Channel，再到 Sink，其本身为一个字节数组，并可携带 headers 的头信息。Event 代表一个数据的最小完整单元，从外部数据源来，向外部目的地去。一个完整的 Event 包括 Event Headers、Event Body、Event 信息，其中 Event 信息就是 Flume 收集到的日志记录。

图 11.2 展示了 Flume 的体系架构。

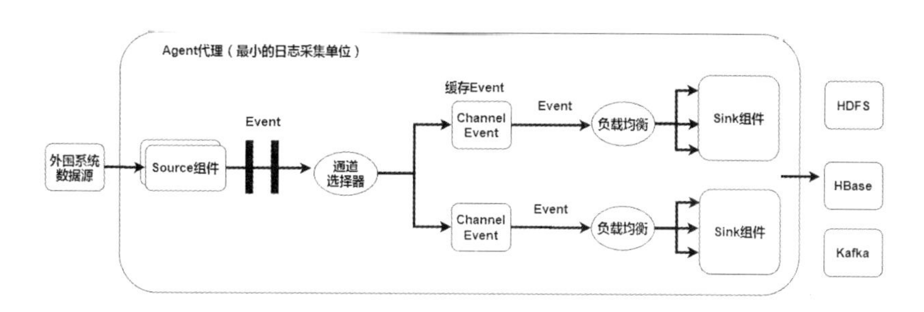


图 11.2　Flume 的体系架构

下面列出了一些常见的组件类型。

### 1．Source 组件

Source 组件主要用于采集外围系统的数据，如日志信息等。表 11.1 是一些常见的 Source 组件类型。

表 11.1　常见的 Source 组件

| | |
|---|---|
| Avro Source | 这种类型的 Source 监听 Avro 端口，从 Avro 客户端接收 Events。Avro 是一个数据序列化系统，设计用于支持大批量数据交换的应用 |
| Thrift Source | Thrift Source 与 Avro Source 基本一致。Thrift 是一种 RPC 常用的通信协议，它可以定义 RPC 方法和数据结构，用于生成不同语言的客户端代码和服务器端代码 |
| Exec Source | Exec Source 的配置就是设定一个 UNIX（Linux）命令，然后通过这个命令不断输出数据 |
| JMS Source | 从 JMS（Java 消息服务）系统中读取数据 |
| Spooling Directory Source | 该类型的 Source 监测配置目录下新增的文件，并将文件中的数据读取出来 |
| Kafka Source | 接收 Kafka 发送来的消息数据。这种类型的 Source 将作为 Kafka 的消费者使用 |

续表

| | |
|---|---|
| NetCat TCP Source<br>NetCat UDP Source | NetCat 是一个非常简单的 UNIX 工具，可以读、写 TCP 或 UDP 网络连接中的数据。在 Flume 中的 NetCat 支持 Flume 与 NetCat 整合，Flume 可以使用 NetCat 读取网络中的数据 |
| HTTP Source | 有些应用程序环境可能不会部署 Flume，此时可以使用 Http Source 将数据接收到 Flume 中。Http Source 可以通过 Http Post 接收 Event |
| Custom Source | Flume 允许开发人员自定义 Source |

### 2. Channel 组件

Chanel 组件主要用于缓存 Source 组件采集到的数据信息。表 11.2 是一些常见的 Channel 组件类型。

表 11.2 常见的 Channel 组件

| | |
|---|---|
| Memory Channel | 将 Source 接收到的 Event 保存在 Java Heap 的内置中。如果允许数据小量丢失，则推荐使用 |
| JDBC Channel | 将 Source 接收到的 Events 存储在持久化存储库中，即存储在数据库中。这是一个支持持久化的 Channel，对于可恢复性非常重要的流程来说是理想的选择 |
| Kafka Channel | 将 Source 接收到的 Events 存储在 Kafka 消息系统的集群中。Events 能及时被其他 Flume 的 Sink 使用，即使当 Agent 或 Kafka Broker 崩溃的时候，也能为 Kafka 提供高可用性和高可靠性的特性 |
| File Channel | 类似 JDBC Channel，二者的区别是 File Channel 将 Source 接收到的 Events 存储在文件系统中 |
| Spillable Memory Channel | 将 Source 接收到的 Events 存储在内存队列和磁盘中。该 Channel 目前正在试验中，不建议在生产环境中使用 |
| Pseudo Transaction Channel | 只用于单元测试，不用于生产环境使用 |
| Custom Channel | Flume 允许开发人员自定义 Channel |

### 3. Sink 组件

Sink 组件主要用于将 Channel 组件缓存的数据信息写到外部的持久化存储介质上，如 HDFS、HBase、Kafka 等。表 11.3 是一些常见的 Sink 组件类型。

表 11.3 常见的 Sink 组件

| | |
|---|---|
| HDFS Sink | 此 Sink 将事件写入 Hadoop 分布式文件系统 HDFS 中。它支持创建文本文件和序列化文件，对这两种格式都支持压缩 |
| Hive Sink | 该类型的 Sink 将包含分割文本或 JSON 数据的 Events 直接传送到 Hive 表或分区中。当一系列 Events 提交到 Hive 时，它们马上可以被 Hive 查询到 |

续表

| | |
|---|---|
| Logger Sink | 记录指定级别的日志，通常用于调试 |
| Avro Sink | 和 Avro Source 配置使用，是实现复杂流动的基础 |
| Thrift Sink | 该类型的 Sink 将 Event 转换为 Thrift Events，从配置好的 Channel 中批量获取 Events 数据，并发送到配置好的主机地址上 |
| IRC Sink | IRC Sink 从 Channel 中获取 Event 消息，并将 Event 消息推送到配置好的 IRC 目的地。IRC 是 Internet Relay Chat 的英文缩写，是一种互联网中继聊天的协议 |
| File Roll Sink | 在本地文件系统中存储事件，每隔指定时长生成文件保存这段时间内收集到的日志信息 |
| Null Sink | 这种类型的 Sink 将直接丢弃 Channel 中接收到的所有 Events |
| HBase Sinks | 通过 HBase Sink 可以将 Channel 中的 Event 写到 HBase 中 |
| ElasticSearch Sink | 通过 ElasticSearch Sink 可以将 Channel 中的 Event 写到 ElasticSearch 搜索引擎中 |
| Kafka Sink | Flume Sink 实现可以将 Channel 中的 Event 数据导出到一个 Kafka Topic 中。这时 Kafka Sink 将作为 Kafka 集群消息的生产者使用 |
| HTTP Sink | 该类型的 Sink 将会从 Channel 中获取 Events 数据，并使用 Http 协议的 Post 请求将这些 Event 数据发送到远端的 Http Server 上。Event 的数据内容将作为 Post 请求的 body 发送 |
| Custom Sink | 如果以上内置的 Sink 都不能满足需求，Flume 允许开发人员自己开发 Sink |

在后续的章节中将详细介绍每种类型组件的详细用法。这里我们先给出一个简单的示例。具体的 Agent 配置信息如下所示。

```
01  #定义 agent 名称为 a1，同时指定 source 为 r1、channel 为 c1、sink 为 k1
02  a1.sources = r1
03  a1.channels = c1
04  a1.sinks = k1
05
06  #具体定义 source
07  a1.sources.r1.type = netcat
08  a1.sources.r1.bind = localhost
09  a1.sources.r1.port = 1234
10
11  #具体定义 channel
12  a1.channels.c1.type = memory
13  a1.channels.c1.capacity = 1000
14  a1.channels.c1.transactionCapacity = 100
15
16  #具体定义 sink
17  a1.sinks.k1.type = logger
18
19  #组装 source、channel、sink
20  a1.sources.r1.channels = c1
21  a1.sinks.k1.channel = c1
```

在这个 Agent 中，我们定义了一个 Netcat Source 从本机的 1234 端口上接收 Event 消息，并将其缓存到 Memory Channel 中，最终通过 Logger Sink 将接收的 Event 输出到终端。将配置文件保存到/root/training/flume/myagent 目录下的 a1.conf 文件中。这里的/root/training/flume/目录是 Flume 的安装目录。在 11.1.2 节中将会介绍 Flume 的安装和部署。

### 11.1.2 Apache Flume 的安装和部署

从 Flume 的官方网站上安装介质，这里我们使用的版本是 apache-flume-1.9.0-bin.tar.gz。读者可以按照以下步骤进行安装和配置，总体上比较简单。

（1）直接在 kafka101 的主机上进行配置，由于之前已经配置好了 JDK 的环境。这里直接将 Flume 的安装包解压到/root/training/目录下。

```
tar -zxvf apache-flume-1.9.0-bin.tar.gz -C ~/training/
```

（2）重命名解压缩的文件夹为 Flume，方便以后更新维护。

```
cd /root/training
mv apache-flume-1.9.0-bin/ flume/
```

（3）进入 Flume 下的 conf 文件夹，将文件 flume-env.sh.template 重命名为 flume-env.sh。

```
cd /root/training/flume/conf/
mv flume-env.sh.template flume-env.sh
```

（4）修改 flume-env.sh 中的 JAVA HOME 配置参数。

```
vi flume-env.sh
修改
export JAVA_HOME=/root/training/jdk1.8.0_181
```

（5）保存退出，并验证 Flume 的版本。

```
cd /root/training/flume
bin/flume-ng version
```

图 11.3 展示了 Flume 配置完成后的版本信息。

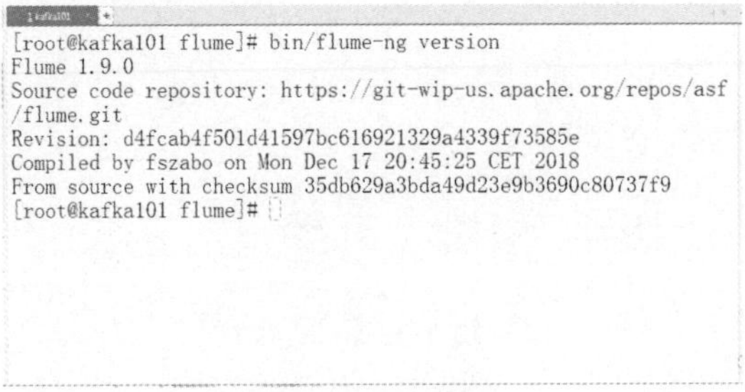

```
[root@kafka101 flume]# bin/flume-ng version
Flume 1.9.0
Source code repository: https://git-wip-us.apache.org/repos/asf/flume.git
Revision: d4fcab4f501d41597bc616921329a4339f73585e
Compiled by fszabo on Mon Dec 17 20:45:25 CET 2018
From source with checksum 35db629a3bda49d23e9b3690c80737f9
[root@kafka101 flume]#
```

图 11.3　Flume 的版本信息

Flume 的安装和部署完成后，我们就可以演示在前面章节中配置好的 Agent（a1.conf）的运行效果了。

（1）进入 Flume 的安装目录，执行下面的语句命令，启动 Agent，如图 11.4 所示。

```
bin/flume-ng agent -n a1 -f myagent/a1.conf -c conf -Dflume.root.logger=INFO,
console
```

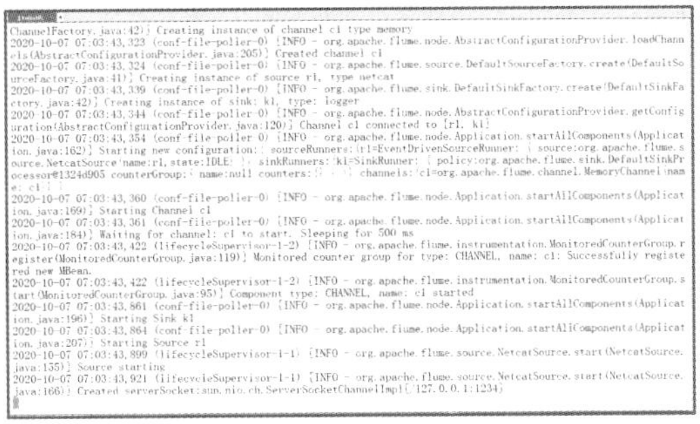

图 11.4　启动 Agent

打印出来的日志的最后一行如下。

```
Created serverSocket:sun.nio.ch.ServerSocketChannelImpl[/127.0.0.1:1234]
```

可以看到 Flume 已经成功地在本机的 1234 端口上创建了 Socket Server。这时候只要有消息从本机的 1234 端口上发送过来，就可以被 Flume 的 Source 捕获。

（2）单独启动一个 Netcat 命令终端，运行在本机的 1234 端口上，如图 11.5 所示。

```
nc 127.0.0.1 1234
```

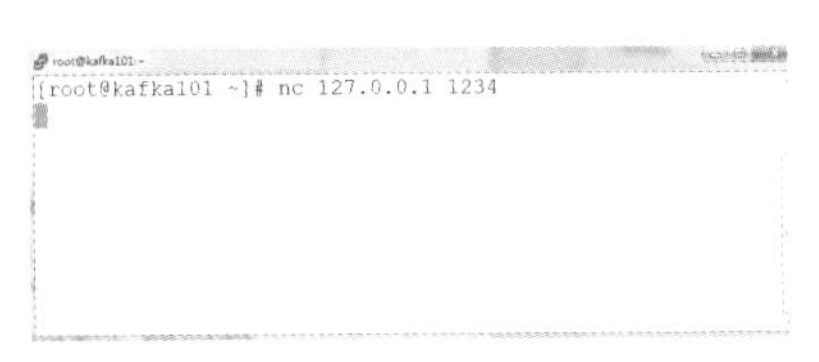


图 11.5　启动 Netcat

（3）在 Netcat 中输入一些内容，这里我们输入的是“Hello Flume”，并按 Enter 键；观察 Flume 命令行窗口的变化，如图 11.6 所示。

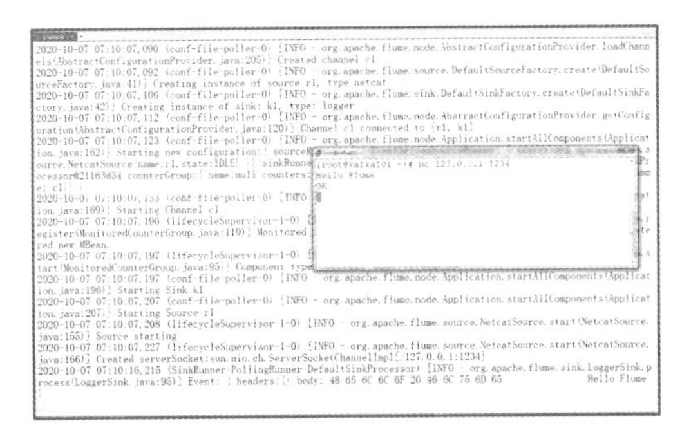

图 11.6　Flume 采集 Netcat 的数据

可以看到在 Flume 终端的日志中输出了如下信息。

```
[INFO - org.apache.flume.sink.LoggerSink.process(LoggerSink.java:95)]
Event: { headers:{} body: 48 65 6C 6C 6F 20 46 6C 75 6D 65 Hello Flume }
```

Flume 通过 Netcat Source 采集到了相应的数据信息，并直接打印在 Flume 的命令行终端中。

## 11.2 Flume 的 Source 组件

前面的内容已经简要介绍了 Flume 中的一些 Source 组件的类型。下面我们对这些 Source 组件进行详细的讨论，并给出相应的配置信息。

### 1. Avro Source

这种类型的 Source 将监听 Avro 端口，内部启动一个 Avro 服务器，用于接收来自 Avro Client 的请求，并且将接收数据存储到 Chanel 中。

下面是一个 Avro Source 的配置信息。

```
01  a1.sources = r1
02  a1.channels = c1
03  a1.sources.r1.type = avro
04  a1.sources.r1.channels = c1
05  a1.sources.r1.bind = 0.0.0.0
06  a1.sources.r1.port = 4141
```

### 2. Thrift Source

这种类型的 Source 将监听 Thrift 端口，接收来自 Thrift Client 的请求。当与另一 Flume Agent 上的内置 Thrift Sink 配对时，它可以创建分层集合拓扑。

```
01  a1.sources = r1
02  a1.channels = c1
03  a1.sources.r1.type = thrift
04  a1.sources.r1.channels = c1
05  a1.sources.r1.bind = 0.0.0.0
06  a1.sources.r1.port = 4141
```

### 3. Exec Source

可以使用这种类型的 Source 采集一个文件最新写入的内容。Exec Source 在启动时，通过给定的 UNIX 命令，同时当该 UNIX 命令进程连续产生标准输出数据时，执行数据的采集。如果该进程退出，则 Source 也退出，不再产生数据。

```
01  a1.sources = r1
02  a1.channels = c1
03  a1.sources.r1.type = exec
04  a1.sources.r1.command = tail -f /root/training/kafka_2.11-2.4.0/logs/server. log
05  a1.sources.r1.channels = c1
```

#### 4. JMS Source

JMS（Java Message Service）Source 能从 Queue 或 Topic 这样的 JMS 模型中读取消息。JMS Source 提供了可配置的批处理数目、消息选择器、用户名/密码及从消息到 Event 的转换器。

```
01   a1.sources = r1
02   a1.channels = c1
03   a1.sources.r1.type = jms
04   a1.sources.r1.channels = c1
05   a1.sources.r1.initialContextFactory = org.apache.activemq.jndi.
ActiveMQInitialContextFactory
06   a1.sources.r1.connectionFactory = GenericConnectionFactory
07   a1.sources.r1.providerURL = tcp://mqserver:61616
08   a1.sources.r1.destinationName = BUSINESS_DATA
09   a1.sources.r1.destinationType = QUEUE
```

#### 5. Spooling Directory Source

可以使用这种类型的 Source 监听一个目录的变化，如系统的日志目录/logs 等。因为 Spooling Directory Source 可以监控一个指定的目录，如果出现新的文件时，就会解析文件中的 Event，当文件被完全写入 Channel 以后，文件会被重命名表示已经完成。如果出现下列问题，该 Source 组件会出错并停止进程。

（1）文件已经被放入 Source 组件中，又被写入。

（2）文件名被重用。

为防止出现以上错误，文件被放入 spooling 目录时最好加上特定的标记，如时间戳。

```
01   a1.channels = ch-1
02   a1.sources = src-1
03   a1.sources.src-1.type = spooldir
04   a1.sources.src-1.channels = ch-1
05   a1.sources.src-1.spoolDir = /root/training/kafka_2.11-2.4.0/logs
06   a1.sources.src-1.fileHeader = true
```

#### 6. Kafka Source

这种类型的 Source 将作为 Kafka 消息的消费者使用。Kafka Source 是从 Kafka Topics 中读取消息的消费者。如果有多个 Kafka Sources，可以将它们配置为同一个消费组，每个 Source 会读取 Topics 中唯一的一组分区。

#### 7. NetCat TCP Source 与 NetCat UDP Source

类似于 Netcat 的 Source，当我们执行命令 nc -k -l [port]时，该类型的 Source 监听给定端口号并将每一行文本转换为 Event。

11.1.1 节最后就是一个 Netcat Source 的例子，这里就不单独举例。

### 8. HTTP Source

使用 HTTP Source 采集数据时，Flume 首先启动一个 Web 服务来监听指定的 IP 和端口。常用的使用场景：在有些应用环境中，不能部署 Flume 及其依赖项，可以在代码中通过 HTTP 协议，而不是 Flume 的 PRC 发送数据的情况，此时 HTTP Source 可以用来将数据接收到 Flume 中。

```
01   a1.sources = r1
02   a1.channels = c1
03   a1.sources.r1.type = http
04   a1.sources.r1.port = 5140
05   a1.sources.r1.channels = c1
06   a1.sources.r1.handler = org.example.rest.RestHandler
07   a1.sources.r1.handler.nickname = random props
08   a1.sources.r1.HttpConfiguration.sendServerVersion = false
09   a1.sources.r1.ServerConnector.idleTimeout = 300
```

### 9. Custom Source

Flume 允许开发人员自定义 Source 组件的类型。一个用户自定义 Source 的程序代码及其相关的依赖，都必须包含在 Agent 的 Classpath 中。这样当 Flume Agent 启动时，就可以找到相关的运行信息。

## 11.3 Flume 的 Channel 组件

Flume 中的 Channel 组件用于缓存 Source 组件采集的数据信息，下面我们对这些 Channel 组件进行详细的讨论，并给出相应的配置信息。

### 1. Memory Channel

基于内存的 Channel，实际就是将 Events 存放于内存的一个固定大小的队列中，其优点是速度快，其缺点是可能丢失数据。

```
01   a1.channels = c1
02   a1.channels.c1.type = memory
03   a1.channels.c1.capacity = 10000
04   a1.channels.c1.transactionCapacity = 10000
05   a1.channels.c1.byteCapacityBufferPercentage = 20
06   a1.channels.c1.byteCapacity = 800000
```

### 2. JDBC Channel

将 Events 存放于一个支持 JDBC 连接的数据库中，目前官方推荐的是 Derby 库，其优点是数据可以恢复。

```
01  a1.channels = c1
02  a1.channels.c1.type = jdbc
```

### 3. Kafka Channel

Kafka Channel 可以被多个场景使用。

- 当有 Flume Source 和 Flume Sink 时，Kafka Channel 为 Events 提供可靠和高可用的 Channel。
- 当有 Flume Source 和 Interceptor 拦截器，但是没有 Flume Sink 时，Kafka Channel 允许写 Flume Events 到 Kafka topic。
- 当有 Flume Sink，但是没有 Flume 的 Source 时，这是一种低延迟容错的方式。将从 Kafka 发送 Events 到 Flume Sinks，例如，HDFSHBase 或 Solr。

```
01  a1.channels.channel1.type = org.apache.flume.channel.kafka.KafkaChannel
02  a1.channels.channel1.kafka.bootstrap.servers = kafka101:9092,kafka102:9092
03  a1.channels.channel1.kafka.topic = channel1
04  a1.channels.channel1.kafka.consumer.group.id = flume-consumer
```

### 4. File Channel

在磁盘上指定一个目录用于存放 Events，同时也可以指定目录的大小，其优点是数据可以恢复，相对 Memory Channel 来说，其缺点是要频繁读取磁盘，速度较慢。

```
01  a1.channels = c1
02  a1.channels.c1.type = file
03  a1.channels.c1.checkpointDir = /mnt/flume/checkpoint
04  a1.channels.c1.dataDirs = /mnt/flume/data
```

### 5. Spillable Memory Channel

将 Event 存放在内存和磁盘上，内存作为主要存储，当内存达到一定临界点的时候会溢写到磁盘上。

```
01  a1.channels = c1
02  a1.channels.c1.type = SPILLABLEMEMORY
03  a1.channels.c1.memoryCapacity = 10000
04  a1.channels.c1.overflowCapacity = 1000000
05  a1.channels.c1.byteCapacity = 800000
06  a1.channels.c1.checkpointDir = /mnt/flume/checkpoint
07  a1.channels.c1.dataDirs = /mnt/flume/data
```

## 11.4 Flume 的 Sink 组件

Flume 通过 Source 组件采集数据源的数据，并缓存到 Channel 中，最后通过 Flume 的 Sink 组件将采集到的 Event 数据存入目的地。Flume Sink 支持多种类型，下面我们对这些 Sink 组件进行详细的讨论，并给出相应的配置信息。

### 1. HDFS Sink

HDFS Sink 将事件写入 Hadoop 分布式文件系统 HDFS 中。目前它支持创建文本文件和序列化文件，对这两种格式都支持压缩。使用这个 Sink 要求 Hadoop 必须已经安装好，以便 Flume 可以通过 Hadoop 提供的 jar 包与 HDFS 进行通信。

```
01  #具体定义 sink
02  a1.sinks.k1.type = hdfs
03  a1.sinks.k1.hdfs.path = hdfs://hadoop101:8020/flume/%Y%m%d
04  a1.sinks.k1.hdfs.filePrefix = events-
05  a1.sinks.k1.hdfs.fileType = DataStream
06
07  #不按照条数生成文件
08  a1.sinks.k1.hdfs.rollCount = 0
09  #HDFS 上的文件达到 128MB 时生成一个文件
10  a1.sinks.k1.hdfs.rollSize = 134217728
11  #HDFS 上的文件达到 60 秒生成一个文件
12  a1.sinks.k1.hdfs.rollInterval = 60
```

### 2. Hive Sink

Hive Sink 将包含分割文本或 JSON 数据的 Events 直接传送到 Hive 表或分区中。使用 Hive 事务写 Events。当一系列 Events 提交到 Hive 时，它们马上可以被 Hive 查询到。

```
01  a1.channels = c1
02  a1.channels.c1.type = memory
03  a1.sinks = k1
04  a1.sinks.k1.type = hive
05  a1.sinks.k1.channel = c1
06  a1.sinks.k1.hive.metastore = thrift://127.0.0.1:9083
07  a1.sinks.k1.hive.database = logsdb
08  a1.sinks.k1.hive.table = weblogs
09  a1.sinks.k1.hive.partition = asia,%{country},%y-%m-%d-%H-%M
10  a1.sinks.k1.useLocalTimeStamp = false
11  a1.sinks.k1.round = true
12  a1.sinks.k1.roundValue = 10
13  a1.sinks.k1.roundUnit = minute
14  a1.sinks.k1.serializer = DELIMITED
15  a1.sinks.k1.serializer.delimiter = "\t"
16  a1.sinks.k1.serializer.serdeSeparator = '\t'
17  a1.sinks.k1.serializer.fieldnames =id,,msg
```

需要注意的是，使用这种类型的 Sink 需要在 Hive 中先建立下面的表结构。

```
01  create table weblogs ( id int , msg string )
02  partitioned by (continent string, country string, time string)
03  clustered by (id) into 5 buckets
04  stored as orc;
```

### 3. Logger Sink

Logger Sink 记录指定级别（比如 INFO、DEBUG、ERROR 等）的日志，通常用于调试，将数据直接输出到终端。它在--conf（-c )参数指定的目录下有 log4j 的配置文件。

#### 4. Avro Sink

Avro Sink 和 Avro Source 配合使用，是实现复杂流动的基础。

```
01  a1.channels = c1
02  a1.sinks = k1
03  a1.sinks.k1.type = avro
04  a1.sinks.k1.channel = c1
05  a1.sinks.k1.hostname = 10.10.10.10
06  a1.sinks.k1.port = 4545
```

#### 5. Thrift Sink

Flume 采集的数据发送到 Thrift Sink，将其转换为 Thrift Events，并发送到配置好的主机地址和端口。配置好的 Channel 按照大小批量获取 Events。

```
01  a1.channels = c1
02  a1.sinks = k1
03  a1.sinks.k1.type = thrift
04  a1.sinks.k1.channel = c1
05  a1.sinks.k1.hostname = 10.10.10.10
06  a1.sinks.k1.port = 4545
```

#### 6. IRC Sink

IRC Sink 从链接的 Channel 获取消息并将消息推送到配置的 IRC 目的地。

```
01  a1.channels = c1
02  a1.sinks = k1
03  a1.sinks.k1.type = irc
04  a1.sinks.k1.channel = c1
05  a1.sinks.k1.hostname = irc.yourdomain.com
06  a1.sinks.k1.nick = flume
```

#### 7. File Roll Sink

File Roll Sink 在本地系统中存储事件，每隔指定时长生成文件，保存这段时间内收集到的日志信息。

```
01  a1.channels = c1
02  a1.sinks = k1
03  a1.sinks.k1.type = file_roll
04  a1.sinks.k1.channel = c1
05  a1.sinks.k1.sink.directory = /var/log/flume
```

#### 8. Null Sink

当接收到 Channel 的 Events 数据时，将直接丢弃所有 Events。

```
01  a1.channels = c1
02  a1.sinks = k1
```

```
03  a1.sinks.k1.type = null
04  a1.sinks.k1.channel = c1
```

### 9. HBase Sink

HBase Sink 写数据到 HBase。

```
01  a1.channels = c1
02  a1.sinks = k1
03  a1.sinks.k1.type = hbase
04  a1.sinks.k1.table = foo_table
05  a1.sinks.k1.columnFamily = bar_cf
06  a1.sinks.k1.serializer = org.apache.flume.sink.hbase.RegexHbaseEventSerializer
07  a1.sinks.k1.channel = c1
```

### 10. ElasticSearch Sink

ElasticSearch Sink 写数据到 ElasticSearch 集群。

```
01  a1.channels = c1
02  a1.sinks = k1
03  a1.sinks.k1.type = elasticsearch
04  a1.sinks.k1.hostNames = 127.0.0.1:9200,127.0.0.2:9300
05  a1.sinks.k1.indexName = foo_index
06  a1.sinks.k1.indexType = bar_type
07  a1.sinks.k1.clusterName = foobar_cluster
08  a1.sinks.k1.batchSize = 500
09  a1.sinks.k1.ttl = 5d
10  a1.sinks.k1.serializer = org.apache.flume.sink.elasticsearch.
ElasticSearchDynamicSerializer
11  a1.sinks.k1.channel = c1
```

### 11. Kafka Sink

Flume Sink 实现可以导出数据到一个 Kafka Topic，将在 11.6 节中单独介绍。

### 12. HTTP Sink

HTTP Sink 从 Channel 获取 Events，并使用 HTTP POST 请求将这些 Events 发送到远程服务。Events 内容作为 POST body 发送。

```
01  a1.channels = c1
02  a1.sinks = k1
03  a1.sinks.k1.type = http
04  a1.sinks.k1.channel = c1
05  a1.sinks.k1.endpoint = http://localhost:8080/someuri
06  a1.sinks.k1.connectTimeout = 2000
07  a1.sinks.k1.requestTimeout = 2000
08  a1.sinks.k1.acceptHeader = application/json
09  a1.sinks.k1.contentTypeHeader = application/json
10  a1.sinks.k1.defaultBackoff = true
```

```
11  a1.sinks.k1.defaultRollback = true
12  a1.sinks.k1.defaultIncrementMetrics = false
13  a1.sinks.k1.backoff.4XX = false
14  a1.sinks.k1.rollback.4XX = false
15  a1.sinks.k1.incrementMetrics.4XX = true
16  a1.sinks.k1.backoff.200 = false
17  a1.sinks.k1.rollback.200 = false
18  a1.sinks.k1.incrementMetrics.200 = true
```

#### 13. Custom Sink

如果以上内置的 Sink 都不能满足需求，可以自己开发 Sink。按照 Flume 要求写一个类，实现相应接口，将类打成 jar 包放置到 Flume 的 lib 目录下，在配置文件中通过类的全路径名加载 Sink。

## 11.5　集成 Kafka 与 Flume

Apache Flume 的主要功能就是收集同步数据源的数据，并将数据保存到持久化系统中，适合数据来源比较广、数据收集结构比较固定的场景。Apache Kafka 主要以一个中间件系统的方式存在，适合高吞吐量和负载的情况，可以作为业务系统中的缓存、消息通知系统、数据收集等场景。

在实际生产环境中，Kafka 生产的数据是由 Flume 的 Sink 提供的，Flume 将 Agent 的日志通过 Sink 组件收集分发到 Kafka；Flume 的 Source 也可以作为 Kafka 的消费者来采集 Kafka 的消息。

#### 1. Kafka 接收 Flume Sink 的数据

下面给出这种方式完整的 Agent 配置信息。

```
01  #定义 agent 名， source、channel、sink 的名称
02  a2.sources = r1
03  a2.channels = c1
04  a2.sinks = k1
05
06  #具体定义 source
07  a2.sources.r1.type = spooldir
08  a2.sources.r1.spoolDir = /root/training/logs
09
10  #定义拦截器，为消息添加时间戳
11  a2.sources.r1.interceptors = i1
12  a2.sources.r1.interceptors.i1.type = org.apache.flume.interceptor.
TimestampInterceptor$Builder
13
14  #具体定义 channel
```

```
15  a2.channels.c1.type = memory
16  a2.channels.c1.capacity = 10000
17  a2.channels.c1.transactionCapacity = 100
18
19  a2.sinks.k1.channel = c1
20  a2.sinks.k1.type = org.apache.flume.sink.kafka.KafkaSink
21  a2.sinks.k1.kafka.topic = mytopic1
22  a2.sinks.k1.kafka.bootstrap.servers = kafka101:9092
23  a2.sinks.k1.kafka.flumeBatchSize = 20
24  a2.sinks.k1.kafka.producer.acks = 1
25  a2.sinks.k1.kafka.producer.linger.ms = 1
26  a2.sinks.k1.kafka.producer.compression.type = snappy
27
28  #组装 source、channel、sink
29  a2.sources.r1.channels = c1
30  a2.sinks.k1.channel = c1
```

下面我们来验证当 Flume 采集到数据后，是否能够发送到 Kafka 中。最终将在 Kafka Consumer 的终端打印相应的数据信息。

（1）创建目录/root/training/logs。如果在这个目录下有新的数据产生，数据将被 Flume 的 Source 采集。

```
mkdir /root/training/logs
```

（2）创建 a2.conf 配置文件，将上面的配置内容输入，保存并退出。

```
vi /root/training/flume/myagent/a2.conf
```

（3）启动 Flume 的 Agent，执行下面的语句。

```
cd /root/training/flume
bin/flume-ng agent -n a2 -f myagent/a2.conf -c conf -Dflume.root.logger=INFO,
console
```

图 11.7 展示了 Flume Agent 的启动过程。

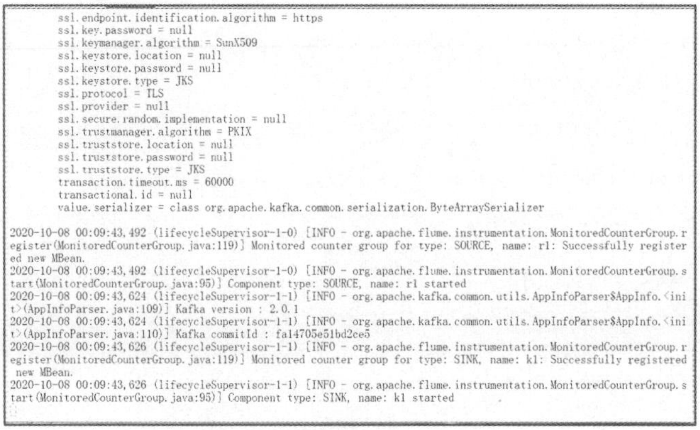

```
        ssl.endpoint.identification.algorithm = https
        ssl.key.password = null
        ssl.keymanager.algorithm = SunX509
        ssl.keystore.location = null
        ssl.keystore.password = null
        ssl.keystore.type = JKS
        ssl.protocol = TLS
        ssl.provider = null
        ssl.secure.random.implementation = null
        ssl.trustmanager.algorithm = PKIX
        ssl.truststore.location = null
        ssl.truststore.password = null
        ssl.truststore.type = JKS
        transaction.timeout.ms = 60000
        transactional.id = null
        value.serializer = class org.apache.kafka.common.serialization.ByteArraySerializer

2020-10-08 00:09:43,492 (lifecycleSupervisor-1-0) [INFO - org.apache.flume.instrumentation.MonitoredCounterGroup.r
egister(MonitoredCounterGroup.java:119)] Monitored counter group for type: SOURCE, name: r1: Successfully register
ed new MBean.
2020-10-08 00:09:43,492 (lifecycleSupervisor-1-0) [INFO - org.apache.flume.instrumentation.MonitoredCounterGroup.s
tart(MonitoredCounterGroup.java:95)] Component type: SOURCE, name: r1 started
2020-10-08 00:09:43,624 (lifecycleSupervisor-1-1) [INFO - org.apache.kafka.common.utils.AppInfoParser$AppInfo.<ini
t>(AppInfoParser.java:109)] Kafka version : 2.0.1
2020-10-08 00:09:43,624 (lifecycleSupervisor-1-1) [INFO - org.apache.kafka.common.utils.AppInfoParser$AppInfo.<ini
t>(AppInfoParser.java:110)] Kafka commitId : fa14705e51bd2ce5
2020-10-08 00:09:43,626 (lifecycleSupervisor-1-1) [INFO - org.apache.flume.instrumentation.MonitoredCounterGroup.r
egister(MonitoredCounterGroup.java:119)] Monitored counter group for type: SINK, name: k1: Successfully registered
 new MBean.
2020-10-08 00:09:43,626 (lifecycleSupervisor-1-1) [INFO - org.apache.flume.instrumentation.MonitoredCounterGroup.s
tart(MonitoredCounterGroup.java:95)] Component type: SINK, name: k1 started
```

图 11.7　Flume Agent 的启动过程

（4）启动 Kafka 集群，并打开一个 Kafka Consumer 的终端。

```
bin/kafka-console-consumer.sh --bootstrap-server kafka101:9092 --topic mytopic1
```

（5）重新打开一个新的 Linux 命令终端，并复制一个文件到/root/training/logs 中。观察 Kafka Consumer 命令行终端的变化，如图 11.8 所示。

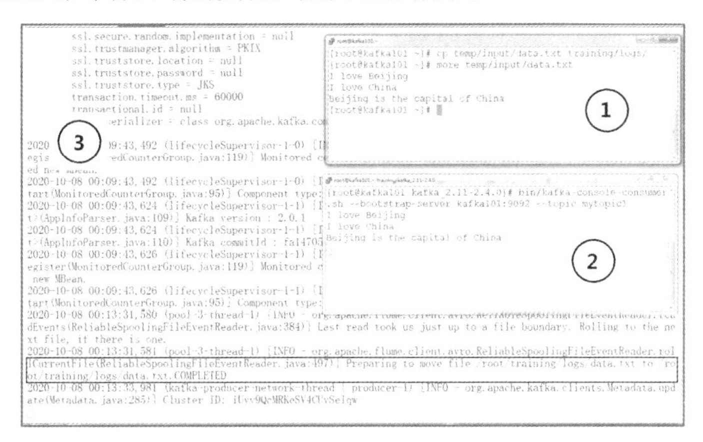

图 11.8　将 Flume 采集的数据发送给 Kafka

其中，

- 窗口 1 用来模拟数据的产生，这里我们使用的测试数据是下面的代码。

```
I love Beijing
I love China
Beijing is the capital of China
```

- 窗口 2 是 Kafka Consumer 的命令行终端。
- 窗口 3 是 Flume 后台输出的日志信息。

当在窗口 1 中将测试数据复制到目录/root/training/logs 下，模拟数据的产生时，可以在窗口 3 中看到 Flume 成功采集到数据，如图中方框所示；最后在窗口 2 中输出采集到的内容。

### 2．将 Kafka 的数据发送给 Flume Source

下面给出这种方式完整的 Agent 配置信息。

```
01    #定义 agent 名， source、channel、sink 的名称
02    a3.sources = r1
03    a3.channels = c1
04    a3.sinks = k1
05
06    #具体定义 source
07    a3.sources.r1.type = org.apache.flume.source.kafka.KafkaSource
08    a3.sources.r1.channels = c1
09    a3.sources.r1.batchSize = 100
```

```
10   a3.sources.r1.batchDurationMillis = 2000
11   a3.sources.r1.kafka.bootstrap.servers = kafka101:9092
12   a3.sources.r1.kafka.topics = mytopic1
13   a3.sources.r1.kafka.consumer.group.id = mygroupID
14
15   #具体定义channel
16   a3.channels.c1.type = memory
17   a3.channels.c1.capacity = 1000
18   a3.channels.c1.transactionCapacity = 100
19
20   #具体定义sink
21   a3.sinks.k1.type = logger
22
23   #组装source、channel、sink
24   a3.sources.r1.channels = c1
25   a3.sinks.k1.channel = c1
```

下面我们来验证这里的配置是否能够正常工作。

（1）创建 a3.conf 配置文件，将上面的配置内容输入，保存并退出。

```
vi /root/training/flume/myagent/a3.conf
```

（2）启动 Flume 的 Agent，执行下面的语句。

```
cd /root/training/flume
bin/flume-ng agent -n a3 -f myagent/a3.conf -c conf -Dflume.root.logger=INFO,
console
```

（3）启动Kafka的生产者命令行终端，并输入测试的数据。这里的数据是“Hello Kafka”。图 11.9 展示了这个过程。

```
bin/kafka-console-producer.sh --broker-list kafka101:9092 --topic mytopic1
```

```
adata.java:285)] Cluster ID: iUyv9QcMRKeSV4CUvSelqw
2020-10-08 00:27:38,881 (PollableSourceRunner-KafkaSource-r1) [INFO - org.apache.kafka.clients.consumer.internals.
AbstractCoordinator$FindCoordinatorResponseHandler.onSuccess(AbstractCoordinator.java:677)] [Consumer clientId=con
sumer-1, groupId=mygroupID] Discovered group coordinator kafka101:9092 (id: 2147483647 rack: null)
2020-10-08 00:27:38,883 (PollableSourceRunner-KafkaSource-r1) [INFO - org.apache.kafka.clients.consumer.internals.
ConsumerCoordinator.onJoinPrepare(ConsumerCoordinator.java:472)] [Consumer clientId=consumer-1, groupId=mygroupID]
 Revoking previously assigned partitions []
2020-10-08 00:27:38,883 (PollableSourceRunner-KafkaSo
AbstractCoordinator.sendJoinGroupRequest(AbstractCoor
roupID] (Re-)joining group
2020-10-08 00:27:38,905 (PollableSourceRunner-KafkaSo
AbstractCoordinator$1.onSuccess(AbstractCoordinator.
uccessfully joined group with generation 3
2020-10-08 00:27:38,907 (PollableSourceRunner-KafkaSo
ConsumerCoordinator.onJoinComplete(ConsumerCoordinat
] Setting newly assigned partitions [mytopic1-2, myto
2020-10-08 00:27:38,907 (PollableSourceRunner-KafkaSo
eListener.onPartitionsAssigned(KafkaSource.java:648)] topic mytopic1 - partition 2 assigned.
2020-10-08 00:27:38,907 (PollableSourceRunner-KafkaSource-r1) [INFO - org.apache.flume.source.kafka.SourceRebalanc
eListener.onPartitionsAssigned(KafkaSource.java:648)] topic mytopic1 - partition 1 assigned.
2020-10-08 00:27:38,907 (PollableSourceRunner-KafkaSource-r1) [INFO - org.apache.flume.source.kafka.SourceRebalanc
eListener.onPartitionsAssigned(KafkaSource.java:648)] topic mytopic1 - partition 0 assigned.
2020-10-08 00:27:38,944 (PollableSourceRunner-KafkaSource-r1) [INFO - org.apache.kafka.clients.consumer.internals.
Fetcher.resetOffsetIfNeeded(Fetcher.java:601)] [Consumer clientId=consumer-1, groupId=mygroupID] Resetting offset
for partition mytopic1-2 to offset 876.
2020-10-08 00:27:38,945 (PollableSourceRunner-KafkaSource-r1) [INFO - org.apache.kafka.clients.consumer.internals.
Fetcher.resetOffsetIfNeeded(Fetcher.java:601)] [Consumer clientId=consumer-1, groupId=mygroupID] Resetting offset
for partition mytopic1-0 to offset 1742.
2020-10-08 00:27:47,708 (SinkRunner-PollingRunner-DefaultSinkProcessor) [INFO - org.apache.flume.sink.LoggerSink.p
rocess(LoggerSink.java:95)] Event: { headers:{topic=mytopic1, partition=2, offset=876, timestamp=1602131267143} bo
dy: 48 65 6C 6C 6F 20 4B 61 66 6B 61                Hello Kafka }
```

```
[root@kafka101 kafka_2.11-2.4.0]# bin/kafka-console-producer
.sh --broker-list kafka101:9092 --topic mytopic1
>Hello Kafka
>
```

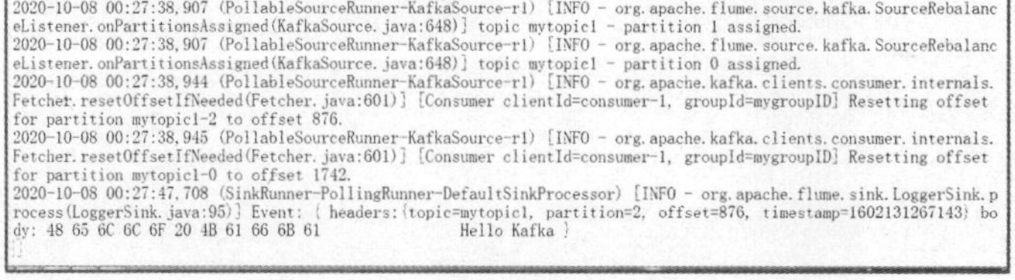

图 11.9　Flume 接收 Kafka 的数据